CHS 12-17

HUMAN FACTORS IN ENGINEERING AND DESIGN

FOURTH EDITION

HUMAN FACTORS IN ENGINEERING AND DESIGN

ERNEST J. McCORMICK
Professor of Psychology
Purdue University

McGRAW-HILL BOOK COMPANY
New York   St. Louis   San Francisco   Auckland
Düsseldorf   Johannesburg   Kuala Lumpur   London
Mexico   Montreal   New Delhi   Panama   Paris
São Paulo   Singapore   Sydney   Tokyo   Toronto

This book was set in Times Roman by Textbook Services, Inc.
The editors were Richard R. Wright and M. E. Margolies;
the cover was designed by Nicholas Krenitsky;
the production supervisor was Leroy A. Young.
The drawings were done by J & R Services, Inc.
Fairfield Graphics was printer and binder.

Library of Congress Cataloging in Publication Data

McCormick, Ernest James.
    Human factors in engineering and design.

    First published in 1957 under title: Human engineering; 2d-3d editions have title: Human
factors engineering.
    Bibliography: p.
    Includes index.
    1. Human engineering. I. Title. [DNLM: 1. Human engineering. T175 M131h]
TA166.M3 1976   620.8   75-15905
ISBN 0-07-044886-8

To Emily, Wynne, and Jan

CONTENTS

PREFACE

One of the important factors that characterizes the "industrial" or "developed" countries is that many of the people live in very much of a "man-made" world (as contrasted with living in more of a "natural" environment). This man-made world includes such varied items as buildings, equipment, machines, tools and other devices, appliances, vehicles, highways, various aspects of the environment such as illumination and atmospheric control, many personal items, and a range of "services" such as health and recreation services. This text deals with some of the problems and processes that are involved in man's efforts to so design these products and environments that they optimally serve their intended use by human beings. These objectives, of course, are as old as man; human beings have always endeavored to adapt the things they make and their environments to their own use. It is only in recent years, however, that systematic, concerted action has been directed toward these objectives. This general area of human endeavor (and its various facets) has come to be known as *human factors engineering*, or simply *human factors, biomechanics, engineering psychology*, or (in most European countries) *ergonomics*.

The previous edition of this text was *Human Factors Engineering;* however, the title of this edition was broadened to *Human Factors in Engineering and Design* since the design and development of many of the physical facilities and services people use are not entirely within the domain of engineering but involve other disciplines as well (as mentioned below). For the sake of semantic simplicity the term "human factors" will be used frequently in this text.

The attention to this endeavor in the very recent decades reflects a significant shift in emphasis; this shift has been toward placing greater reliance upon systematic research, and reduced emphasis upon human experience, as the basis for developing principles and data to be applied in human factors design processes. The impetus of the current attention to human factors originated from the pressing requirement in the development of certain new types of systems that their suitability for their intended use be ensured. Although the initial focus of such attention was on certain items of military, aircraft, and electronics equipment and space vehicles, some inroads have since been made in the design of other items such as automobiles, transportation equipment, machine tools, computers, agricultural equipment, certain types of consumer products, and some features of the physical ambient environment. In addition there have been slight stirrings of interest in giving more systematic attention to the human factors implications of architectural design, city planning, and a range of civic and community facilities and services. It should be added, however, that the human factors inroads in the case of many areas have been very limited. As a frame of reference (and one that will be followed in the text), it is suggested that virtually all the things people use (ranging from can openers to huge industrial facilities and transportation systems), the environments in which they live and work (including their houses and total communities), and many types of services to the public (whether offered by communities, government agencies, or private enterprises) be viewed in terms of their possible human factors implications.

Several professions are concerned with human factors in one way or another. It is essentially an interdisciplinary activity rather than a separate profession as such. While the interested professions overlap in their roles in this field, they generally fall into two groups. On one side of the coin are those professions that are concerned with creating the facilities that human beings use, the procedures and manners in which such facilities are used, the environments in which people live and work; working within this area are engineers of all sorts (mechanical, electrical, industrial, illuminating, air conditioning, acoustical, civil, construction, traffic, etc.), industrial designers, architects, and city planners, and probably others; these are the potential *appliers* of human factors data and principles. They have it within their hands to facilitate or, conversely, to complicate the subsequent use by human beings of those facilities which they create. On the other side of the coin are the disciplines that are the *producers* of human factors information; working within this area are primarily the behavioral scientists (such as psychologists, sociologists, and anthropologists) and the biological scientists (biologists, physiologists, etc.). Their dominant roles in the human factors area are those of carrying out research to generate new information about human beings that might be relevant to the human factors field, and of serving as consultants or advisors on matters relating to their areas of specialization.

The objectives of human factors are those of achieving functional effectiveness of whatever physical equipment or facilities people use and of maintaining or enhancing human welfare (such as health, safety, and satisfaction) by appropriate design of the equipment, facilities, and environments they use. The central theme of this text is that of illustrating how these objectives can better be fulfilled by concentration on human factors during the design stage. This will be done primarily by illustrating, by relevant research, the relationships between various "design" features and functional effectiveness or some aspect of human welfare.

In line with this objective, the text deals with several of the most important areas of human factors and is therefore something of a survey of human factors. In the case of each topic it has been the intent to delineate it, to characterize its major dimensions and related concepts, and to present examples of research that are relevant to the area. In the selection of research material dealing with any given topic, it was of course necessary to choose from a wide variety of sources. Although practical requirements have made it necessary to be selective in the inclusions of illustrative material, usually material was selected that would be broadly relevant rather than highly situational; however, situational material is included in some instances, this being done in part to represent the varied spectrum of the human factors area. The materials are drawn from a number of disciplines that are concerned with the human factors area, but especially from the behavioral and biological sciences; in this connection the bulk of the material comes from the field of psychology. This predominance is the consequence of psychology having undoubtedly contributed more to the human factors area than any other discipline, and is also due to its being the author's own field of work.

A book such as this should be dedicated to the many individuals whose research has formed its basis. Their names are referred to throughout the text and are listed in the chapter references and in the authors' index, and so they will not be repeated here. If their data and conclusions have been misinterpreted, the present author assumes the responsibility; the reader should not blame the investigator. Appreciation is expressed to the publishers who have granted permission to reproduce or to adapt the many tables and figures.

I would like to express appreciation to Mrs. Bernice Ezra and Mrs. Charlene Wamer for typing the manuscript.

Ernest J. McCormick

Part One

Introduction

Chapter 1

Human Beings in a Man-Made World

Our ancestors of bygone millenniums lived in an essentially "natural" environment, their existence virtually depending on what they could do directly with their hands. Over time, of course, they developed simple tools and utensils and constructed shelter for themselves toward the end of aiding and abetting the process of keeping alive and of making life a bit more tolerable. The more recent centuries, decades, and years have seen the production of physical accouterments and facilities that simply could not have been imagined by our ancestors in their wildest dreams. In at least many civilizations of our present world, the majority of the "things" people use are man made. In other words, many people live in very much of a man-made world. Even those engaged in activities close to nature—the fishermen, farmers, campers, and bird watchers —use many man-made devices. An abbreviated listing of man-made facilities could include hand tools, kitchen utensils, vehicles, highways, machinery, houses and other buildings, TV sets, telephones, and space capsules (plus the binoculars for the bird watchers). This text is concerned essentially with the "human factors" aspects of the design of the man-made facilities of our civilization on the presumption that the particular design of such facilities can enhance or degrade their use by people. What we will here refer to as *human factors* addresses itself to this problem.

HUMAN FACTORS DEFINED

Definitions sometimes represent treacherous exercises in semantics, but, for better or worse, they probably are necessary exercises. We shall approach the definition of human factors in three stages, as follows:

- The central *focus* of human factors relates to the consideration of human beings in the design of the man-made objects, facilities, and environments that people "use" in the various aspects of their lives.
- The *objectives* of human factors in the design of these man-made objects, facilities, and environments are twofold, as follows: (1) to enhance the *functional effectiveness* with which people can use them; and (2) to maintain or enhance certain desirable *human values* in the process (e.g., health, safety, and satisfaction); this second objective is essentially one of human welfare.
- The central *approach* of human factors is the systemic application of relevant information about human characteristics and behavior to the design of the man-made objects, facilities, and environments that people use.

In capsule form, the nub of human factors can be considered as the process of designing for human use. Aside from the emphasis on designing the "things" people use, however, the human factors discipline can be viewed as embracing certain related functions (when relevant to the equipment or facility in question) as: operational methods and procedures; testing and evaluating such items in terms of their human factors aspects; job design; development of job aids and training materials; and the selection and training of personnel who would be involved in the use of such items. Further, the human factors discipline usually is viewed as embracing relevant supportive research that provides some of the guidelines for the design and related processes.

MAN'S INVOLVEMENT WITH WHAT HE USES

An inventory of the man-made objects, facilities, and environments that man "uses" would comprise a mixed bag, but in general terms it would include the three classes discussed below.

Man-Machine Systems

We can consider a man-machine system as a combination of one or more human beings and one or more physical components interacting to bring about, from given inputs, some desired output. In this frame of reference, the common concept of "machine" is too restricted, and we should rather consider a "machine" to consist of virtually any type of physical object, device, equipment, facility, thing, or "what have you" that people use in carrying out some activity that is directed toward achieving some desired purpose or in performing some function. In a relatively simple form a man-machine system (or what we will sometimes refer to as a "system") can be a man with a hoe, a hammer, a hod, or hair clippers. Going up the scale of complexity, one can regard as systems the family automobile, an office machine, a lawn mower, and a roulette wheel, each equipped with its operator. More complex systems include aircraft, bottling machines, conveyor systems, telephone systems, and automated oil refineries, along with their personnel. Some systems are less delineated and more "amorphous" than these, such as the servicing system of a gasoline station, the operation of an amusement park or a highway and traffic system, and the rescue operations for locating an aircraft downed at sea.

The essential nature of man's involvement in a system is an active one, interacting with the system to fulfill the function for which the system is designed.

Physical Environment

The physical environments which people "use" include two general categories. The first consists of the physical space and related facilities which people use, ranging from the immediate environment (such as a work station, a lounge chair, or a typing desk) through the intermediate (such as a home, an office, a factory, a school, or a football stadium) to the general (such as a neighborhood, a community, a city, or a highway system). The second category consists of the various aspects of the ambient environment, such as illumination, atmospheric conditions (including pollution), and noise. It should be noted that some aspects of the physical environment in which we live and work are part of the natural environment and may not be amenable to modification (although one can provide protection from certain undesirable environmental conditions such as heat or cold). Although the nature of man's involvement with his physical environment is essentially passive, the environment tends to impose certain constraints on his behavior (such as limiting the range of his movements or restricting his field of view) or to predetermine certain aspects of his behavior (such as stooping down to look into a file cabinet, wandering through a labyrinth in a supermarket to find where the bread is, or trying to see the edge of the road when driving on a rainy night).

Personal and Protective Items

The third class of man-made things people use consists of many types of personal items (such as apparel and handbags) and protective equipment and gear (such as safety shoes and hats, safety goggles, astronaut suits, gloves, and earplugs). The human involvement with such items is typically passive, although their design can also impose certain constraints on behavior or predetermine the nature of certain aspects of behavior.

Discussion

Aside from these three categories of the man-made accouterments and trappings of civilization, there are some other odds and ends that tend to defy nice, neat categorization, such as morning newspapers, playing cards, and postage stamps. Aside from the complications of classification, however, we would like to reinforce the central point we have been alluding to, namely, that whatever the nature of the human involvement with the man-made features of civilization, the specific design features thereof can influence, for better or worse, their functional utility or some relevant human value.

THE CASE FOR HUMAN FACTORS

Since man has somehow survived for these many thousands of years without people specializing in human factors, one might wonder why—at the present stage of history—it has become desirable to have human factors experts who specialize in worrying about these matters. In reflecting about this it should be recognized that the objectives of human factors are not new; the history of man is filled with evidence of his efforts, both successful and unsuccessful, to create tools and equipment which satisfactorily serve his purposes and to control more adequately the environment within which he lives and works. But during most of the centuries of man's history, the development of tools and equipment depended in large part on the process of evolution, of trial and error. Through the use of a particular device—an axe, an oar, a bow and arrow—it was possible to identify its deficiencies and to modify it accordingly, so that the next "generation" of the device would better serve its purpose.

The industrial revolution brought about major changes in the tools, equipment, devices, and environments people used; although the evolutionary process still was important as a basis for improvement in terms of human considerations, it is probable that the increased tempo of technological developments may have strained the evolutionary process as the basis for adapting devices for human use. The tempo has continued to increase in more recent years, the time during and since World War II having witnessed an epidemic of entirely new and markedly altered types of equipment and devices for human use. It has been found, often through unhappy experiences, however, that some of these devices were not designed appropriately for human use. It was found, for example, that some items of military equipment, such as high-speed aircraft, radar, and fire-control systems, could not be managed effectively by their operators, that human errors were excessive, and that many accidents occurred because of human mistakes which were attributed to design deficiencies. Similar deficiencies have been documented for certain types of civilian equipment.

The upshot of all of this is that it is becoming increasingly important in the design of equipment, facilities, etc., to consider human factors aspects early in the design game and in a systematic manner. A few questions that illustrate some of the types of considerations that might be taken into account during the design stage could include the following: Should a mail-sorting system have an optical scanner that automatically activates mechanisms for sorting mail by zip code or should human operators scan the mail and activate keying devices for sorting by zip code? Should a particular warning signal be visual or auditory?

How much "feel" should be built into the power-assisted-steering mechanism of a car? Would the information load of a given (tentative) assortment of visual displays be within reasonable human limits? How much illumination should be provided for a given operation?

In effect, then, the increased complexities of the things people use (as the consequence of technology) place a premium on having assurance that the item in question will fulfill the two objectives of functional effectiveness and human welfare. The need for such assurance requires that human factors be taken into account early during the (usually long) design and development process.

COVERAGE OF THIS TEXT

Since a comprehensive treatment of the entire scope of human factors would fill a small library, this text must be restricted to a rather modest segment of the total human factors domain. The central theme intended is that of illustrating the way in which the fulfillment of the two primary objectives of human factors (i.e., functional effectiveness and human welfare) can be influenced by the extent to which relevant human considerations have been taken into account during the design of the object, facility, or environment in question. Further, this theme will be followed as it relates to at least some of the more commonly recognized human factors "content" areas (such as the design of displays for presenting information to people, human control processes, and physical environment). It is suggested that pursuing this theme across the several subareas would have the further advantage of an overview of the content areas of human factors.

The implications of various perceptual, mental, and physical characteristics as they might affect or influence the objectives of human factors probably can best be reflected by the result of relevant research investigations and of documented operational experience. Therefore the theme of the text will generally be carried out by presenting and discussing the results of illustrative research and by bringing in generalizations or guidelines that are supported by research or experiences that have relevance to the design process in terms of human factors considerations. Thus, in the various subject or content areas much of the material in this text will consist of summaries of research that reflect the relationships between design variables on the one hand and criteria of functional effectiveness or human welfare on the other hand.

It is recognized that the illustrative material that will be brought in to carry out this theme will be in no way comprehensive, but it is hoped that it will represent in most content areas some of the more important facets of those areas.

Although the central theme will, then, deal with the human factors aspects of the design of the many things people use, there will be some modest treatment of certain related topics, such as how human factors fits in with the other phases of design and development processes, and of certain of the personnel-related functions, such as selection and training.

THE NATURE OF MAN-MACHINE SYSTEMS

Since the involvement of human beings with man-machine systems represents the dominant aspect of human factors, we should take an overview of such systems with particular reference to the roles of people in such systems.

Especially in fairly complex systems, one can identify systems within systems within systems. In this connection Linvill [2][1] suggests that some systems can be viewed as coming in layers. An exterior system consists of a set of component interior systems, which, in turn, may be composed of component interior systems that are interior to them, etc. We could view a complete telephone system as an exterior system, with a given telephone exchange as an interior system to it, and a specific switchboard as an interior system to the telephone exchange, etc. Interior systems frequently are referred to as *subsystems* or *components* (a component being itself something of a system). Within a given complex system, one can also envision a sequence of subsystems, such as the various sequential production processes in an industrial plant. Recognizing the various possible relationships among systems or subsystems, in system development processes one may, operationally, consider a particular entity—big or little, exterior or interior—as the system to be dealt with.

Characteristics of Systems

Most systems, at whatever level of detail, have certain general characteristics, or properties, in common.

Purposes of the System   First, every system has some purpose or objective. This should be clearly understood and in most situations should be made a matter of record, including a definitive statement of the specifications that are to be fulfilled. In contemplating the design of an electric car for home-to-office and around-the-town use, for example, one should set forth in as precise terms as possible such specifications as speed, range, and maneuverability and set about designing a vehicle that will fulfill these specifications. Some of the specifications may be of a strictly engineering nature (e.g., the horsepower), whereas others may have strong human factors overtones (e.g., the physical dimensions for the passengers or the ratio of the vehicular turn to steering-wheel turn).

Operational Functions and Components   In order for a system to fulfill its purposes certain operational functions need to be performed. For example, in a postal operation it is necessary to perform such functions as mail collection, stamp cancellation, mail sorting, and delivery. Each such operational function, in turn, needs to be performed by an individual or by a physical component. In system-design processes it is sometimes the practice to specify these opera-

[1]The numbers in brackets refer to references that are listed at the end of each chapter.

tional functions and to set them forth in blocks in a block diagram, with the tentative expectation that each function can be *allocated* to a corresponding physical component or to a human being. More will be said about this allocation process later. But it should be noted here, as Jones [1] points out, that this assumption of a one-to-one correspondence between functional blocks and separate physical (and presumably human) components is most applicable to *flow* systems, in which there is a sequence of clearly discriminable operations. But although all systems do involve the execution of functions by human or physical components, or by both, in some systems the functions are, as Jones puts it, less "determinate" and may not be clearly discriminable from each other. In such instances the assumption of a one-to-one correspondence between functions and system components simply may not apply; Jones cites, as an example, the operation of an automobile by its driver, indicating that there is no quick and unambiguous way of knowing that any particular block diagram (of functions and components) represents the constantly changing interaction between the driver, the vehicle, and the environment. The moral of these observations seems to be that the function-component frame of reference may serve a useful purpose in the development of some systems but may be of less value for other systems.

The execution of any operational function, in turn, typically involves a combination of four more basic functions, as follows: sensing (information receiving), information storage, information processing and decision, and action functions; they are depicted graphically in Figure 1-1. Since information

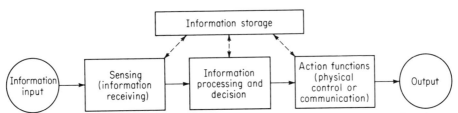

Figure 1-1 Types of basic functions performed by man or machine components of man-machine systems.

storage interacts with all the other functions, it is shown above the others. The other three functions occur in sequence.

1 *Sensing (information receiving):* One of these functions is sensing, or information receiving. Some of the information entering a system is from outside the system, for example, airplanes entering the area of control of a control-tower operator, an order for the production of a product, the heat that sets off an automatic fire alarm, various cues regarding the presence of schools of fish, and telegraph communications. Some information, however, may originate from inside the system itself. Such information can be of a feedback nature (such as the reading on the speedometer of the action on the accelerator or the feel of a control lever), or it can be information that is stored in the system.

The sensing, if by a human being, would be through the use of the various sense modalities, such as vision, audition, and touch. There are various types of machine sensing devices, such as electronic, photographic, and mechanical. Sensing by a machine in some cases is simply a substitute for the same sensing function by a man. The electronic device in an automated post office which identifies the location of a stamp on an envelope is simply doing the same thing that a man would otherwise do to place the envelope in proper position for canceling the stamp. The sonar used for detecting schools of fish, however, involves "sensing" the fish in a manner that a man is not capable of.

2 *Information storage:* For human beings, information storage is synonymous with memory of learned material. Information can be stored in physical components in many ways, as on punch cards, magnetic tapes, templates, records, and tables of data. Most of the information that is stored for later use is in coded or symbolic form.

3 *Information processing and decision:* Information processing embraces various types of operations performed with information that is received (sensed) and information that is stored. When human beings are involved in information processing, this process, simple or complex, typically results in a decision to act (or in some instances, a decision *not* to act). When mechanized or automated machine components are used, their information processing must be programmed in some way in order to cause the component to respond in some predetermined manner to each possible input. Such programming is, of course, readily understood if a computer is used. Other methods of programming involve the use of various types of schemes, such as gears, cams, electrical and electronic circuits, and levers.

4 *Action functions:* What we shall call the *action* functions of a system generally are the operations which occur as a consequence of the decisions that are made. These functions fall roughly into two classes. The first of these is some type of physical control action or process, such as the activation of certain control mechanisms or the handling, movement, modification, or alteration of materials or objects. The other is essentially a communication action, be it by voice (in human beings), signals, records, or other methods. Such functions also involve some physical actions, but these are in a sense incidental to the communication function.

Input and Output Other essential features of man-machine systems are their inputs and outputs. The input into a system consists of the ingredients that are necessary in order to achieve the desired outcome. The input may consist of physical objects or materials, such as lumber in a sawmill, crude petroleum in a refinery, or beef for the broth to be brewed. It may be information in some form, for example, account records, telegraphic messages, or the presence of objects such as aircraft in an area. In communication systems, the primary input would be information in some form. Further, the input may consist of energy, such as electric power, heat, or other types. In any given system, the input may consist of any or all of these. The output is of course the result or outcome of the system, such as a change in a product, a communication transmitted, or a service rendered. When the system in question has various components, the output of one component frequently serves as the input to another.

Communication Links Most systems involve some form of communications and communication links to make this possible. Some systems (such as telephone systems) exist for the sole purpose of sending communications. In some other systems the end action is a communication (as in computers). Although such systems have dominant communication objectives, there is some form of internal communication inherent in many systems. In the most obvious form, it consists of voice or written communications. In less obvious form, it may be the transmission of a control signal (mechanical or otherwise) from an operator to the machine through the activation of his control device; in turn, the machine may "talk back" through, say, visual instruments that provide information or by direct cues such as sound and sight. By a bit of extrapolation, one can consider one component, whether man or machine, as "communicating" with another. Every such communication must be provided for by some type of communication link. In systems design, then, it is necessary to anticipate who (or what) is to "talk" to whom (or what) and to provide an appropriate link, human or mechanical, to make this possible.

Procedures Another attribute of most systems is the set of procedures or practices that are followed in their operation. Such procedures may have been formally set forth as the system was developed, or they may have evolved with the use of the system (and thus be simply "ways of doing things").

Discussion In discussing the characteristics of systems particular mention should be made of the relationship between what we referred to as the "basic" functions involved in system operations. Three of these (sensing, information processing and decision, and action) represent what is conventionally referred to by psychologists as the S-O-R (stimulus-organism-response) paradigm, these three functions being part and parcel of every human activity in the sense that a *stimulus* acts upon an *organism* (the human being) to bring about a *response*. As Meister points out [3], it may appear superficially that system inputs and outputs are somewhat the same as stimuli and responses. He goes on to express the point of view, however, that inputs and outputs are linked directly to system requirements (thus mediating system processes), whereas stimuli and responses mediate the behavioral processes of the operator. One might argue with the notion that stimuli and responses are the exclusive functions of operators, since physical components also can sense "stimuli" and (with appropriate programming) make certain "responses." But aside from this point, we should recognize the essential distinction that Meister does make, namely, that inputs/outputs are "different" entities from stimuli/responses. Inputs/outputs implement system requirements, in primarily an operational sense, whereas stimuli/responses implement the inputs/outputs. In other words, the ability of the human beings and/or physical components to sense stimuli and to make appropriate responses is necessary to effect the input-output processes of a system.

Fundamental Man-Machine System Assumptions

In discussing systems Meister [3] differentiates between man-machine systems (MSS) and non-MSS systems by emphasizing a couple of points mentioned above, namely, that the MSS is a man-made entity and that it is specifically developed to satisfy specified requirements (that is, to fulfill specific purposes). In addition, in characterizing man-machine systems, he sets forth the following fundamental assumptions, most of which have been touched on or implied in the previous discussion:

   1   The man-machine relationship forms a system (MMS) composed of the man, the machine, and the system environment.
   2   The MMS is an artificial entity purposefully directed (required) to produce specified outputs on the basis of specified inputs.
   3   The MMS and its subsystems function in space and time and at various levels of size and complexity.
   4   Overall MMS requirements define subsystem, input, and output requirements.
   5   MMS subsystems interact with and influence each other.
   6   The MMS is most effective when inputs and outputs required to accomplish the system requirement are appropriately balanced.
   7   Failure to accomplish MMS requirements causes a change in MMS activity.
   8   Accomplishment of the subsystem requirement involves a continuing comparison between that requirement and the condition of the subsystem.

TYPES OF SYSTEMS

The nature of a system of course predetermines the nature of the human involvement in it. Before discussing various major types of systems, however, let us clarify the distinction between closed- and open-loop systems. A closed-loop system is continuous, performing some process which requires continuous control (such as in vehicular operation), and requires continuous feedback for its successful operation. The feedback provides information about any error that should be taken into account in the continuing control process. An open-loop system is one which, when activated, needs no further control or at least cannot be further controlled. In this type of system the "die is cast" once the system has been put into operation, and no further control can be exercised, such as in firing a rocket that has no guidance system. Although feedback with such systems obviously cannot serve continuous control, feedback can, if provided, serve to improve subsequent operations of the system. It should also be noted that in most open-loop systems there is almost inevitably some internal feedback within the operator, even if not provided for outside the operator.

In a very gross sense, systems can be characterized by the degree of human versus machine control. In a fairly extensive classification of systems of various types, Jones [1] includes four types that he refers to as machine systems. Three of these are listed in Table 1-1 along with their mode of operation and the nature of their physical components and couplings.

Table 1-1 Certain Machine Systems Classified according to Mode of Operation and the Physical Nature of Their Components and Couplings

Kind of system and mode of operation	Components	Couplings between components	Examples
1. Manual system, operator-directed, flexible	Hand tools or aids	One human operator	Cook plus utensils Craftsman plus tools Singer plus amplifier
2. Mechanical system,* operator-controlled and inflexible	Highly interdependent physical parts forming indistinguishable components and couplings		Engine Automobile Machine tool
3. Automatic system, preset, programmed, or adaptive	Powered mechanical systems*	Cables, pipes, conduits, levers, etc., forming a control circuit	Process plant Automatic telephone exchange Digital computer

*The original source refers to these as subsystems. For our purpose, however, we shall refer to them as systems.
Source: Adapted from Jones [1], Table 1.

Manual Systems

Manual systems consist of hand tools and other aids which are coupled together by the human operator who controls the operation, using his own physical energy as a power source. The operator (usually a craftsman) transmits to, and receives from, his tools a great deal of information, typically operates at his own speed, and can readily exploit his ability to act as a "high variety" system.

Mechanical Systems

These systems (also referred to as *semiautomatic* systems) consist of well-integrated physical parts, such as various types of powered machine tools. They are generally so designed as to perform their functions with little variation. The power typically is provided by the machine, and the operator's function is then essentially one of control, usually by the use of control devices. (Jones observes that these systems, or what he refers to as "mechanical subsystems," are the components of what he refers to as "mechanized systems"; the components typically are linked together by tracks, cables, conduits, etc., and by human operators to form the much larger ensemble of a mechanized system

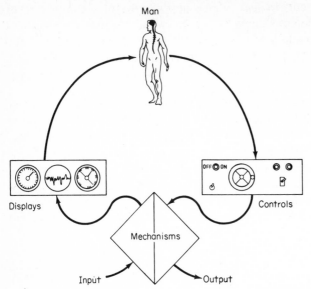

Man

Displays

Controls

Mechanisms

Input   Output

Figure 1-2 Schematic representation of a man-machine system as depicted by Taylor [4]. This represents essentially a mechanical or semiautomatic system.

such as a railway or an assembly line.) The basic ingredients of mechanical systems are well represented by Taylor [4], as shown in Figure 1-2. In that figure the man receives information about the state of affairs of the system via displays, performs essentially an information-processing and decision function, and implements the decisions by the use of control devices—or as Taylor expresses it, the human operator is represented as an organic data-transmission and -processing link inserted between the mechanical or electronic displays and the controls of a machine. In some mechanical systems, however, the information about the state of affairs is sensed directly, not by the use of intervening displays.

Automatic Systems

When a system is fully automated, it performs all operational functions, including sensing, information processing and decision making, and action. Such a system needs to be fully programmed in order to take appropriate action for all possible contingencies that are sensed. Most automatic systems are of a closed-loop nature. If such a system were perfectly reliable, it conceivably could offer the possibility of taking over all functions and leaving men to twiddle their thumbs or take off for the golf course. No one would have to stay to tend the store. Since perfectly reliable automated systems are not a likely possibility, however, at least in our lifetime, it is probable that certain primary human functions in such systems will be those of monitoring, programming, and maintenance.

The distinctions made between manual, mechanical, and automatic systems are not really clear-cut. In fact, within any given system, the different components (which can be considered subsystems) can vary in the degree of their manual versus automatic features.

DISCUSSION

The technological developments of recent decades have (at least in the developed countries of the world) resulted in people living more and more in a man-made world in which many or most of the things they deal with and the features of their environment are "artificial" (i.e., man made) rather than "natural." The major theme to which this text is addressed is that further systematic attention should be given to the human factors aspects of the design of such things and environments, toward the ends of enhancing the functional effectiveness of whatever things people use and of maintaining or enhancing desirable human values.

A WORD ABOUT INDIVIDUAL DIFFERENCES

In the design of physical systems and accouterments in terms of human considerations there is an easy reference to the "typical" or "average" human being. However, three words of caution are in order in conceptualizing the "model" of the human beings for whom the designer is designing whatever he is designing: (1) Human beings come in assorted sizes, shapes, and varieties; although there are circumstances in which it is appropriate to design for the somewhat mythical "typical" or "average" individual, the designer should be ever mindful of the fact of individual differences. (2) Some things are to be designed for special groups, such as infants, children, teenagers, the elderly, or the infirm; in such instances the designer obviously should consider such groups as a "model." (3) When certain things or facilities are to be designed for "the public," the designer should provide for almost the entire gamut of human beings; for example, in public places the various features such as entry ways, doors, ramps, escalators, and signs should be suitable for a doddering grandmother or a kid who is tagging along behind his mother or father, as well as for the hale and hardy businessman or shopper.

REFERENCES

1 Jones, J. C.: The designing of man-machine systems, *Ergonomics*, 1967, vol. 10, no. 2, pp. 101–111.
2 Linvill, W. K.: "Sensitivity of system design to system objectives," in *Proceedings of the National Symposium on Human Factors in Systems Engineering*, IRE Professional Group on Military Electronics, Dec. 3–4, 1957.
3 Meister, D.: Toward a distinctive theoretical structure for human factors, *JSAS Catalogue of Selected Documents in Psychology* (APA), vol. 2, Winter, 1972, p. 41.
4 Taylor, F. V.: Psychology and the design of machines, *The American Psychologist*, 1957, vol. 12, pp. 249–258.

Chapter 2

The Data Base of Human Factors

As indicated above, the central approach of the human factors discipline is the application of relevant information about human characteristics and behavior to the design of the man-made objects, facilities, and environments that people use. There are essentially two sources of the information that are relevant for this purpose, namely, experience and research. Because of the importance of research in this regard, and because of the emphasis on the research basis of human factors in this text, this chapter will deal with a few facets of human research as related to human factors. In addition, the chapter will cover the somewhat related matters of performance reliability and human performance, since data of these types also serve as inputs in human factors design processes.

AN OVERVIEW OF RESEARCH METHODS

Although this text is not intended as a text in research methods, an overview of certain aspects of research methodology may be in order.[1]

---

[1] For an excellent discussion of research techniques applicable to human factors, see Chapanis [5].

Types of Variables Used in Research

There are two types of variables that must ordinarily be dealt with in most research projects. One of these is the factor that is being investigated, such as illumination, designs of instruments, information channels, or gravitational forces. Such factors are referred to as the *independent variable*. The second type of variable, the *dependent variable*, is a measure of the possible "effects" of the independent variable (frequently a measure of performance, such as reaction time). The dependent variable usually is called a *criterion*, and will be discussed later.

Laboratory versus Real-Life Research

Frequently, if an investigator has a particular research purpose in mind, he may have some freedom to choose the situational context of his research—that is, whether to carry out his research in a laboratory setting or in the real world. There are other circumstances, however, in which the nature of the research problem virtually dictates the locale. For example, one cannot realistically study in the laboratory the extent to which different types of speed-zone signs affect driver behavior; the laboratory would remove the spontaneous aspect of driver reactions to such signs. The experimenter is such a case obviously would set up shop along the highways and streets. On the other hand, if an investigator wished to study the differential thresholds for discriminating amplitude of vibration, he would have to do so in a laboratory, since there probably are no situations in the real world where such discriminations can be studied. Where the investigator has the opportunity for choice, there are certain considerations that become pertinent in making such a decision, such as control of variables (which usually is easier within the pristine walls of the laboratory); the realism (if this is relevant), which is clearly greater in the real world; the differential motivation of subjects; the safety of subjects; and perhaps others.

Simulation is used in human factors research to "represent" some item of equipment, some operation or procedure, or some environment. It is usually used when it is impractical or impossible to carry out the research in a real-world situation. Physical simulation can range from very simple items (such as mockups of an item of equipment) to extremely complex forms such as those used in space research programs. Computer simulation is also used in some circumstances.

Sampling

Practically every study or experiment includes a sample from some theoretical population. Depending on the nature of the research, the population from which a sample is drawn can consist of people, events, and even objects. Thus, a psychologist might consider the population in a study to be comprised of males with automobile driving permits in a given state, or of men undergoing a specific training course. A population of events could consist of traffic at a particular location during rush hours, or of flights of commercial airlines. A population of objects could consist of physical components of a given type (such as radio transmitters), or frozen meat pies.

The data from a sample of the population serve as the basis for extrapolating to, or drawing inferences about, the parent population. The sample should be of such a size and such a nature that it could be considered representing the population.

Statistical Analysis of Data

Once an experiment or study has been carried out and the data have been gathered, the experimenter is in a position to analyze the data to see what relationships there are between or among the independent and dependent variables using appropriate statistical analyses. It is not intended in this text to deal extensively with statistics, or to discuss elaborate statistical methods. Probably most readers are already familiar with most of the statistical methods and concepts that will be touched on later, especially frequency distributions and different measures of central tendency (the mean, median, and mode). For those readers who are not familiar with the concepts of the standard deviation, correlations, and statistical significance, these will be described very briefly.

Standard Deviation The quantitative values of cases within a sample— values such as errors made, height of people, or scores on tests—naturally vary from each other. Certain statistical indexes can be used to quantify the degree of variability among the cases. One such measure is the *standard deviation*.[2] This is expressed in terms of the original numerical values of the data and reflects the variability in the distribution of the cases from the mean. In relatively normal distributions approximately two-thirds of all the cases fall within one standard deviation above or below the mean, and over 99 percent of the cases fall within three standard deviations above or below the mean. If the standard deviation of the height of one group of boys is, say, 1 in, they would be more homogeneous (i.e., less variable) than another group whose standard deviation is 2 in.

Correlations A *coefficient of correlation* is a measure of the degree of relationship between two variables. The magnitude of correlations may range from +1.00 (which is a perfect positive correlation) down through various intermediate positive values to zero (which is the absence of any relationship), down through various negative values to −1.00 (which is a perfect negative correlation).

Statistical Significance To evaluate his results, an experimenter should determine to what extent the results are *statistically significant*, such as the dif-

---

[2]The standard deviation, mathematically, is the square root of the average of the squares of the deviations of the individual cases from the mean. Formulas for computing the standard deviation may be found in most elementary statistical texts.

ference between two means or the size of a correlation. Statistical significance refers to the *probability* that the results (whatever they may be) could have occurred by *chance* (as opposed to being brought about by the experimental variables under investigation). It is common practice to use either the "5 percent" or the "1 percent" level as the acceptable level of statistical significance. If a difference is significant at the 1 percent level, this means that the obtained difference is of such a magnitude that it could have occurred *by chance* only 1 time out of 100.

CRITERIA IN RESEARCH AND SYSTEM DEVELOPMENT

The criterion (or dependent variable) as used in research is a measure of the possible "effects" of the independent variable (that is, the factor being investigated). Thus, we might compare the speed and accuracy with which people can read different road signs, or the heart rates of people working under two conditions of temperature. Aside from the use of criteria in research undertakings, however, criteria are also used in the testing and evaluation of systems or components such as in testing the performance of a dishwasher, a tractor, or a telephone system.

Types of Criteria

In general terms the criteria in which we will be interested fall into two broad classes, namely, those associated with systems (and components) and those of a human nature.

System Criteria   Basically, system criteria are those which relate to system performance or output, or in other words those which reflect something about the degree to which the system achieves its intended use. For example, a computer keyboard might be evaluated in terms of such criteria as number and accuracy of data entries made per unit of time, and an earth-moving vehicle might be evaluated in terms of the amount of earth moved per unit of time. Chapanis [4] lists other examples of system criteria, as: the anticipated life of a system; ease of operation or use; maintainability; reliability; operating cost; and manpower requirements. Some such criteria are rather strictly mechanistic, in the sense that they reflect essentially engineering performance (e.g., the maximum rpm of an engine), whereas others reflect more the performance of the system as it is used by the people involved in it (such as errors in cards punched).

Human Criteria   There are four relatively different types of human criteria, namely, human performance measures, physiological indices, subjective responses, and accident frequency. In a strict sense human performance must be considered in terms of various sensory, mental, and motor activities. In specific work situations, however, it is usually difficult, if not impossible, to measure human performance in strictly human activity terms, since such performance usually is inextricably intertwined with the performance characteristics of the physical equipment being used. Thus, the typing performance of a typist is not entirely a function of the typist but is also in part the consequence of the typewriter (its make, condition, etc.).

For some purposes indices of various physiological conditions are pertinent criteria. Such possible indices include heart rate, blood pressure, composition of the blood, galvanic skin response, brain waves, respiration rate, skin temperature, blood sugar, and many other measures. Some of these and other physiological variables are used as indices of the physiological effects on people of various methods of work, of work performed with equipment of various designs, of work periods, and of work performed under various environmental situations (such as heat and cold).

For some purposes the subjective responses of people can serve as appropriate criteria; examples are ratings of the performance of individuals, of alternative design features of a system, of the judged importance of different types of information for use in a system, and of the comfort of seats.

For still other purposes accident or injury frequency may serve as appropriate criteria. For example, the number of injuries or deaths per million miles traveled gives a comparison (in terms of this criterion) of various types of transportation systems, such as commercial airlines, railways, buses, and automobiles.

Discussion of Types of Criteria   It should be noted that some "human" criteria tend to reflect something about the systems involved, such as accident rates and ratings of comfort. Thus, the two classes of criteria do not form nice, neat dichotomies, but rather tend to comprise a continuum, ranging from strictly mechanistic system criteria at one end to strictly behavioral criteria at the other end. In the case of both system and human criteria, there are many specific types that can be used. Examples of various types will be given in the later chapters.

Requirements of Satisfactory Criteria

Criteria used in research investigations generally should fulfill three requirements, namely, relevance, freedom from contamination, and reliability.

Relevance   "Relevance" refers to the judged appropriateness of the criterion for the intended purpose. If, for example, the purpose at hand is that of evaluating the performance of a system or a component, one would agree with Chapanis [4] in his argument for use of appropriate system criteria. In this regard he states that it is difficult to see what some "experimental," essentially human, criteria (such as certain physiological indices, reaction time, etc.) have to do with more strictly "system" criteria. Although one must agree that such relationships sometimes are not at all manifest, there are human factors investigations in which distinctly human criteria could definitely be relevant, such as in comparing the physical energy costs of different work methods, or the ratings by secretaries of the comfort of different typing chairs.

Freedom From Contamination A criterion measure should not be influenced by variables that are extraneous to the variable that is being measured. Contamination would occur, for example, if the laboratory equipment for measuring oxygen consumption was faulty, thus producing erroneous values.

Reliability of Criterion Measure As the term "reliability" is used here, it refers to the *reliability* of the *measure* being used (as opposed to the concept of system or component reliability, to be discussed later). A measure is reliable to the extent that the criterion values will be about the same under similar circumstances, such as the heart rate measured twice under similar work conditions. If two "sets" of criterion values are available, reliability frequently is determined by computing a coefficient of correlation between the two.

PERFORMANCE RELIABILITY

Akin to the question of performance criteria (either system or behavioral) is the matter of performance reliability. Having just discussed reliability of criterion *measures*, let us hasten to distinguish between that use of the term "reliability" and the use of the term as a measure of the *dependability* of *performance* of the system or individual in carrying out an intended function. This use of the term stems from engineering practice[3] in which it refers to quantitative values that characterize the dependability of system or component performance. Actually, there are different measures of system reliability, each being relevant to certain types of systems or situations. However, most of these measures of reliability can be applied to human performance as well as to system performance. Certain variations on the reliability theme include: (1) probability of successful performance (this is especially applicable when the performance consists of discrete events, such as detecting a defect or starting a car); (2) mean time to failure (abbreviated MTF; there are several possible variations on this, but they all relate to the amount of time a system or individual performs successfully, either until failure or between failures; this index is most applicable to continuous types of activities). These are other variations that could also be mentioned. For our present discussion, let us consider reliability in terms of the probability of successful performance.

If a system includes two or more components (machine or human, or both), the reliability of the composite system will depend on the reliability of the individual components and how they are combined within the system. Components can be combined within a system either in *series* or in *parallel*.

Components in Series

In many systems the components are arranged in series (or sequence) in such a manner that successful performance of the total system depends upon success-

[3]For texts on reliability in engineering practice, the reader is referred to G. H. Sandler, *System reliability engineering*, Prentice-Hall, Inc., Englewood Cliffs, N. J., 1963, and W. G. Ireson, *Reliability handbook*, McGraw-Hill Book Company, New York, 1966.

ful performance of each and every component, man or machine. By taking some semantic liberties, we could assume components to be *in series* that may, in fact, be functioning concurrently and interdependently, such as a human operator using some type of equipment. In analyzing reliability data in such cases, two conditions must be fulfilled: (1) failure of any given component results in system failure, and (2) the component failures are independent of each other. When these assumptions are fulfilled, the reliability of the system for error-free operation would be the product of the reliabilities of the several components. To estimate system reliability quantitatively then, it is necessary to express the reliability of the individual components quantitatively, specifically as the probability of error-free performance.

If, for example, there are two components in a system, each of which has a reliability of .90 (meaning that it "works" 90 percent of the time), the reliability of the system would be the product of these two, or .81. The basic reliability formula for sequential situations as presented by Lusser [9], but with changed notations, is

$$R_{\text{system}} = R_1 \times R_2 \times R_3 \cdots R_n$$

in which R_1, R_2, $R_3 \cdots R_n$ are the reliabilities of the individual components, expressed as percentages of successful functioning.

It can thus be seen that each component *decreases* system reliability by its own factor. If a system consists of, say, 100 components, each with a reliability R of .99, the system reliability would be only .365. If the system consisted of 400 such components, the system reliability would be only .03, which indicates almost certain failure.

When successful system performance depends upon successful performance of two components in series, for example, a physical component and a human operator, we can estimate the system reliability from the reliabilities of the two components, as shown in Figure 2-1. That figure is equally applicable for two physical components, or two human beings, operating independently, when system performance depends on performance of both.

Components in Parallel

The reliability of a system in which the components are in parallel is entirely different from the situation in which they are in a series. With parallel components there are two or more which in some way are performing the same function. This is sometimes referred to as a "back-up" or "redundancy" arrangement—one in which one component backs up another, increasing the probability that the function will be performed. In such a case the system reliability is estimated by combining the probabilities of success (the reliabilities) of the several components, using the following formula [Gordon, 6]:

$$R_{\text{system}} = [1 - (1 - r)^m]^n$$

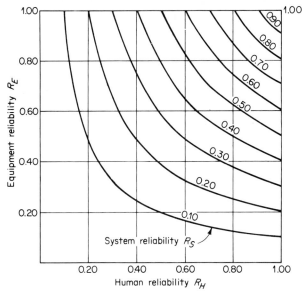

Figure 2-1 Effect of human and equipment reliabilities on system reliability. This same relationship applies to two independent physical components or human beings. [From Meister and Rabideau, 12, p. 16.]

where m = number of components in parallel for each function
n = number of functions
r = unit reliability

This formula applies only where the number of components in parallel for each of the n functions is the same and the unit reliability of each component is the same. Where these conditions do not apply, the derivation becomes somewhat more complex.[4]

In a strictly parallel arrangement, identical components would be used independently, either physical components or people as the case may be. If the reliabilities of individual components are, say, .90, the joint probability for two components (which would be the system reliability) would be .99. Even with relatively low component reliabilities such as .70, system reliability of

---

[4]Where these conditions do not exist, the reliability can be built up by an iterative technique, as follows: For the components in parallel, select the first. If the system included only that one, the system reliability R_s would equal that of the component R_1. The reliability of the system that included both components 1 and 2 (call it $R_{s(1,2)}$) would be as follows:

$$R_{s(1,2)} = R_1 + (1 - R_1)R_2$$

and, in turn,

$$R_{s(1,2,3)} = R_{s(1,2)} + (1 - R_{s(1,2)})R_3$$

and

$$R_{s(1,2,3,\ldots,n)} = R_{s(1,2,\ldots,n-1)} + (1 - R_{s(1,2,\ldots,n-1)})R_n$$

99.2 can be achieved with four parallel components. In some instances the parallel components can be of very different types, such as a man backing up a machine component.

Although physiological measures and subjective judgments have their appropriate and important places as criteria in human factors affairs, much of the raw data about human beings that should be used in this business deal with some aspect of human performance. For example, in various human factors contexts one might want to obtain answers to such questions as: What is the effect on speech intelligibility of various filters in a communication system? What are the effects on visual acuity of various types and levels of vibration? What is the probability of identification of a signal light at various angles from the operator? How many defects might occur in the task of soldering electric wires? The gamut of human performance information that might be required to answer these and scads of other questions ranges from very basic data about human beings (and their sensory, perceptual, mediation, and physical performance abilities) to data that are very task-oriented in an operational sense and are specific to a very limited range of situations (such as a soldering task). This broad spectrum of data has come to be referred to as *human performance data* and, at the next order of abstraction, *human performance theory*. Much of the illustrative data presented later will consist of an assortment of human performance data. Task-oriented performance data are particularly relevant to the above matter of system reliability and will warrant a bit of elaboration.

Human Performance of System Tasks

As human beings become involved in systems, their abilities and limitations are manifested in their performance of the tasks that are required. Considering the human being as an essential component of a system, for some purposes it may be necessary, as Meister [10] puts it, to measure his potentiality for failure (which, in behavioral terms, is error). This is, of course, a facet of the human reliability theme of a few paragraphs back. (Error is, of course, the complement of reliability.) If we wish to estimate the effect of human performance on system reliability, we need to do so in units of human performance that are system related, these units of behavior being *tasks*. The steps proposed by Swain [19] to do this are:

1 Define the system or subsystem failure which is to be evaluated.
2 Identify and list all the human operations performed and their relationships to system tasks and functions.
3 Predict error rates for each human operation or group of operations pertinent to the evaluation.

4 Determine the effect of human errors on the system.

5 Recommend changes necessary to reduce the system or subsystem failure rate as a consequence of the estimated effects of the recommended changes.

The particularly sticky wicket of these operations is item 3, the prediction of error rates for human operations (tasks), which would require something other than a Ouija board—it may be hoped, some reasonably quantitative data. Before discussing the collection of such data, however, let us scan the gamut of human errors in systems.

Types of Human Errors in System Tasks   People are quite inventive in the kinds of bloopers they perpetrate, but it can serve some purposes (such as trying to reduce their frequency or severity) to figure out what kinds of mistakes people do make. As a first step in this direction, classification may be helpful. Rook [18] developed one scheme, to be used in the classification of errors in an operating system, that provides for a cross classification of errors in terms of two bases, as illustrated below:

Conscious level or intent in performing act	Behavior component		
	Input I	Mediation M	Output O
A Intentional	AI	AM	AO
B Unintentional	BI	BM	BO
C Omission	CI	CM	CO

In this formulation the intentional act obviously does not refer to an intentional error, but rather to an act that the individual performed intentionally, thinking he was doing the right thing when in fact he was not (such as pushing the "door close" button in an elevator instead of the "door open" button, thereby clobbering an entering passenger). The input-mediation-output model of this classification system corresponds to a common sequence of psychological functions that is basic to all behavior, namely, S (stimulus), O (organism), R (response). As Meister [11] points out, human error occurs when any element in this chain of events is broken, such as failure to perceive a stimulus, inability to discriminate among various stimuli, misinterpretation of meaning of stimuli, not knowing what response to make to a particular stimulus, physical inability to make a required response, and responding out of sequence. Identifying the *source* of errors in terms of input-mediation-output behaviors is a first step in developing inklings about how to reduce the likelihood of errors.

Rook's classification system above applies primarily to discrete (i.e., individual, separate, distinguishable) acts. Some variations in classifying errors in discrete acts, as well as in certain other tasks, have been proposed by Altman

[1]. An abbreviated recap of his formulation is given below, with types of errors listed in each category:

Discrete acts	Continuous actions	Monitoring
Omissions	Failure to achieve end state in avail-	False detections
Insertions	able time	Failure to detect
Sequence	Displacement from target condition	
Unacceptable performance	over time	

Besides defining the nature of the task itself, it is sometimes useful, for analytical purposes, to classify the task in terms of its degree of *revocability* as proposed by Altman [2], such as: immediate correction; correction only after intervening steps; no correction within a given "mission" (e.g., within the present operation); and irrevocable consequences.

Collection of Task Data

The types of task information that may be useful in at least some system-development operations cover primarily such parameters as reliability and time, although in some instances other parameters may be relevant, such as task criticality. In general there are two kinds of sources of such information, namely, empirical observations and human estimates.

Empirical Task Data   Aside from sets of industrial engineering time-study data, there are two particular sources of essentially empirical task data that should be mentioned. One of these, the Data Store, was developed by The American Institute for Research [Munger, 13; Munger, Smith, and Payne, 14; and Payne and Altman, 15] particularly for tasks in the operation of electronic equipment. This Data Store was pulled together from many different sources, supplemented by special laboratory studies. Without going into the tale of how the data were developed, an example is given in Table 2-1 specifically related to the use of a joy stick. Note that the data include performance time and reliability in performance under the variations of the illustrated dimensions. In using the Data Store, the time that would be estimated for a given instance would consist of the base time (in this case 1.93 s) plus the times to be added for the particular dimension characteristics that would apply (such as 1.50 s for a joy stick of 24 in, plus 0.50 s if it is to be moved 50°, etc.).

The second source is referred to—in this day of acronyms—as SHERB (Sandia Human Error Rate Bank) [Rigby, 17], which is a compilation of error-rate data based on the THERP (Technique for Human Error Rate Prediction) [Swain, 19, and Rook, 18]. This body of data consists of human error rates for many industrial tasks based on large numbers of observations. Following are a few examples [Rook, 18]:

Type of error	Probability of error, P	Reliability, 1 − P
Two wires which can be transposed are transposed	.0006	.9994
A component is omitted	.00003	.99997
A component is wired backward	.001	.999

Judgmental Task Data   Where empirical data on tasks are not lying around loose, it may be useful to resort to human estimates about certain task parameters for use in a system development process. To illustrate such a process, in the estimation of the error rates of tasks, Irwin, Levitz, and Freed [7] asked experienced missile engineers, technicians, and mechanics to rate 60 tasks by sorting them into 10 piles ranging from those that would be performed with the least error to those with the greatest error. By a subsequent manipulation (using empirical reliability data that were available for some of the tasks and were extrapolated to others) it was possible to derive quantitative estimates of the reliability of all 60 tasks. A few illustrations are listed below:

Task	Mean rating, 10-point scale (10 = greatest error)	Derived reliability
Read time (brush recorder)	8.2	.9904
Install gasket	6.0	.9945
Inspect reducing adapter	4.9	.9958
Open hand valves	3.8	.9968
Remove drain tube	2.6	.9976

Although data on human reliability in performing tasks would be useful in

Table 2-1   Time and Reliability of Operation of Joy Sticks of Various Lengths When Moved Various Distances

Dimension	Time to be added to base time,* s	Reliability
Stick length, in		
6–9	1.50	.9963
12–18	0.00	.9967
21–27	1.50	.9963
Stick movement, degrees		
5–20	0.00	.9981
30–40	0.20	.9975
40–60	0.50	.9960

*Base time, 1.93 s.
Note: Other data (not shown here) are given for variations in other dimensions, specifically, control resistance, presence or absence of arm support, and time lag between movement of control and corresponding movement of display.
Source: Munger [13].

estimating total system reliability, it cannot be assumed that errors in performance of all tasks are necessarily equal in their effects on system performance.
(To draw from Gilbert and Sullivan, it seems doubtful that to have "polished up
the handle of the big front door" was a particularly critical task in the British
Admiralty system.) Thus, another task parameter might be criticality, and it is
this parameter that was used by Pickrel and McDonald [16] in one phase of
their elaboration of human performance reliability. Specifically, they elicited
ratings of task criticality, setting ranges of rating values for the following classification:

1   Safe (Error will not result in degradation, damage, hazard, or injury.)
2   Marginal
3   Critical
4   Catastrophic (Error will produce severe degradation—loss of system or
death or multiple deaths or injuries.)

Using such data as a springboard, these investigators urge that a concentration of efforts for failure reduction be made for those errors which are most
likely to occur and most likely to have the more serious consequences along the
"criticality" scale.

Discussion

The above discussion of task performance, reliability, and error simply opens
the door to a broad area of increasing importance to the whole human factors
domain.[5] This door opener may be relevant at this time, however, to alert the
reader to the importance of human performance and human error as they may
relate to later chapters.

REFERENCES

1   Altman, J. W.: Improvements needed in a central store of human performance data,
    Human Factors, 1964, vol. 6, no. 6, pp. 681–686.
2   Altman, J. W.: Classification of human error, in W. B. Askren (ed.), *Symposium on
    reliability of human performance in work*, AMRL, TR 67–88, May, 1967.
3   Askren, W. B. (ed.): *Symposium on reliability of human performance in work*,
    AMRL, TR 67–88, May, 1967.
4   Chapanis, A.: Plenary discussion: Relevance of physiological and psychological
    criteria to man-machine systems: The present state of the art, *Ergonomics*, 1970,
    vol. 13, no. 3, pp. 337–346.
5   Chapanis, A.: *Research techniques in human engineering*, The Johns Hopkins
    Press, Baltimore, 1959.
6   Gordon, R.: Optimum component redundancy for maximum system reliability,
    Operations Research, 1957, vol. 5, pp. 229–243.

[5]For additional discussion the reader is referred to other sources such as Altman [2],
Askren [3], Meister [10], Leuba [8], and Swain [19].

7   Irwin, I. A., J. J. Levitz, and A. M. Freed: "Human reliability in the performance of maintenance," in *Proceedings, Symposium on quantification of human performance, Aug.* 17–19, 1964, *Albuquerque, New Mexico,* pp. 143–198, M-5.7 Subcommittee on Human Factors, Electronic Industries Association.

8   Leuba, H. R.: Quantification in man-machine systems, *Human Factors,* 1964, vol. 6, no. 6, pp. 555–583.

9   Lusser, R.: The notorious unreliability of complex equipment. *Astronautics.* February, 1958, vol. 3, no. 2, pp. 26, 74–78.

10 Meister, D.: Methods of predicting human reliability in man-machine systems, *Human Factors,* 1964, vol. 6, no. 6, pp. 621–646.

11 Meister, D.: "Human factors in reliability," in W. G. Ireson, *Reliability handbook,* sec. 12, McGraw-Hill Book Company, New York, 1966.

12 Meister, D., and G. F. Rabideau: *Human factors evaluation in system development,* John Wiley & Sons, Inc., New York, 1965.

13 Munger, S. J.: *An index of electronic equipment operability: evaluation booklet,* The American Institute for Research, Pittsburgh, 1962.

14 Munger, S. J., R. W. Smith, and D. Payne: *An index of electronic equipment operability: data store,* The American Institute for Research, Pittsburgh, 1962.

15 Payne, D., and J. W. Altman: *An index of electronic equipment operability: report of development,* The American Institute for Research, Pittsburgh, 1962.

16 Pickrel, E. W., and T. A. McDonald: Quantification of human performance in large, complex systems, *Human Factors,* 1964, vol. 6, no. 6, pp. 647–663.

17 Rigby, L.V.: *The Sandia Human Error Rate Bank (SHERB),* paper presented at Symposium on Man-Machine Effectiveness Analysis: Techniques and Requirements, Human Factors Society, Los Angeles Chapter, Santa Monica, Calif., June 15, 1967.

18 Rook, L. W., Jr.: *Reduction of human error in industrial production,* Sandia Corporation, Albuquerque, N.M., SCTM 93–62(14), 1962.

19 Swain, A. D.: "THERP (Technique for Human Error Rate Prediction)," in *Proceedings, Symposium on quantification of human performance, Aug.* 17–19, 1964, *Albuquerque, New Mexico,* pp. 109–117, M-5.7 Subcommittee on Human Factors, Electronic Industries Association.

Information Input and
Mediation Processes

Chapter 3

Information Input and Processing

We mortals are continually being bombarded with stimuli from our environment, these stimuli consisting of various forms of energy to which our sense organs are sensitive. The interpretation of such stimuli (that is, the information they convey) is generally a function of our perceptual processes and of our learned associations (such as learning the alphabet). Our common notion of *information* is reflected by such everyday examples as what we read in the newspapers or hear on TV, the bill for automobile repairs, the gossip over the backyard fence, and the directions on road signs. However, we can view "information" in a much broader frame of reference as embracing the transfer of energy that has meaningful implications in any given situation, such as a driver "communicating" with his car via the control mechanisms; the sensing of air temperature by people or by thermometers; and mechanical, hydraulic, and servo linkages in various types of equipment. Thus, we can envision information flowing along two-way streets involving man, the various physical components with which he interacts, and the environment, as follows:

Man ↔ Man
Man ↔ Physical components
Man ↔ Environment
Physical components ↔ Physical components
Physical components ↔ Environment

SOURCES AND PATHWAYS OF STIMULI

The input to people in these interactions is of course the "information" received through the sense organs. Actually we do not receive information as such through the senses. Rather, our sensory mechanisms are sensitive to certain stimuli, which in turn convey meaning to us. The stimuli are various forms of energy, such as light, sound, heat, and mechanical pressure. In considering human information input and processing, however, let us first take an overview of the types of original sources of information (actually stimuli), the pathways of such information, and the variations in the form of the information that may occur between the original source and the receiver. Perhaps most typically the original source (the *distal* stimuli, if we want to use the long-haired term) is some object, event, or environmental condition. Information from these original sources may come to us *directly* (such as by *direct* observation of an airplane), or it may come to us *indirectly* through some intervening mechanism or device (such as a radar or telescope). In either case, the distal stimuli are sensed by the individual only through the energy that they generate (directly or indirectly) through *proximal* stimuli (light, sound, mechanical energy, etc.). In the case of *indirect* sensing, the *new* distal stimuli may be of two types. In the first place, they may be *coded* stimuli, such as visual or auditory displays. In the second place, they may be *reproduced* stimuli, such as those presented by TV, radio, or photographs or through such devices as microscopes, microfilm viewers, binoculars, and hearing aids; in such cases the reproduction may be intentionally or unintentionally modified in some way, as by enlargement, miniaturization, amplification, filtering, or enhancement. With either coded or reproduced stimuli, the new, or converted, stimuli become the actual distal stimuli to the human sensory receptors. Figure 3-1 illustrates in a schematic way these various pathways of information reception to the individual, both direct and indirect.

The human factors aspect of design enters into this process in those circumstances in which *indirect* sensing applies, for it is in these circumstances that the designer can design displays for presenting information to people. *Display* is a term that applies to virtually any man-made method of presenting information, such as a highway traffic sign, a family radio, or a page of Braille print. Human information input and processing operations depend, of course, upon the sensory reception of relevant stimuli (such as from displays), but they also depend upon perceptual and learning processes. In other words, it is necessary that sensory stimuli be correctly recognized by the receiver and that their "meaning" be understood. If the stimulus is a code symbol of some type (such as a red light), it is necessary that its meaning be understood (in this case, to stop).

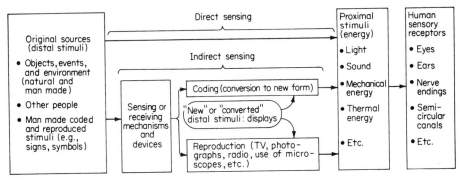

Figure 3-1 Schematic illustration of pathways of information from original sources to sensory receptors. (Although typically the original, basic source is an object, event, condition, the environment, etc., in some situations the effective original source to an individual consists of some man-made coded or reproduced stimuli; office personnel, for example, usually deal with recorded symbols which, for practical purposes, are their "original" distal stimuli.)

In the design of some systems the circumstance may require a determination as to whether a given information input or processing function can best be performed by an individual or by some physical component of a system.

This chapter will deal with certain human factors that may be relevant to the design considerations that relate to human information-receiving and processing functions, but before discussing these matters let us sidetrack ourselves to discuss briefly the topic of information theory.

INFORMATION THEORY

Although the relevance of information theory to human factors is rather nominal, at least a brief exposure to information theory is in order since some of the human factors research does involve reference to it, especially to the measurement of information that is central to it, namely, the *bit*. (This term is a boiled-down version of *binary unit*.) The bit has been defined as the amount of information we obtain when one of two equally likely alternatives is specified [Abramson, 1, p. 12]. When in fact the various alternatives are equally probable, the total amount of information (usually symbolized by the letter *H*) is derived from the following formula:

$$H = \log_2 n$$

where n is the number of equally probable alternatives. This, in turn, can be expressed in terms of the probabilities of each alternative, that probability being the reciprocal of n. Thus,

$$H = \log_2 \frac{1}{p}$$

where p is the probability of each such alternative.

When the probabilities of various alternatives are equal, the amount of information in bits is measured by the logarithm, to the base 2, of the number of such alternatives. With only two alternatives, the information, in bits, is equal to the logarithm of 2 to the base 2, which is then 1. When Paul Revere was to receive a signal from the Old North Church, he was to see 1 signal (1 lantern) if the enemy came by land and 2 if by sea. Assuming that these two alternatives were equally probable, the amount of information available would be 1 bit; a discrimination was to be made between 2 alternatives.

Let us now take four alternatives, such as four lights on a panel, only one of which may be on at a time. In this case we should have 2 bits of information ($\log_2 4 = 2$). If we had eight such lights, we should have 3 bits of information ($\log_2 8 = 3$), etc.

If we were playing Twenty Questions, it would be possible to determine the correct answer out of 1,048,576 possible alternatives if the questions were properly framed and if all the information were used. The amount of information that would be so obtained with 20 questions would be 20 bits ($\log_2 1,048,576 = 20$). When the probabilities of the individual events are not equal, the computation of the bit is more complicated, but the basic concept is still applicable.

INFORMATION INPUT PROCESSES

If we view "information" in a broad frame of reference, it becomes manifest that there are many forms of human involvement in different facets of life in receiving, storing, retrieving, processing, and transmitting information. These intertwined functions lead up to virtually every human activity. In view of this, let us consider the nature of the human sensorimotor system to see how information "flow" is related to this.

The Human Sensorimotor System

In many types of human activities the physical responses people make have (or at least should have) a direct and clear-cut relationship to some input, as when dialing a phone number (the input being the number in the directory) or typing from prepared copy. In more complex tasks, however, there may be more information processing (including making judgments and decisions) between the information input stage and the actual response, as in driving an automobile in heavy traffic. Granting such variations in complexity, however, the same generalized representation of the various intervening processes applies. One such representation is that shown in Figure 3-2 [Welford, 33]. This hypothetical block diagram shows the functions of: sensing by the sense organs; perception; short-term store (memory); the translation from perception to action (the basis of response); control of response; the action of the effectors; and related functions of long-term store (memory) and certain feedback loops. Needless to say, these various functions cannot be neatly differentiated one from another, but this may serve as a representation of the ensemble of blending functions.

Although Figure 3-2 is a reasonably straightforward representaton of the

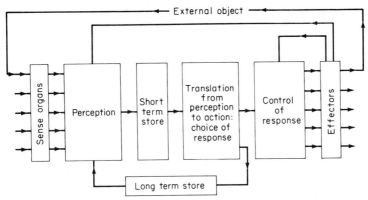

Figure 3-2 Representation of the human sensorimotor system. One can view this system as representing the sequence of information input, processing, and output on the part of the human being. [From Welford, 33, p. 6, in Welford, A. T., and Birren, J.E., *Behavior, ageing and the nervous system,* 1965. Courtesy Charles C. Thomas, Publisher, Springfield, Illinois.]

various "stages" of information processing when there is only one source of information input, one would wonder what happens when several sensory inputs occur at or about the same time. In this regard Broadbent [6, p. 297] poses the theory that the nervous system acts to some extent as a single communication channel, thus having a limited capacity. Further, he suggests that the individual makes some "selection" from all the sensory inputs he receives, this selection being based on some combination of the nature of the stimuli (e.g., their intensity) and the state of the individual (e.g., his drive). Although this theory implies that we give our attention to only one aspect of our environment at a time, this does not preclude rapid shifting of attention to, or alternation of attention between, two or more aspects.

In all of this, it is apparent that the bottleneck is the cortex of the brain, not the sensory mechanisms. We can still "sense" the tremendous variety of stimuli in our environment, such as traffic patterns, landscapes, TV scenes, and football games, even though—at any one moment—we "tune in" only one aspect. In this connection, Steinbuch [27] has summarized the information reduction that occurs from the initial reception by the sense organs through the intermediate processes to permanent storage (memory) and presents the following estimates:

Process	Maximum flow of information, bits/s
Sensory reception	1,000,000,000
Nerve connections	3,000,000
Consciousness	16
Permanent storage	0.7

He postulates certain as yet unexplained intermediate reduction processes between the neural connections of the sensory organs and the conscious perception of the stimuli. Granting that the above estimates are rough, we can indeed see that the central-nervous-system processes of consciousness and storage are capable of handling only a fraction of the potentially tremendous information input to the sensory receptors.

Stimulus Dimensions

The stimulus inputs that people receive from their environment via any sensory modality (vision, audition, etc.) differ in terms of their characteristics. For example, we make visual discriminations in terms of shape, configuration, size, position, color, etc., and auditory discriminations in terms of frequency, intensity, etc. The "natural" environment we sense is of course very complex. But as information is presented to people via displays, the nature of the stimuli used usually is simplified, typically consisting of variations in a given "class" of stimuli which can be considered as a *stimulus dimension*. Thus, the positions of the hands of a clock are used to tell time (especially when we cannot read the numbers); the number and time spacing of Morse code signals represent letters; and the shapes and designs of road signs have different meanings. The utility of any given stimulus dimension to convey information, however, depends upon the ability of people to make the sensory and perceptual distinctions that are required in differentiating one stimulus of a given class from another (such as telling one color from another). Such discriminations, however, usually have to be made on an *absolute* basis rather than on a *relative* basis. A relative judgment is one which is made when there is an opportunity to compare two or more stimuli; thus, one might compare two or more sounds in terms of loudness or two or more lights in terms of brightness. In absolute judgments, there is no opportunity to make comparisons, such as identifying a given note on the piano (say, middle C) without being able to compare it with any others, or identifying a given color out of several possible colors when it is presented by itself.

As one might expect, people are generally able to make fewer discriminations on an absolute basis than on a relative basis. For example, it has been estimated that most people can differentiate as many as 100,000 to 300,000 different colors on a relative basis when comparing two at a time (taking into account variations in hue, lightness, and saturation). On the other hand, the number of colors that can be identified on an absolute basis is limited to no more than a dozen or two.

Absolute judgments can be required in either of two types of circumstances. In the first place, several discrete positions (levels or values) along a stimulus dimension might be used as codes, each position representing a different item of information. If the stimuli consist of tones of different frequencies, the receiver is supposed to identify the *particular* tone. In the second place, the stimulus may be of any value along the stimulus dimension, and the individual needs to make some judgment regarding its value or position along the dimension. Such judgments might be used in various operational ways. A radio operator, for example, might use his judgment of loudness of a radio signal to adjust the gain (volume) up or down to some subjective standard. On the other hand an inspector might use his judgment to classify all items as either "pass" or "fail," or possibly by grades, such as A, B, and C.

The ability of people to make absolute discriminations among individual stimuli of most stimulus dimensions is really not very large, being generally in the range from 4 to 9 or 10, with corresponding bits from about 2.0 to 3.0 or 3.4, as illustrated by the following examples:

Stimulus dimension	Average number of discriminations	Number of bits
Pure tones	5	2.3
Loudness	4–5	2 to 2.3
Size of viewed objects	5–7	2.3 to 2.8
Brightness	3–5	1.7 to 2.3

In this connection Miller [33] refers to the "magical number seven, plus or minus two," meaning that the range of such discriminations is somewhere around 7 ± 2 (5 to 9); for some dimensions the number is greater, and for some less, than this specific range. Seven discriminations would transmit 2.8 bits. With *combinations* of dimensions, however, the information that can be transmitted sometimes is noticeably greater.

Information in Human Memory

We saw above that the amount of information that can enter "permanent storage" (i.e., memory) per unit of time is very modest, being about 0.7 bits/s. However, the amount that can be retained in such storage is tremendous, with there being marked individual differences. While the specific processes of learning and storage are not entirely understood, it is generally recognized that they are based on certain changes within the nerve cells (the neurons) of the brain. It has been estimated that there are something like 10 billion of these. Assuming the efficient utilization thereof, it has been estimated that the overall storage capacity of the human memory is somewhere between 100 million and 1 million billion bits [Geyer and Johnson, 16]. This range, of course, is far greater than the storage capacity of any computer now in existence or likely to be developed within any reasonable time. To extend the analogy between human information storage and that of computers, it has been pointed out that storage components of electronic systems are of two types, namely, static (consisting of special patterns of binary data unchanging in time) and dynamic (information in the form of electrical or mechanical impulses). In turn, there is some evidence that the brain also combines both of these schemes [Geyer and Johnson, 16]. One of the schemes is provision for storage of "old" information; the other has been described as a "circulatory" conception, accounting for the recording of current or recent information (short-term memory).

Information in Human Responses

Following the previous discussion of information, we can view human responses as "conveying" information. This notion is most clear-cut in instances in which the output responses are intended to correspond with input stimuli, as in the case of a keypunch operator (who is to "reproduce" the input information by pressing corresponding keys), or in the case of a lathe operator (who is to "reproduce" in the wood or metal being turned the design given on a blueprint). The efficiency with which people can "transmit" information through their responses depends upon the nature of the initial information input and upon the type of response that is required. As an example of the influence of the type of response on the amount of information that can be transmitted, in one study a comparison was made of the verbal versus motor (i.e., key-pressing) responses of subjects who were presented with Arabic numerals (2, 4, or 8 under a number of experimental conditions) at specified rates (1, 2, or 3 per second) and were asked to repeat them *verbally* or to respond by *pressing keys* which corresponded to the numerals to be presented [Alluisi, Muller, and Fitts, 2]. The maximum rate for verbal responses was 7.9 bits per second and for the motor (key-pressing) responses, 2.8 bits per second. In a somewhat more complex task, an analysis was made of the information of placing pegs in holes with varying degrees of tolerance (the difference in diameter of the peg and of the hole) and varying amplitudes (distance of movement) [Fitts, 13]. Information in bits was measured by a special formula, the number of bits for the various experimental conditions ranging from 3 to 10.

Although the amount of information that can be "transmitted" by people through their physical responses thus varies with the situation, it has been estimated that a reasonable ceiling value is of the order of about 10 bits/s [Singleton, 26].

DISPLAYS USED IN INFORMATION INPUT

There are of course many sources of information (i.e., distal stimuli) that people can sense directly without any problem. There are, however, many circumstances in which the information that people need in performing some activity can best be presented indirectly by the use of some type of display. Some such circumstances are given below:

1 When distal stimuli are of the type that humans generally can sense, but cannot sense adequately under the circumstances because of such factors as:
   a Stimuli at or below threshold values (e.g., too far, too small, or not sufficiently intense) that need to be amplified by electronic, optical, or other means
   b Stimuli that require reduction for adequate sensing (e.g., large land areas converted to maps)
   c Stimuli embedded in excessive noise that generally need to be filtered or amplified

   d   Stimuli far beyond human sensing limits that have to be converted to another form of energy for transmission (e.g., by radio and TV) and then reconverted to the original form or converted to another form
   e   Stimuli that need to be sensed with greater precision than people can discriminate (e.g., temperature, weights and measures, and sound)
   f   Stimuli that need to be stored for future reference (e.g., by photograph and tape recorder)
   g   Stimuli of one type that probably can be sensed better or more conveniently if converted to another type in either the same sensory modality (e.g., graphs to represent quantitative data) or a different modality (e.g., auditory warning devices)
   h   Information about events or circumstances that by their nature virtually require some display presentations (such as emergencies, road signs, and hazardous conditions)
   2   When distal stimuli are of the type that humans generally cannot sense or that are beyond the spectrum to which humans are sensitive, and so have to be sensed by sensing devices and converted into coded form for human reception (e.g., certain forms of electromagnetic energy and ultrasonic vibrations)

   In these and other types of circumstances, it may be appropriate to transmit relevant information (stimuli) *indirectly* by some type of display. For our purposes we shall consider a display to be any method of presenting information indirectly, in either *reproduced* or *coded* (symbolic) form. If a decision is made to use a display, there may be some option regarding the sensory modality and the specific type of display to use, since the method of presenting information can influence, for better or worse, the accuracy and speed with which information can be received.

Types of Displays

Displays can be described as either *dynamic* or *static*. Dynamic displays are those that continually change or are subject to change through time, and include the following types: displays that depict the status or condition of some variable, such as temperature and pressure gauges, speedometers, and altimeters; certain CRT displays such as radar, sonar, TV, and radio-range signal transmitters; displays that present intentionally transmitted information, such as record players, TV, and movies; and those that are intended to aid the user in the control or setting of some variable, such as the temperature control of an oven. (It might be observed, incidentally, that there are some devices that do double duty as both displays and controls; this is especially the case with devices used for making settings, such as oven controls.) Static displays, in turn, are those that remain fixed over time, such as signs, charts, graphs, labels, and various forms of printed or written material.

Types of Information Presented by Displays

Some of the major types of information presented by displays are described below. Although most displays can be considered as falling into categories that parallel these, there is not a perfect one-to-one relationship because certain specific displays present two or more "types" of information, or because the user "uses" the displayed information only for certain purposes.

- *Quantitative information:* Display presentations which reflect the quantitative value of some variable, such as temperature or speed. Although in most instances the variable is dynamic, some such information may be static (such as that presented in monographs and tables).
- *Qualitative information:* Display presentations which reflect the approximate value, trend, rate of change, direction of change, or other aspect of some changeable variable. Such information usually is predicated on some quantitative parameter, but the displayed presentation is "used" more as an indication of the change in the parameter than for obtaining a quantitative value as such.
- *Status information:* Display presentations which reflect the condition or status of a system, such as: on-off indications; indications of one of a limited number of conditions, such as stop-caution-go lights; and indications of independent conditions of some class, such as a TV channel.
- *Warning and signal information:* Display presentation used to indicate emergency or unsafe conditions or to indicate the presence or absence of some object or condition (such as aircraft or lighthouse beacons). Displayed information of this type can be static or dynamic.
- *Representational information:* Pictorial or graphic representations of objects, areas, or other configurations. Certain displays may present dynamic images (such as TV or movies) or may present symbolic representations (such as heartbeats shown on an oscilloscope or blips on a cathode-ray tube). Others may present static information (such as photographs, maps, charts, diagrams, blueprints, and graphic representations such as bar graphs and line graphs).
- *Identification information:* Display presentations used to identify some (usually) static condition, situation, or object, such as the identification of hazards, traffic lanes, and color-coded pipes. The identification usually is in coded form.
- *Alphanumeric and symbolic information:* Display presentations of verbal, numerical, and related coded information in many forms, such as signs, labels, placards, instructions, music notes, printed and typed material including Braille, and computer printouts. Such information usually is static, but in certain circumstances it may be dynamic, as in the case of news bulletins displayed by moving lights on a building.
- *Time-phased information:* Display presentations of pulsed or time-phased signals, e.g., signals that are controlled in terms of duration of the signals and of intersignal intervals, and of their combinations, such as the Morse code and blinker lights.

The kinds of displays that would be preferable for presenting certain types of information are virtually specified by the nature of the information in question, but for presenting most types of information there are options regarding the kinds of displays and certainly about the specific features of the displays. Examples of some visual and auditory displays will be given in the next couple of chapters, along with examples of some of the research that has been carried out regarding the design of displays. Since the coding of stimuli has rather general implications in connection with information input and processing, however, some aspects of coding will be covered in this chapter.

Coding of Sensory Inputs

Many displays involve the coding of information (actually, of stimuli), as opposed to some form of direct representation or "reproduction" of the total stimulus. Any such coding implies first the selection of the particular sensory modality, and second the selection of the particular stimulus (or coding) dimension to use within that sensory modality.

Selection of the Sensory Modality   In selecting or designing displays for transmission of information in some situations, the selection of the sensory modality (and even the stimulus dimension) is virtually a foregone conclusion, as in the use of vision for road signs and the use of audition for many various purposes. Where there is some option, however, the intrinsic advantages of one over the other may depend upon any of a number of considerations, such as those given in Table 3-1. The advantages of one over the other, as given in the

Table 3-1   When to Use the Auditory or Visual Form of Presentation

Use auditory presentation if:	Use visual presentation if:
1 The message is simple.	1 The message is complex.
2 The message is short.	2 The message is long.
3 The message will not be referred to later.	3 The message will be referred to later.
4 The message deals with events in time.	4 The message deals with location in space.
5 The message calls for immediate action.	5 The message does not call for immediate action.
6 The visual system of the person is overburdened.	6 The auditory system of the person is overburdened.
7 The receiving location is too bright or dark-adaptation integrity is necessary.	7 The receiving location is too noisy.
8 The person's job requires him to move about continually.	8 The person's job allows him to remain in one position.

Source: Deatherage [11, Table 4-1], p. 124.

table, are based on considerations of substantial amounts of research and experience relating to these two sensory modalities.

Selection of the Stimulus Dimension   When the stimulus dimension is not clearly dictated by the nature of the situation at hand, the selection can be based on the relative pros and cons associated with the various dimensions that logically could be considered. Discussion of some of these will be included in the following chapters. In considering various alternatives, however, it should be pointed out that sometimes two or more stimulus dimensions can be used in combination. Most codes are undimensional, which means that a given code symbol has a given meaning, such as red, yellow, and green colors of traffic lights, the unique "You're out!" arm signal of the baseball umpire, or the $, c, ?, &, =, and ! signs of the printed page. On the other hand, in some contexts two or more coding dimensions are used in combination, such combinations varying in the degree of *redundancy*. A completely redundant code is one in which each thing to be identified has two (or more) unique attributes (such as a unique color *and* a unique shape). In turn, a completely nonredundant code is one in which each and every unique combination of specific stimuli of each of two (or more) stimulus dimensions represents a specific item. For example, each combination of a color *and* a shape might be used to code items in a warehouse. In some circumstances a partially redundant coding scheme might be used.

General Guidelines in the Use of Coding Systems

Generalizations in an area such as the use of codes are pretty treacherous, especially since this ball park is a very complicated one. (We will see later a few instances of apparent inconsistencies in research findings.) Further, in any given situation it may be necessary to effect a trade-off of one advantage for another. There may be, however, a few guidelines that clearly stick out, and others that might be teased out, of the available evidence, that may have some generality.

Detectability of Codes   To begin with, any stimulus used in coding information needs to be detectable; specifically, it has to be of such a nature that it can be sensed by the sensing mechanism in the situation at hand. In the lingo of the psychologists this is an "absolute threshold." But "absolute" thresholds vary from individual to individual, and even for the same individual at various times, and vary with the situation.

Discriminability of Codes   Further, every code symbol, even though detectable by the sensory mechanism, needs to be discriminable from other code symbols. This may be a problem where differences along some stimulus dimension are to be used, such as sound frequency and brightness. As discussed above, stimulus dimensions vary in the number of levels that can be identified on an absolute basis. In this connection, the *degree* of difference between adjacent stimuli may have some influence on the effectiveness of a coding system, even if all differences used are reasonably discriminable. The results of one study will illustrate this [Mudd, 23]. In that study three different levels of interstimulus difference were used with each of the four auditory cueing dimensions, the differences between adjacent stimuli being varied. In the case of certain dimensions—especially intensity—the mean response times were related to the magnitude of the differences between stimuli as follows: largest stimulus difference: 4.6 s; average difference: 5.7 s; and smallest difference: 6.8 s. The effectiveness of such "separation" of specific stimuli of some class, however, presumably is a function of the nature of the stimulus dimension; in this study, for example, increasing the difference between frequency stimuli did not reduce response time appreciably.

Compatibility of Codes *Compatibility* is a very generalized concept that has substantial applicability to human factors engineering. For our general purposes it might be defined thus: Compatibility refers to the spatial, movement, or conceptual relationships of stimuli and of responses, individually or in combination, which are consistent with human expectations. This concept is most straightforward in the case of stimulus-response (S-R) compatibility, but the concept of compatibility is also applicable where there is simply information transfer (as in the use of certain coding systems) in the absence of any corresponding physical response. In this connection, the degree of compatibility of coding systems generally tends to be maximum when the task at hand requires a minimum amount of information transformation—encoding or decoding. Thus, we should take advantage of associations that are already built into the repertoire of people; these may be either natural or learned associations. To refer again to traffic lights, if someone were foolish enough to change the system from red, yellow, and green to, say, violet, chartreuse, and azure, we would really be loused up—*because* of the need to recode from one system to another, at least until we learned the new one.

Meaningfulness of Codes The use of any code implies that its meaningfulness to the user should be manifest. This can be predicated on either of two bases: (1) the fact that the code is actually symbolic of that which it represents (such as a road sign showing a curve rather than saying "turn right"); and (2) the learning of the association between the display and what it represents.

Standardization of Codes When coding systems are to be used by different people in different situations (as in the case of traffic signs), their standardization will facilitate their use when people shift from one situation to another.

Use of Multidimensional Codes In very general terms, the use of two or more coding dimensions in combination tends to facilitate the transfer of information to human beings, probably especially so where there is complete redundancy, and perhaps less so with partial redundancy.

Discussion of Coding System

As indicated before, any coding system is based on a particular stimulus dimension. Chapters 4 and 5 will include, respectively, discussions of visual codes and of auditory and tactual codes, along with examples.

Organization of Information in Displays

As pointed out earlier, a display does not transmit information as such, but rather presents stimuli which may be meaningful to the receiver. Thus, when we discuss the *organization* of information, this is really a euphemistic way of referring to the organization of stimuli—such as their temporal characteristics and their number and type. It should be noted here, again, that human abilities to deal with a barrage of stimulus inputs depend very much on perceptual and mediation processes rather than exclusively on sensory processes.

Speed and Load of Stimuli   *Load* refers to the variety of stimuli (in type and number) to which the receiver must attend. Thus, if an individual has several different types of visual instruments, or several of the same type, to which he must give his attention, the load on the visual system will be greater than if there were fewer. On the other hand, *speed*, when used in this context, relates to the number of stimuli per unit of time or, conversely, to the time available per stimulus. Incidentally, one could contemplate speed and load for a single sensory modality (such as vision or hearing) or for combinations of the sensory modalities.

The effects of load and speed have been investigated in a number of experiments. In one study, for example, Mackworth and Mackworth [18] had a visual search task which utilized a display panel with 3 clocks at the top and 50 columns with numbers in certain positions of each. The clocks presented a three-digit number that increased its value one number at a time, this change being identified by a loud click. The chore of the subject was a rather complicated one that need not be described in detail, but basically consisted of a continuous comparison of the changing number of the clocks with the numbers in the columns; he then reported the column (i.e., A, B, C, etc.) in which the most nearly corresponding number appeared. Load was varied by controlling the number of columns to be used (5, 10, 15, 20, 30, 40, or 50). Speed was controlled by the rate at which the clocks kept changing. Subjects were scored on the number of errors made.

Some of the results are shown in Figure 3-3. In particular, this showed for

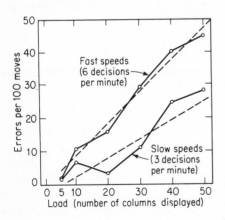

Figure 3-3 Effects of increasing load (number of channels used) upon errors. These data come from the study using a panel of various numbers of columns of comparison numbers. The fast and slow speeds were, respectively, 6 and 3 decisions per minute. [From Mackworth and Mackworth, 18.]

fast speeds (6 decisions per minute) and slow speeds (3 decisions) the relationship between load and errors. The dotted lines (which reflect the general relationships) indicate that errors increase with load (even though the number of decisions per minute remains constant) and that speed also is related to errors.

Interestingly enough, in studies of load and speed it has been found that the arithmetic product of these two variables typically results in a linear relationship with performance. This was illustrated, for example, by one of a series of studies by Conrad [8], with a clock-watching task in which the subjects were to press a key as a pointer approached 12 or 6 o'clock on any of the clock dials used. In the experiment in question, under different circumstances two, three, or four dials were used, and speed was varied. The relationship between the product of speed and load on the one hand, and omissions on the other, is shown in Figure 3-4. It can be seen that this relationship is essentially linear over the primary range of speed times load.

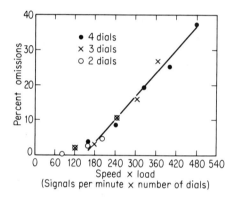

Figure 3-4 Relationship between speed × load and "omissions" in a clock-watching study. The load is the number of sources (in this case two, three, or four clock dials). The relationship is essentially linear over the primary range of speed × load values. [From data from Conrad, 8, as adapted by Mackworth and Mackworth, 18.]

Speed Stress and Load Stress The patterns of results illustrated above have been confirmed in a number of studies. Such results have led Conrad to postulate the existence of *speed stress* and of *load stress* in situations such as these [9]. He suggests that *speed stress* is essentially a reaction on the part of a person working on a task that has the effect of worsening his performance beyond what might be expected from the physical characteristics of the display. *Load stress,* on the other hand, changes the character of the task. As the number of signal sources (visual displays) is increased, more time is needed to make judgments simply because of the greater scanning coverage required.

Time-Phasing of Signals When stimuli are presented briefly and in close temporal sequence, perceptual failure may occur because of speed or load stress. Some of the aspects of timing of auditory signals have been investigated by Conrad in a series of studies using the multidial displays mentioned earlier. In one of these studies [Conrad, 10], for example, four of his dials were used, signal speed being varied from 40 to 160 signals per minute. A detailed analysis was made of the number of occasions that a response was late (pressing a key corresponding to the dial *after* the pointer had passed the 12 or 6 o'clock positions). Further, this analysis was made in relation to the timing of adjacent stimuli and responses. Considering two stimuli and their responses, one could envision this usual sequence of events (S = stimulus; R = response):

$$S_1, R_1, S_2, R_2$$

An analysis of the timing of the intervals between these events indicated that if stimuli are close together, or if they come in bunches, the responses to them frequently are missed, delayed, or otherwise affected. While absolute limits of minimum desirable interstimulus intervals are a bit hard to come by, it has been pointed out that where such an interval is shorter than, say, 0.5 s, the stimuli are likely to be confused; in fact, an individual may then respond to the two as though they were one.

On the basis of one phase of his study Conrad [10] makes the point that, when individuals have some control over the timing of input signals (i.e., self-pacing), their task performance tends to be better than when the timing of signals is entirely outside their control.

Discussion   The implication for equipment design of some of the research relating to speed, load, and timing of frequently occurring visual signals is relatively clear. Where visual signals would likely occur fairly rapidly, it would seem desirable (where possible) to provide that the information be presented over a limited number of sources (channels) rather than over many. In addition, the rate of presentation should be within acceptable bounds as far as human response performance is concerned. Further, where the spacing of signals to be presented can be controlled, one should avoid short intersignal intervals, the bunching of signals, and short intervals between signals following previous responses. Where feasible, it seems desirable to permit the individual to control the rate of signal input.

Multiple-Sensory Inputs

The above discussion of load, speed and related variables generally was restricted to frequently occurring visual signals of some general class. Sometimes, however, workers are subject to several inputs through the same sensory channel and to inputs through the various senses, especially vision and hearing. The relationships of the multiple inputs can be of several kinds, including the following: (1) *time-sharing* (potentially meaningful and relevant information may be coming from different sources through the same sensory channel or through two or even more sensory channels); (2) *use of redundant sensory channels* (two or more senses may be used to transmit identical or supporting information, usually simultaneously, or at least in very close temporal sequence); and (3) *noise* ("noise" generally refers to some irrelevant, usually undesirable, stimuli; although we usually think of noise as auditory in nature, the term is also applied to other irrelevant stimuli such as *visual* noise, or clutter).

There have been many studies of the effects of different inputs into the same sensory channel and of inputs from one sensory modality upon inputs from another. It is not feasible here to discuss these effects extensively, although a few points may be made.[1]

Time-Sharing

"Time-sharing" refers to situations in which the human being has two or more chores to which he has to alternate his attention. In a strict sense an individual cannot give simultaneous attention to two or more aspects of a situation. In performing various functions simultaneously, such as steering a car, controlling the accelerator, and keeping our eyes peeled on the traffic, we actually keep shifting our conscious attention from one to another, sometimes very rapidly. Time-sharing can take many forms, such as receiving two or more sensory inputs, taking various physical actions, or combinations of these. Where these demands press the limits of individuals, load or speed stress, or both, may occur. In effect, then, time-sharing can induce a form of load and speed stress.

Time-Sharing of Visual Inputs   The investigations of speed and load discussed above dealt with relatively homogeneous tasks and therefore did not involve time-sharing. Many activities of the real world, however, do not conform to the pristine patterns of many laboratory tasks. This disparity led Weisz and McElroy [32] to study the effects of task-induced stress (really speed stress) on actually time-shared tasks that would be more representative of human operator activities in systems, including tasks that required storage and integration of information over time (which most laboratory tasks in this domain have not required). Since the details of their experimental procedures might serve as an overload to the reader, we shall try to boil the study down to its major features. The experiment involved five different response tasks to be carried out in response to different stimuli presented simultaneously by frames with a CRT. The range of stimuli consisted of variations of each of four geometric forms (i.e., rectangle, trapezoid, triangle, and parallelogram); the stimuli were generated on a CRT by the use of a computer, and appeared in the quadrants of the CRT. Each form had seven variations, such as very tall, thin rectangles ranging down to low, wide rectangles. The five types of tasks (events) to be responded to were:

- *Mean*: Estimate of the *population mean* of rectangles in upper right quadrant, with the seven variations in rectangles considered as forming a scale. Every frame.
- *Form*: Identification of *form value* (i.e., the specific variation) of trapezoid in lower right quadrant. Every frame.
- *Pair*: Detection of pairs (and triplets) of identical forms in frame.

---

[1] For a more extensive discussion of this topic the reader is referred to Mowbray and Gebhard [21].

• *Run*: Remembering of location of pairs of stimuli and checking for appearance of a similar pair on the next frame in the same quadrant location (short-term memory required).

• *Line*: Searching the lower two quadrants for *extreme* variations of trapezoid or triangle in dotted-line form (rather than solid line).

These tasks were time-shared, and after a breaking-in period the subjects performed the composite tasks for a series of 18 trials (9 on each of two days), these being varied at three rates, namely, 10 s per frame, 7 s per frame, and 4 or 5 s per frame. The subjects recorded their responses by the use of special push buttons. Performance was measured by errors, including a composite (total) of errors in all tasks. Let us now see what gleanings there are from this study. Figure 3-5 shows the percent of all errors (*a*) and the percent of omissions (*b*) for

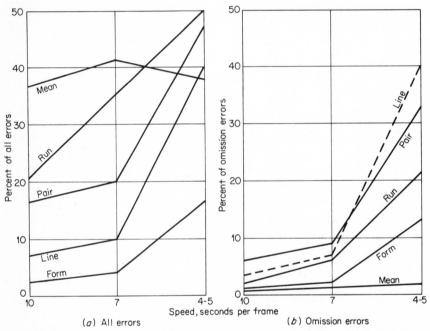

(*a*) All errors

Speed, seconds per frame

(*b*) Omission errors

Figure 3-5 Percent of all errors *a* and omission errors *b* in tasks as a function of speed (in seconds per frame as presented to subjects). See text for description of tasks. [From Weisz and McElroy, 32.]

each of the tasks. Certain points stand out from these figures. Obviously, the highest level of speed stress was accompanied by an increase in total error. (Although not shown in the figure, the total composite error index, as computed, increased from 18.4 to 22.4 and 32.2 percent.) But the interesting thing is that percent of all errors (*a*) on the *mean* task, despite the difficulty of the task, did not increase like the others, and the *form* task did not increase as much as the others. These two tasks also had the lowest percent of omission indexes (*b*), especially the *mean* task, its omission rate being virtually unaffected by speed stress.

Some speculations about the results by the investigators may cast some light on the effects of speed stress on time-shared visual-input tasks. To begin with, it should be noted that the *mean* and *form* tasks required a response for each frame, but the others did not; the others involved some temporal or spatial uncertainty which required search of the display to identify significant events, and were more "probabilistic" in the sense that an event calling for a response did *not* occur in every frame. Further, the *run* task (the one that suffered most from speed stress) also depended on short-term memory from one frame to the next. Thus, it appears that in time-sharing of tasks such as these, speed stress does not affect all tasks equally, but rather affects particularly the tasks with greater uncertainty and those which depend on short-term memory.

Another relevant study is that conducted by Olson [24]. The experimental task was sharing visual attention between two tasks. One of these was a tracking task, consisting of a moving "road" that was curved and a wheel that controlled a pointer on the road; the object was to keep the pointer on the road. The other task was the identification of dial pointers (on anywhere from 6 to 18 dials) which deviated randomly from a neutral position. While the primary concern was more with other variables (dial arrangement, load, speed, etc.), one aspect of the study was related to the present topic of time-sharing. Performance on the two tasks was correlated only to the extent of .20, which suggests the hypothesis that individuals tended to adopt their own priority strategies of giving primary attention to one task or the other. Whether the priority strategies are fairly common to most subjects because of the intrinsic nature of the task (as in the case of the study by Weisz and McElroy) or self-selected (as presumably in the study by Olson), it nonetheless appears evident that when the pressures of speed stress tax the capacities of people, something has to give, specifically performance on some of the time-shared tasks.

Time-Sharing of Auditory Inputs   Essentially the same adverse effect of time-sharing (i.e., of something having to give) is apparent in the case of auditory inputs, such as when two or more inputs occur simultaneously, overlap each other to some degree, or occur very close together in time. If an individual is listening, say, for verbal messages, and two messages occur at the same time, only one of them usually will get through. If, however, there is a slight lag in one, the first typically is identified more accurately than the second [Webster and Thompson, 31]. But if there is a distinct intensity difference, the second being the more intense, it will tend to have priority on the receiver's attention, even though it may lag after the first by as much as 2 s.

The adverse effect of simultaneous messages occurs even when only one is relevant and needs to be attended to [Broadbent, 5]. There is also some evidence that when a competing (and irrelevant) message is relatively similar in content and words to a relevant message, the interference effects on the relevant message are greater than if the irrelevant message is rather different [Peters, 25].

Time-Sharing of Auditory and Visual Sensory Channels   Since vision and audition are the most important senses for receiving information from displays, it would be particularly in order to compare these two senses in the degree to which they are influenced by interference from each other and from other possible work activities. Present inklings suggest that when both visual and auditory inputs are being time-shared, the auditory channel is more resistant to interference effects than the visual channel [Mowbray, 20].

Discussion   It is fairly evident that there are bounds beyond which the time-sharing of sensory inputs typically results in some degradation of performance. When circumstances permit (and they sometimes do not), efforts should, of course, be made to so manipulate the situation that degradation will at least be minimized if not eliminated. As in other contexts, we need to accept generalizations with caution. From this point of view, research evidence—some cited above and some not—seems to suggest a few general guidelines, although these may not be applicable across the board:

1   Where possible, the number of potentially competing sources should be minimized.

2   Where time-sharing is likely to impose speed or load stress, the receiver should be provided with some inklings about priorities, so his strategy in giving attention to first things first can take these into account.

3   Where possible, the requirements for use of short-term memory and for dealing with low-probability events should be minimized.

4   Where possible, input stimuli which require individual responses should be separated temporally and presented at such a rate that they may be responded to individually. Extremely short intervals (say, less than 0.5 or 0.25 s) should be avoided if at all possible. Where possible, the receiver should be permitted to control the input rate.

5   Where a choice of sensory modalities is feasible in a situation where a sensory input has competition, the auditory sense is generally more durable and is less influenced by other inputs.

6   Some means of directing attention to relevant and more important sources will increase the likelihood of their priority in the receiver's attention; for example, in some situations visual stimuli (such as lights) might be used as advance cues to the location of relevant auditory sources, or vice versa.

7   Where two or more auditory inputs might have to be time-shared, it would be desirable to schedule relevant messages or signals so they do not occur simultaneously, to separate physically the sources (such as speakers) of relevant versus irrelevant messages, in order to filter out (if possible) any irrelevant messages, and where they cannot be filtered out, to make them as different as possible from those that are relevant such as by making the relevant stimuli more intense or using clearly distinct spectral characteristics.

8 Especially when repetitive manual tasks are time-shared with nonrelated sensory inputs, the greater the learning of the manual task, the less will be its possible effect on the reception of the sensory input.

Use of Redundant Sensory Channels

Two or more sensory channels, such as vision and audition, can be used in a redundant fashion to present the same information. While the evidence about some aspects of human behavior is a bit inconclusive, there is virtually no question but that the use of redundant visual and auditory coding (the simultaneous presentation of identical information to both senses) increases the odds of reception of the information. Different specific studies could be used to illustrate this point. We shall use a vigilance study to illustrate this [Buckner and McGrath, 7].

Three of the conditions in the study were as follows:

- *Visual task:* Subjects detected an increment in the brightness of an intermittent light that was on for 1 s and off for 2.
- *Auditory task:* Subjects detected an increment in the loudness of an intermittent 750 Hz tone that was on for 1 s and off for 2.
- *Combined visual and auditory task:* The visual and auditory tasks were combined, and both visual and auditory signals occurred simultaneously.

In this task the subjects were supposed to detect 24 signals during 60-min vigilance watches; during any watch, only visual or auditory or combined signals were used. Figure 3-6 shows the percent of signals of each type that were

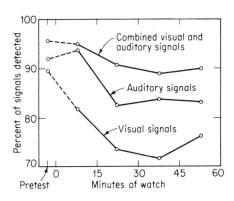

Figure 3-6 Performance on a visual, auditory, and combined visual-auditory vigilance task. The vertical scale shows the percentage of signals that were detected during certain periods of a 1-h watch. [From Buckner and McGrath, 7.]

detected. The consistent advantage of the combined signal is evident throughout the hour watch.

In another study that reinforces this point, subjects were to press one of three keys (left, center, or right) in response to a visual signal (red, orange, or green, respectively), an auditory signal (100, 700, or 5000 Hz), or a combined signal [Klemmer, 17]. The percentages of correct responses were as follows:

visual signal, 89 percent; auditory signal, 91 percent; and combined visual and auditory signal, 95 percent.

Noise and the Theory of Signal Detection (TSD)

In some circumstances meaningful stimuli may occur in the presence of "noise" that may interfere with the reception of the stimuli whatever they may be, such as warning bells, foghorns, Morse code signals, radar blips, signal lights against their backgrounds, or defects in products that are being inspected. The possible effects of the noise (auditory, visual, or otherwise) on the detection of stimuli have given rise to the formulation of the theory of signal detection (TSD) [Swets et al., 29; Swets, 28].

Basis of TSD To illustrate this theory, let us assume the case of an ambient noise in a factory with an intensity that varies randomly over time, with the occasional occurrence of a warning signal of a crane that increases the total intensity. The probabilities (for individual points in time) of the noise might be like that at the left of Figure 3-7, and the distribution of the combined intensity

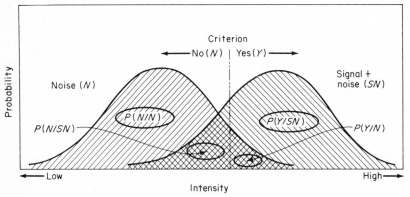

Figure 3-7 Illustration of certain of the concepts of the theory of signal detection (TSD), using sound intensity as the parameter. The two distributions reflect the probabilities (at points in time) at which the intensity of the noise (N) or of the signal plus noise (SN) might occur. The figure illustrates a criterion point that might be selected for making a "yes" decision. Other parameters can be viewed in the same manner. See text for discussion, including the meaning of the probability (P) symbols shown.

(noise plus signal) might be like that at the right. In the case of very low or very high intensity values, there is no appreciable problem in determining whether only the noise is present or whether the signal is present. It is in the overlapping areas that confusion can occur. The actual probabilities of one or the other are reflected in the figure by the relative proportion of overlap of the two distributions at the overlapping intensity values.

In making a determination about the presence or absence of a signal, there are four response alternatives, as follows, with a probability (P) associated with each:

Response	Stimulus	
	Noise (N)	Signal + noise (SN)
Yes (Y)	P (Y/N)	P (Y/SN)
No (N)	P (N/N)	P (N/SN)

We can see in the example in Figure 3-7 that no observer would be able to detect signals 100 percent of the time [P (Y/SN)] and have no "false alarms" [P (Y/N)]. However, the "criterion" selected by the observer in making his yes-or-no decision has an obvious effect on the frequency of the types of errors he would make. If he is "lenient" in his identification of signals, he would set his criterion level toward the left, with an ensuing increase in the number of "false alarms"; whereas if he were operating under a "set" to be quite "sure" of a signal, he would move his criterion level to the right, with a reduction in the "hit" rate (the number of actual signal detections).

Generality of TSD   Although this discussion has been based on the illustration of an auditory signal against a background of noise (with the sensory continuum being intensity), the same concepts apply in the case of other types of "signals" and "noise," such as the waveform of auditory sounds, lights, and other visual signals, such as blips on radarscopes.

Implications of TSD   The important implications of signal detection theory include at least the following: wherever possible, the signal should be such that, when combined with the noise, the combined values form a distribution that is *clearly* separated from that of the noise by itself; and when some overlap cannot be avoided, a decision needs to be made regarding the type of error that can best be tolerated ("false alarms" or failure to detect signals), since this will influence the criterion level that the observer should establish.

HUMAN INFORMATION PROCESSING

In his discussion of the ergonomics of information presentation, Singleton [26] points up very effectively the need for presenting information to people via displays in such a manner as to enhance the use of the information in operational situations. The "use" of such information in the human factors context implies the involvement of any of many different types of mediation or cognitive processes, including the following:

Information storage
- *Long-term memory*
- *Short-term memory:* Remembering relevant information for short periods of time, such as a message to give to someone

Information retrieval and processing
- *Recognition:* Essentially a perceptual process involving the recognition or detection of relevant stimuli or signals
- *Recall:* Including the recall both of previously learned factual information, procedures, processes, sequences, and other such classes, and of information in short-time storage, mentioned above
- *Information processing:* Categorizing, calculating, coding, computing, interpolating, itemizing, tabulating, translating, etc.[2]
- *Problem solving and decision making:* Analyzing, calculating, choosing, comparing, computing, estimating, planning, etc.[2]
- *Control of physical responses:* The exercise of control over a wide range of physical responses including conditioned responses, selection of responses appropriate to specific stimuli, sequences of responses, and continuous control responses.

Practically every one of these (and perhaps other) mediation processes serves as the subject of a major field of knowledge and research in its own right, and it is not possible in this text to deal with the role of each as related to the human factors context. Because of the central role of learning in this context, however, we will take an overview of the processes of acquisition of knowledge and skill and the related aspect of transfer of learning.

Acquisition of Skills and Knowledge

Learning consists of a relatively permanent change in an individual as manifested by his behavior. Whether an individual has learned something cannot be determined directly, but only by his subsequent performance. The spectrum of human performance abilities is of course rooted in the knowledge and skills that people have learned. Although staggeringly varied in their specific nature, the skills of people probably can be grouped into such classes as: gross bodily skills (walking, maintaining equilibrium, etc.), manipulative skills (including those of continuous, sequential, and discrete types), perceptual skills, and language skills (including conventional communications, mathematics, metaphor, and other representations people use in thinking and problem solving, and in coding computer languages, for example).

Types of Learning   The nature of the performance or material to be learned (the task, operation, activity, content, etc.) virtually predetermines the type of learning involved. Gagné [15] has postulated certain types of learning that form what he refers to as a *cumulative learning sequence*, with each level depending on lower levels in a building-block fashion, somewhat as shown in Figure 3-8. Although Gagné postulates this hierarchical scheme for human development, we can probably see some relevance of each of these levels to cer-

---

[2]These enumerations of behaviors were developed by Berliner. Angell, and Shearer [4] and reflect the consensus of several people as representing reasonably unambiguous mediation processes.

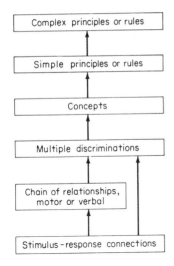

Figure 3-8 A generalized illustration of the cumulative learning sequence proposed by Gagné [15, p. 182]. In this model the learning at any given level depends on relevant learning at the lower levels.

tain types of activities in the operation of systems. For example, the stimulus-response level applies to such straightforward tasks as turning a control switch when a signal is heard; the chain relationship, to following certain specified motions; the multiple discriminations, to the selection of the appropriate response to each of several stimuli (such as traffic lights); and the application of principles, to certain maintenance problems.

Conditions That Contribute to Learning The conditions and methods that contribute most beneficially to learning will not be discussed in detail here. However, a few such conditions will be mentioned briefly: motivation on the part of the learner; knowledge of results (i.e., feedback to the individual regarding his performance); distribution of training periods (usually some spacing of learning periods is desirable, although the optimum duration and spacing of learning periods is unique to the knowledge or skill being learned); and the types of incentives used (usually positive incentives are more effective than negative incentives, and usually *intrinsic* incentives, i.e., those associated with the activity, are more effective than *external* incentives).

Transfer of Learning Much of that which we learn in one situation is *transferred* to another. This is of course the premise on which all education is founded—that what is learned in school will be transferred to relevant contexts in the real world. In our study of human factors we are of course interested in the transfer-of-learning process as it is used in transferring previously learned knowledge and skills to the types of life circumstances to which such transfer is relevant, such as performing a job or driving a car.

Various theories of transfer have been set forth. The first major formulation was the concept of *identical elements* proposed by Thorndike [30], i.e., that transfer from one situation to another occurred to the extent that there were identical elements in the two situations. Another general theory is that of transfer through principles [Bass and Vaughn, 3, p. 40]. This theory postulates that positive transfer results when an individual applies to new situations the principles learned in previous specific situations which have sufficient generality to cover the class of stimuli that the previous and the new situations have in common. Still other theories are predicated on the degree of similarity or difference between the nature of the *stimuli*, and of the *responses*, of the task on which the initial learning takes place, and of the task to which the learning is to be transferred [Deese and Hulse, 12, p. 349; Muckler et al., 22]. Such theories would predict that transfer would be greatest when the stimulus and the response of the transfer task are the same as those of the initial learning task, and that transfer would be negative if a new response had to be associated with a stimulus for which a different response has already been learned (such as learning to "stop" on a green light). Although such theories have an intuitive appeal, it should be added that there is no operational basis for measuring the degree of similarity between different stimuli or responses, or for estimating "how much" transfer would occur in any given situation. At the present time we probably need to acknowledge the bald fact that there is no generally confirmed theory related to transfer of learning that lends itself to practical application.

Compatibility

One should not leave the subject of information input and processing without touching further on the topic of compatibility. Although there are many different manifestations of compatibility, most instances probably can be considered to fall in one of three groups, namely, (1) *spatial* compatibility, i.e., the compatibility of the physical features, or arrangement in space, of certain items, especially displays and controls; (2) *movement* compatibility, the direction of movement of displays, controls, and system responses; and (3) *conceptual* compatibility, the conceptual associations that people have, such as green representing "go" in certain codes. In the context of perceptual-motor activities there is some presumption of compatibility of stimulus and response in combination. The term *stimulus-response compatibility* (S-R compatibility) was first used by Fitts and Seeger [14], following the earlier use of the term *compatibility* by A. M. Small. Fitts and Seeger characterized S-R compatibility as follows: "A task involves compatible S-R relations to the extent that the ensemble of stimulus and response combinations comprising the task results in a high rate of information transfer." In this information-theory context, the concept of compatibility implies a hypothetical process of information transformation, or recoding, in the activity and is predicated on the assumption that the degree of compatibility is at a maximum when the recoding processes are at a minimum.

Origins of Compatibility Relationships Compatibility relationships stem from two possible origins. In the first place, certain compatible relationships are intrinsic in the situation, for example, turning a steering wheel to the right in order to turn to the right. In certain combinations of displays and controls, for example, the degree of compatibility is associated with the extent to which they are isomorphic or have similar spatial relationships. Other compatible relationships are culturally acquired, stemming from habits or associations that are characteristic of the culture in question. For example, in the United States a light switch is usually pushed up to turn it on, but in certain other countries it is pushed down. How such culturally acquired patterns develop is perhaps the consequence of fortuitous circumstances.

The Identification of Compatibility Relationships   If one wishes to take advantage of compatible relationships in designing equipment or other items, it is of course necessary to know *what* relationships are *compatible*. There generally are two ways in which these can be ascertained or inferred. In the first place, certain such relationships are obvious or manifest; this is particularly true with many relationships that are intrinsic in the situation, such as the arrangement of corresponding displays and controls in juxtaposition to each other. In addition, certain culturally acquired relationships are so pervasive that they, too, are obvious, such as the red, yellow, and green symbols of traffic lights. But when the most compatible relationships are not obvious, it is necessary to identify them on the basis of empirical experiments. Certain examples will be cited in later chapters, but in general, such experiments produce information on the proportion of subjects who choose each specific relationship of different possible relationships.

Discussion of Compatibility   Although different versions of compatibility involve the processes of sensation and perception and also response, the tie-in between these—the bridge between them—is a mediation process. Where compatible relationships can be utilized, the probability of improved performance usually is increased. As with many aspects of human performance, however, there are certain constraints or limitations that need to be considered in connection with compatibility relationships. For example, some such relationships are not self-evident; they need to be ascertained empirically. When this is done, it sometimes turns out that a given relationship is not universally perceived by people; in such instances it may be necessary to "figure the odds," that is, to determine the proportion of people with each possible "association" or response tendency and make a design determination on this basis. In addition, there are some circumstances where trade-off considerations may require that one forgo the use of a given compatible relationship for some other benefit.

Discussion

In any operational situation (as in performing some function in a job), the information input from the environment (whether sensed directly, or sensed indirectly via displays) serves, in combination with the information stored in memory, as the grist for any of numerous types of mediation processes as the basis for decisions about actions to be taken. In this regard, there have been efforts to simplify display presentations so as to reduce the required demands made on operating personnel, including training requirements. The ultimate strategy following this approach would, as Singleton [26] points out, lead to the elimination of human beings in such operations, and to develop an automated system to achieve the desired objective. When complete automation is feasible and practical, such automation would seem to be desirable. When this is not the case, however, there is a question as to how "simplified" an operation should be designed to be.

REFERENCES

1 Abramson, N.: *Information theory and coding*, McGraw-Hill Book Company, New York, 1963.
2 Alluisi, E. A., P. F. Muller, Jr., and P. M. Fitts: An information analysis of verbal and motor responses in a forced-paced serial task, *Journal of Experimental Psychology*, 1957, vol. 53, pp. 153–158.
3 Bass, B. M., and J. A. Vaughn: *Training in industry: the management of learning*, Wadsworth Publishing Company, Inc., Belmont, Calif., 1966.
4 Berliner, C., D. Angell, and J. W. Shearer: "Behaviors, measures and instruments for performance evaluation in simulated environments," in *Proceedings, Symposium on quantification of human performance*, Aug. 17–19, 1964, *Albuquerque, New Mexico*, M-5.7, Subcommittee on Human Factors, Electronic Industries Association.
5 Broadbent, D. E.: Listening to one of two synchronous messages, *Journal of Experimental Psychology*, 1952, vol. 44, pp. 51–55.
6 Broadbent, D. E.: *Perception and communication*, Pergamon Press, New York, 1958.
7 Buckner, D. N., and J. J. McGrath: *A comparison of performances on single and dual sensory mode vigilance tasks*, Human Factors Research, Inc., Los Angeles, Calif., TR 8, ONR Contract Nonr 2649(00), NR 153–199, February, 1961.
8 Conrad, R.: Speed and load stress in a sensori-motor skill, *British Journal of Industrial Medicine*, 1951, vol. 8, pp. 1–7.
9 Conrad, R.: Some effects on performance of changes in perceptual load, *Journal of Experimental Psychology*, 1955, vol. 49, pp. 313–332.
10 Conrad, R.: The timing of signals in skill, *Journal of Experimental Psychology*, 1956, vol. 51, pp. 365–370.
11 Deatherage, B. H.: "Auditory and other sensory forms of information presentation," in H. P. Van Cott and R. G. Kinkade (eds.), *Human engineering guide to equipment design*, rev. ed., Chap. 4, U.S. Government Printing Office, Washington, D.C., 1972.
12 Deese, J., and S. H. Hulse: *The psychology of learning*, 3d ed., McGraw-Hill Book Company, New York, 1967.
13 Fitts, P. M.: The information capacity of the human motor system in controlling the amplitude of movement, *Journal of Experimental Psychology*, 1954, vol. 47, pp. 381–391.
14 Fitts, P. M., and C. M. Seeger: S-R compatibility: spatial characteristics of stimulus and response codes, *Journal of Experimental Psychology*, 1953, vol. 46, pp. 199–210.

15 Gagné, R. M.: Contributions of learning to human development, *Psychological Review*, 1968, vol. 75, pp. 177–191.

16 Geyer, B. H., and C. W. Johnson: Memory in man and machines, *General Electric Review*, March, 1957, vol. 60, no. 2, pp. 29–33.

17 Klemmer, E. T.: Time sharing between frequency-coded auditory and visual channels, *Journal of Experimental Psychology*, 1958, vol. 55, pp. 229–235.

18 Mackworth, N. H., and J. F. Mackworth: Visual search for successive decisions, *British Journal of Psychology*, 1958, vol. 49, pp. 210–221.

19 Miller, G. A.: The magical number seven, plus or minus two: Some limits on our capacity for processing information, *Psychological Review*, 1956, vol. 63, pp. 81–97.

20 Mowbray, G. H.: Simultaneous vision and audition: The detection of elements missing from overlearned sequences, *Journal of Experimental Psychology*, 1952, vol. 44, pp. 292–300.

21 Mowbray, G. H., and J. W. Gebhard: *Man's senses as information channels*, Report CM-936, Johns Hopkins University, Applied Physics Laboratory, Silver Spring, Md., May, 1958.

22 Muckler, F. A., J. E. Nygaard, L. I. O'Kelly, and A. C. Williams, Jr.: *Psychological variables in the design of flight simulators for training*, USAF, WADC, TR 56–369, January, 1959.

23 Mudd, S. A.: *The scaling and experimental investigation of four dimensions of pure tone and their use in an audio-visual monitoring problem*, unpublished Ph.D. thesis, Purdue University, Lafayette, Ind., June, 1961.

24 Olson, P. L.: *Display arrangement, number of channels and information speed as related to operator performance*, unpublished Ph.D. thesis, Purdue University, Lafayette, Ind., July, 1959.

25 Peters, R. W.: *Competing messages: the effect of interfering messages upon the reception of primary messages*, USN School of Aviation Medicine, Report NM 001 064.01.27, 1954.

26 Singleton, W. J.: The ergonomics of information presentation, *Applied Ergonomics*, 1971, vol. 2, no. 4, pp. 213–220.

27 Steinbuch, K.: *Information processing in man*, paper presented at IRE International Congress on Human Factors in Electronics, Long Beach, Calif., May, 1962.

28 Swets, J. A. (ed.): *Signal detection and recognition by human observers*, John Wiley & Sons, Inc., New York, 1964.

29 Swets, J. A., W. P. Tanner, Jr., and T. G. Birdsall: Decision processes in perception, *Psychological Review*, 1961, vol. 68, pp. 301–310.

30 Thorndike, E. L.: Mental discipline in high school studies, *Journal of Educational Psychology*, 1924, vol. 15, pp. 1–22, 83–98.

31 Webster, J. C., and P. O. Thompson: Responding to both of two overlapping messages, *Journal of the Acoustical Society of America*, 1954, vol. 26, pp. 396–402.

32 Weisz, A. Z., and L. S. McElroy: *Information processing in a complex task under speed stress*, Decision Sciences Laboratory, Electronics System Division, AFSC, USAF, Report ESD-TDR 64–391, May, 1964.

33 Welford, A. T.: "Performance, biological mechanisms and age: A theoretical sketch," in A. T. Welford and J. E. Birren (eds.), *Behavior, ageing and the nervous system*, Charles C. Thomas, Springfield, Ill., 1965.

Chapter 4

Visual Displays

Since the use of visual displays depends upon the visual capabilities of people, we will first discuss the process of seeing and certain types of visual skills.

THE PROCESS OF SEEING

The eye is very much like a camera in that it has an adjustable lens through which light rays are transmitted and focused, and a sensitive area (the retina) upon which the light falls. Figure 4-1 illustrates the principal features of the eye

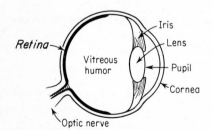

Figure 4-1 Principal features of the human eye in cross section. Light passes through the pupil, is refracted by the lens, and is brought to a focus on the retina. The retina receives the light stimulus and transmits an impulse to the brain through the optic nerve.

in cross section. The lens of the eye is normally flexible, so that it can adjust itself to bring about proper focus on the retina. The image of the object upon the

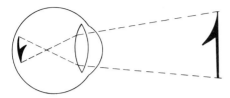

Figure 4-2 Illustration of the manner in which the image of an object is reproduced in inverted form on the retina of the eye.

retina is reversed and inverted, just as it is in a camera, as illustrated in Figure 4-2. The retina consists of two types of sensitive areas, namely, rods and cones. The cones are primarily sensitive to variations in the wavelength of light, which give rise to the subjective sensation of color. There are approximately 6 or 7 million cones in the eye; these generally predominate in the center section of the retina. The rods are primarily sensitive to the amount of light and are not particularly sensitive to differences in wavelength. There are about 130 million rods in the eye; they tend to predominate toward the outer reaches of the retina around the sides of the eyeball. The rods and cones, upon receiving light through the lens, set up nerve impulses which are transmitted through the optic nerve to the brain, where translation then takes place.

Visual Acuity

There are several kinds of visual acuity, but they all deal with the resolution of (i.e., the ability to distinguish visually) black and white detail. The various types of visual acuity depend very largely on the *accommodation* of the eyes, which is the adjustment of the lens of the eye to bring about proper focusing of the light rays on the retina. In normal accommodation, if one is looking at a far object, the lens flattens, and if one looks at a near object, the lens tends to bulge, in order to bring about proper focusing of the image on the retina. This is illustrated in Figure 4-3*a* for far objects, and 4-3*b* for near objects.

In some individuals the accommodation of the eyes is inadequate. This causes the conditions that we sometimes call *nearsightedness* and *farsightedness*. When a person is nearsighted, his lens tends to remain in a bulged condition, so that while he may achieve a proper focus of near objects, he cannot achieve a proper focus of far objects, as shown in Figure 4-3*c*. Farsightedness, in turn, is a condition in which the lens tends to remain too flat. While such a person may see clearly at far distance, he encounters difficulty in seeing properly at near distance, as shown in Figure 4-3*d*. Such conditions can sometimes be corrected by appropriate lenses which change the direction of the light rays before they reach the lens of the eye and thereby bring about proper focusing on the retina.

Types of Visual Acuity

The most commonly used measure of acuity, *minimum separable acuity*, refers to the smallest feature, or the smallest space between the parts of a target, that the eye can detect. The visual targets used in measuring minimum separable acuity include letters and various geometric forms, such as those illustrated in

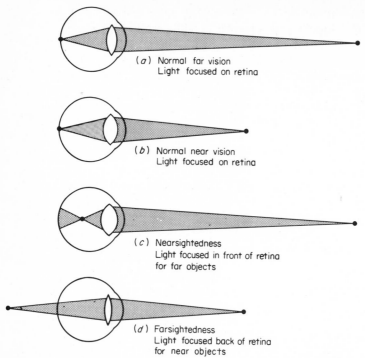

(*a*) Normal far vision
Light focused on retina

(*b*) Normal near vision
Light focused on retina

(*c*) Nearsightedness
Light focused in front of retina
for far objects

(*d*) Farsightedness
Light focused back of retina
for near objects

Figure 4-3   Illustration of normal accommodation at far and near distances, *a* and *b*, and of nearsightedness *c* and farsightedness *d*.

Figure 4-4. Such targets can be varied in size and distance, the acuity of the subject being determined by the smallest target he can properly identify. Such acuity usually is measured in terms of the reciprocal of the visual angle subtended at the eye by the smallest detail that can be discriminated (i.e., the angular subtense of that detail). The reciprocal of a visual angle of 1 minute of arc usually is used as a standard in scoring. This reciprocal, being unity, provides a base with which poorer or better levels of acuity can be compared. If, for example, one individual can identify only a detail that subtends an arc of 1.5 minutes, his acuity score would then be the reciprocal of 1.5 minutes, or 0.67.

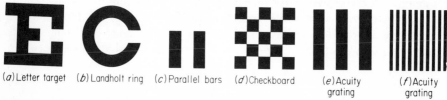

(*a*)Letter target   (*b*)Landholt ring   (*c*)Parallel bars   (*d*)Checkboard   (*e*)Acuity   (*f*)Acuity
grating   grating

Figure 4-4   Illustrations of various types of targets used in visual acuity tests and experiments. The features to be differentiated in targets *a, b, c, d,* and *e* are all the same size and would, therefore, subtend the same visual angle at the eye. With target *a* the subject is to identify each letter; with *c, e,* and *f* he is to identify the orientation (such as vertical or horizontal); and with *b* he is to identify any of four orientations. With target *d* he is to identify one checkerboard target from three others with smaller squares.

On the other hand, if an individual can identify a detail that subtends an arc of 0.8 minute, his score, the reciprocal of 0.8 minute, would be 1.25.

Vernier acuity refers to the ability to differentiate the lateral displacement, or slight offset, of one line from another that, if not so offset, would form a single continuous line (such as in lining up the "ends" of lines in certain optical devices). It is usually measured in terms of the visual angle of the smallest such displacement that can be detected. *Minimum perceptible acuity* is the ability to detect a spot (such as a round dot) from its background, and is typically measured in terms of the visual angle of the smallest spot that can be detected. In turn, *stereoscopic acuity* refers to the ability to differentiate the different images—or pictures—received by the retinas of the two eyes of a single object that has depth. (These two images differ most when the object is near the eyes, and differ least when the object is far away.) Stereoscopic acuity typically is measured by the difference between the parallactic angles (at the eyes) of two similar targets that are at just noticeably different distances from each other.

Convergence (Phoria)

As we direct our visual attention to a particular object, it is necessary that the two eyes converge on the object so that the images of the object on the two retinas are in corresponding positions; in this way we get an impression of a single object. The two images are said to be *fused* if they do so correspond. Convergence is controlled by muscles that surround the eyeball. Normally, as an individual looks at a particular object, these muscles operate automatically to bring about convergence. But some individuals tend to converge too much, and others tend not to converge enough. These conditions are called *phorias*. Since the double images that occur when convergence does not take place are visually uncomfortable, such people usually have compensated for this by learning to bring about convergence; however, muscular stresses and strains can occur in overcoming these muscular imbalances.

Color Discrimination

While people are still arguing about the specific process by which we differentiate between colors, color blindness basically is a deficiency in the ability of the cones to differentiate various wavelengths. True and complete color blindness is very rare. The various degrees and types of partial color blindness include cases in which there is difficulty in differentiating between certain colors, such as between red and green or between blue and yellow.

Dark Adaptation

The adaptation of the eye to different levels of light and darkness is brought about by two functions. In the first place, the pupil of the eye increases in size as we go into a darkened room, in order to admit more light to the eyes; it tends to contract in bright light, in order to limit the amount of light that enters the eye. Another function that affects how well we can see as we go from the light into darkness is a physiological process in the retina in which *visual purple* is built up. Under such circumstances the cones (which are color sensitive) lose much of their sensitivity. Since in the dark our vision depends very largely on the rods, color discrimination is limited in the dark. The time required for complete dark adaptation is usually 30 to 40 min. The reverse adaptation, from darkness to light, takes place in some seconds, or at most in a minute or two. In situations where someone needs to become dark-adapted, such as to go on shipboard watch at night, it is frequently the practice to wear red goggles for a period of time (say, half an hour) before going on watch, to facilitate the dark-adaptation process.

Conditions That Affect Visual Discriminations

The ability of individuals to make visual discriminations is of course dependent upon their visual skills, especially their visual acuity. Aside from individual differences, however, there are certain variables (conditions), external to the individual, that affect visual discriminations. Some of these variables are listed or discussed briefly below.

Luminance Contrast   "Luminance contrast" refers to the difference in luminance of the features of the object being viewed, in particular of the feature to be discriminated by contrast with its background (for example, an arrow on a direction sign against the background area of the sign). The luminance contrast is expressed by the following relationship:

$$\text{Contrast} = \frac{B_1 - B_2}{B_1} \times 100$$

in which B_1 = brighter of two contrasting areas
B_2 = darker of two contrasting areas

The contrast between the print on this page and its white background is considerable; if we assume that the paper has a reflectance of 80 percent and that the print has a reflectance of 10 percent, the contrast would be

$$\frac{80 - 10}{80} \times 100 = \frac{70}{80} \times 100 = 88 \text{ percent}$$

If the printing were on medium-gray paper rather than on white, the contrast would, of course, be much less.

Amount of Illumination   (This will be discussed in Chapter 12.)

Time   Within reasonable limits, the longer the viewing time, the greater is the discriminability.

Luminance Ratio   The luminance ratio is the ratio between the luminance of any two areas in the visual field (usually the area of primary visual attention and the surrounding area).

Movement   The movement of a target object or of the observer (or both) decreases the threshold of visual acuity. The ability to make visual discriminations under such circumstances is called *dynamic visual acuity* (DVA). It is usually expressed in degrees of movement per second [Goodson and Miller, 21]. Acuity deteriorates rapidly as the rate of motion exceeds 60°/s [Burg, 6]. DVA seems not to be strongly related to other visual skills [Burg and Hulbert, 7].

Glare   (This will be discussed in Chapter 12.)

Combinations of Variables   Available evidence indicates that there are interaction effects on visual performance when various combinations of the above variables exist, such as the combined effects of contrast and motion as reported by Petersen and Dugas [46]. Some of these effects will be discussed in Chapter 12.

Discussion

The visual skills people have—especially visual acuity and color discrimination—have a direct bearing upon the design of visual displays, particularly on the ability to *detect* relevant stimuli and to *discriminate* between and among variations thereof (such as positions of pointers on dials, or different letters). But, as discussed in the previous chapter, we can *sense* much more than we can comprehend or remember. The meaningfulness of what we see in visual displays depends in part upon our perceptual processes and the learning of relevant associations (such as learning the alphabet or the shape of road signs). Thus, the appropriate design of various types of visual displays—which we will now discuss—must be predicated in part upon perceptual and learning factors as well as upon the specific visual skills of people.

QUANTITATIVE VISUAL DISPLAYS

Quantitative displays are used to provide information about the quantitative value of some variable, either a dynamic changeable variable (such as temperature or speed) or what is essentially a static variable (such as a measurement of length, as with a rule). In most uses of such displays there is an explicit or implicit level of precision that is required or desired, such as measurement to the nearest millimeter, centimeter, inch, foot, or mile. A great deal of research has been carried out with quantitative displays, directed toward determining the design features that contribute to speed and accuracy of their use. We will bring in the details of only a few such research undertakings, and will otherwise summarize the implications of such research.

Basic Design of Dynamic Quantitative Displays

There are three basic types of dynamic quantitative displays, as follows: (1) fixed scales with moving pointers; (2) moving scales with fixed pointers (or, in some cases, lubber lines); and (3) digital displays or counters (in which the numbers of mechanical counters click into position, as mileage readings on many speedometers). Examples of these three types are given in Figure 4-5.

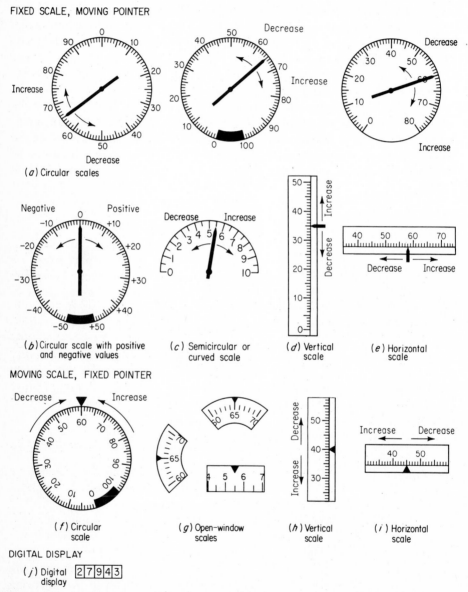

Figure 4-5 Examples of certain types of displays used in presenting quantitative information. (Reference will be made later to certain features of these scales.)

The first two classes are analog indicators in that the position of the pointer relative to the scale is analogous to the value that is represented. It is evident that there are differences in the effectiveness with which people can use these different designs in different types of circumstances.

Comparison of Different Designs Over the years there has been a number of studies in which certain designs of quantitative scales have been compared [Elkin, 18; Graham, 24; Sleight, 59]. Although the results of such studies are somewhat at odds with each other, certain general implications seem to stand out. For example, it has been pointed out [*Applied ergonomics handbook*, 71] that a digital indicator is more suitable than an analog display if precise readings of numerical values are required, since with an analog display the reader has to estimate the position of the pointer with respect to adjacent scale markers. This superiority in precision is supported by Sinclair [58] and is also reflected by the results of a study by Zeff [70] in which there were only 4 reading errors out of 800 readings with a digital display whereas with a circular display there were 50 errors. Further, readings with a digital display were found to be faster, as indicated by the following comparison of mean response times:

Digital display 0.94 s
Circular dial 3.54 s

It should be noted, however, that if the values are changing on a digital indicator, they may not remain visible long enough to be read.

Although digital displays have an advantage in terms of both accuracy and time in indicating specific values, analog displays have advantages of other types. The advantage lies especially with the fixed-scale moving-pointer design, as opposed to the moving-scale fixed-pointer design [Christensen, 9], probably because the position of the pointer (in a fixed scale) adds a perceptual cue that is missing in a moving-scale design. In this connection, many basically quantitative scales are also used in a qualitative manner, such as in noting approximate deviation from a desired value or in noting the rate and direction of change in observing altitude. The use of fixed-scale moving-pointer designs certainly facilitates this purpose. Fixed scales, however, have their limitations, especially when the range of values is so great that it cannot be shown on the face of a relatively small scale. In such a case certain moving-scale fixed-pointer designs, such as rectangular open-window and horizontal and vertical scales, have the practical advantage of occupying a small panel space, since the scale can be wound around spools behind the panel face, with only the relevant portion of the scale exposed.

Further, research and experience tend to favor circular and semicircular scales (*a*, *b*, and *c* in Figure 4-5) over vertical and horizontal scales (*d* and *e* in that figure). However, there are circumstances in which vertical and horizontal scales would have advantages that would argue for their use.

Readability of Altimeters The design of altimeters has posed a special problem for aviation for far too long a time. There have been numerous instances in which aircraft accidents have been attributed to misreading of the commonly used model. That model consists of three pointers representing, respectively, 100, 1000, and 10,000 ft, like the second, minute, and hour hands of a watch. In an early experimental study of altimeters Grether [25] found that that model was read with more errors and took more time than most other (experimental) designs, presumably because the reader had to *combine* three pieces of information. In turn, a combination counter (for thousands of feet) and pointer (for hundreds) proved to be used most accurately and with fewest errors.

Some further inklings about altimeter designs come from a study by Simon and Roscoe [57] as reported by Roscoe [53], in which four types of instruments were compared for use in displaying present altitude, predicted altitude (in 1 min), and command altitude (the altitude at which the plane is supposed to be flying); these are shown in very simplified form in Figure 4-6. The displays were intended to provide comparisons of three design variables, namely, (1) vertical versus circular scales, (2) integrated presentations

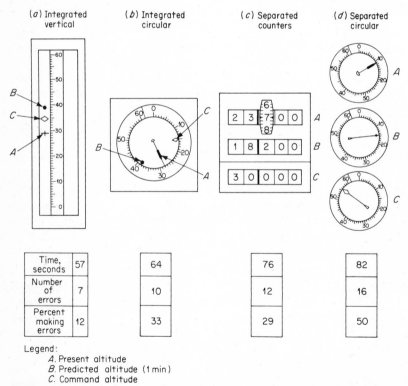

Legend:
    A. Present altitude
    B. Predicted altitude (1 min)
    C. Command altitude

Figure 4-6 Four display designs for presenting (*A*) present altitude, (*B*) predicted altitude (in 1 min), and (*C*) command altitude and three criteria (mean time for 10 trials, number of errors, and percent of 24 subjects making errors). The displays are shown in overly simplified form. [Adapted from Simon and Roscoe, 57, as presented by Roscoe, 53.]

(of the three altitude values mentioned above) versus separate presentations, and (3) spatial-analog presentations versus digital counters. Time and error performance scores of pilots in solving altitude-control decision problems and the percent of pilots making errors in using the four designs are also given in Figure 4-6. The clear and consistent superiority of design *a* (the integrated vertical-scale display) is apparent. The explanation for this given by Roscoe [53] is primarily its pictorial realism in representing relative positions in vertical space by a display in which *up* means *up* and *down* means *down*. (This is again an example of compatibility.) Design *b*, which represents vertical space in a distorted manner (around a circle), did not fare as well as *a*, but it was generally superior to *c* and *d*, both of which consisted of *separate* displays of the three altitude values rather than an *integrated* display. Thus, we can derive a strong hint that integrated displays (where they are indeed appropriate) generally are preferable to displays that have distinctly separate indications for the various values.

As an aside, this example reinforces the point that hard-and-fast generalizations are fairly treacherous. We have seen, for example, that *in this case*, a vertical scale was clearly best and the use of separate counters was manifestly inappropriate, presumably because of the requirement to envision relative positions in vertical space.

Specific Features of Quantitative Scales

The ability of people to make visual discriminations (such as those that are required in the use of quantitative scales) is influenced in part by the specific features that are to be discriminated. Some of the relevant features of quantitative scales are length of scale unit, scale markers (how many and what size), numerical progressions of scales, and the design of pointers.

Length of Scale Unit The length of the scale unit is the length on the scale, that represents the numerical value that is the smallest unit to which the scale is to be read. For example, if a pressure gauge is to be read to the nearest 10 lb, then 10 lb would be the smallest unit of measurement; the scale would be so constructed that a given length (in inches, millimeters, etc.) would represent 10 lb of pressure. (Whether there is, or is not, a marker for each such unit is another matter.)

The length of the scale unit should be such that the distinctions between the values to be read can be made with reasonably optimum reliability in terms of human sensory and perceptual skills. Various investigators have studied this matter with some still unexplained differences. Although certain investigators [Jones et al., 33; Murrell et al., 43] have reported acceptable accuracy in the reading of scales with scale units as low as 0.02 in, most sets of recommendations provide for values ranging from about 0.05 to 0.07, as shown in Figure 4-7. The larger values probably would be warranted when the use of instruments is under less than ideal conditions, such as when used by persons who have below-normal vision, or when used under poor illumination or under

Basic sketches, measurements in inches (parenthetical values in centimeters)

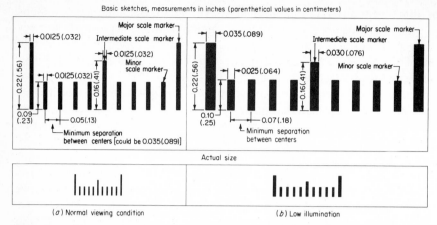

(a) Normal viewing condition   (b) Low illumination

Figure 4-7 Recommended format of quantitative scales, considering length of scale unit and graduation markers. Format a is proposed for normal illumination conditions under normal viewing conditions, and b for low illumination. [Adapted from Grether and Baker, p. 88.]

pressure of time. By implication, there could indeed be favorable circumstances in which smaller scale units could be fully justified, as with individuals with good vision, when the viewing conditions typically would be satisfactory, and perhaps where accuracy requirements would not be stringent.

Scale Markers   There is some conflicting evidence about the desirability of having a scale marker for each scale unit to be read, as opposed to having fewer scale markers (which then requires interpolation of values between the markers). In general, however, experience argues for the inclusion of a scale marker for each scale unit to be read. If the total scale requires that the space for a scale unit needs to be reduced below those normally considered desirable, however, the scale markers would then be too crowded to read accurately and rapidly (especially under low illumination). In such a circumstance it is better to design a scale that does require some interpolation. Actually, people are moderately accurate in interpolation, as reported by Cohen and Follert [10], who report that interpolation of fifths, and even of tenths, may yield satisfactory accuracy in many situations. Even so, where high accuracy is required (as with certain test instruments and fine measuring devices), a marker should be placed at every scale unit, even though this requires a larger scale or a closer viewing distance.

Numerical Progressions of Scales   Every quantitative scale has some intrinsic numerical-progression system that is characterized by the numerical difference between adjacent graduation markers on the scale and by the numbering of the major scale markers. In general, the garden variety of progression by 1s, of 0, 1, 2, 3, etc., is the easiest to use. This lends itself readily to a scale with major markers at 0, 10, 20, etc., with intermediate markers at 5, 15, 25, etc., and with minor markers at individual numbers. Progression by 5s is also satis-

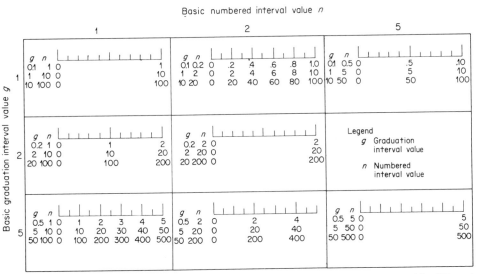

Figure 4-8 Examples of certain generally acceptable quantitative scales with different numerical-progression systems (1s, 2s, and 5s). The values to the left in each case are, respectively, the graduation scale interval g (the difference between the minor markers), and the numbered scale interval n (the difference between numbered markers). For each scale there are variations of the basic values of the system, these being decimal multiples or multiples of 1, 2, or 5.

factory, and by 2s is moderately so. Some examples of scales with these progressions are shown in Figure 4-8. The offbeat progression system by 4s and others by 2.5s, 3s, 6s, etc., usually give trouble and should be avoided except under special circumstances that distinctly justify them. Where large numerical values are used in the scale, the relative readabilities of the scales are the same if they are all multiplied by 10, 100, 1000, etc. Decimals, however, make scales more difficult to use, although for scales with decimals the same relative advantages and disadvantages hold for the various numerical progressions. The zero in front of the decimal point should be omitted when such scales are used [Vernon, 66].

The Design of Pointers The few studies that have dealt with pointer design leave some unanswered questions, but, in general, some of the common recommendations are the following: the use of pointed pointers (with a tip angle of about 20°); having the tip of the pointer meet, but not overlap, the smallest scale markers; having the color of the pointer extend from the tip to the center of the scale (in the case of circular scales); and having the pointer close to the surface of the scale (to avoid parallax).

Combining Scale Features Several of the features of quantitative scales discussed above have been integrated into relatively standard formats for designing scales and their markers, as shown in Figure 4-7. These are based on scale unit lengths of (*a*) 0.05 in for use under normal viewing conditions (with adequate illumination) and (*b*) 0.07 in for low illumination and for viewing at about 28 in. Although the formats in Figure 4-7 are shown in a horizontal scale, the features can of course be incorporated in circular or semicircular scales.

It should be added that the design features shown in this figure should be considered as general guidelines rather than as rigid requirements, and that the advantages of certain features may, in practical situations, have to be traded off for other advantages. The further point should be made that experience and logic in display design suggest the desirability of simple, uncluttered, rather bold designs, as illustrated in Figure 4-9.

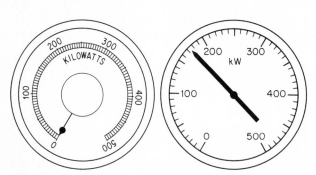

Figure 4-9 Illustration of two designs of a meter. The one at the right would be easier to read because it is bolder and less cluttered than the one on the left. It has fewer graduation markers, and the double arc-line has been eliminated. The scale length is increased by placing the markers closer to the perimeter; although this requires that the numerals be placed inside the scale, the clear design and the fact that the numerals are upright probably would partially offset this disadvantage. [Adapted *from Applied Ergonomics Handbook,* 71, fig. 3.1.]

Scale Size and Viewing Distance   The above discussion of the detailed features of scales is predicated on a normal viewing distance of 28 in.[1] If a display is to be viewed at a greater distance, the features would have to be enlarged in order to maintain, at the eye, the same visual angle of the detailed features. To maintain that same visual angle, the following formula can be applied for any other viewing distance in inches (*x*):

$$\left.\begin{matrix}\text{Dimension}\\ \text{at } x \text{ in}\end{matrix}\right\} = \left(\begin{matrix}\text{Dimension}\\ \text{at 28 in}\end{matrix}\right)\left(\frac{x\text{ in}}{28}\right)$$

QUALITATIVE VISUAL DISPLAYS

In using displays for obtaining qualitative information, the user is primarily interested in the approximate value of some continuously changeable variable (such as temperature, pressure, or speed) or in its trend or rate of change. The basic underlying data used for such purposes usually are quantitative.

---

[1]For a thorough discussion of scale design (based on British practice) taking into account range of values, viewing distance, numerical progression, etc., see Murrell [42, chap. 9].

Quantitative data may be used as the basis for qualitative reading in at least three ways, as follows: (1) for determining the status or condition of the variable in terms of each of a limited number of predetermined ranges (such as determining if the temperature gauge of an automobile is "cold," "normal," or "hot"); (2) for maintaining approximately some desirable range of values (such as maintaining a driving speed between 55 and 65 miles per hour); and (3) for observing trends, rates of change, etc. (such as noting the rate of change in altitude of an airplane). In the qualitative use of quantitative data, however, there is evidence that suggests that a display that is best for a quantitative reading is not necessarily best for a qualitative reading task. Some evidence for support of this contention comes from a study in which open-window, circular, and vertical designs are compared [Elkin, 18]. In one phase of this study, subjects made qualitative readings, as follows:

Pointer setting	Response to be made by subject
Above 60	High
40–60	OK
Below 40	Low

The accuracy of the readings was very high (only 3 errors were made in 1440 readings). The average times taken for the readings, however, are interesting, especially when compared with the *lowest* average reading times for the quantitative reading task (which was with scales graduated to 5s and read to the nearest 5).

Type of scale	Average reading time, s (scales graduated to 5s)	
	Qualitative	Quantitative
Open-window	115	102
Circular	107	113
Vertical	101	118

Thus, while the open-window design took the least time (of the three types) for quantitative reading, it took the longest time for qualitative reading.

The optimum designs of displays for qualitative reading, however, depend on *how* they are to be used, that is, the particular type of qualitative reading. If the entire continuum of values can be sliced up into a limited number of ranges—each of which represents some general "level"—the optimum design would be one in which each range of values is separately coded, such as by color, as illustrated in Figure 4-10. When color coding is not feasible (as under certain illumination conditions or with color-deficient individuals), zones on an instrument can be shape-coded. In this connection, it is desirable (if feasible) to take advantage of any natural associations people may have with designs or shapes. One study was directed toward determining what, if any, such associa-

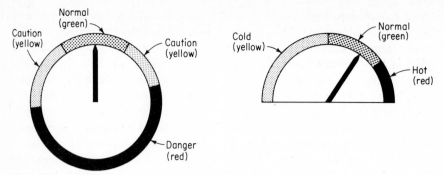

Figure 4-10 Illustration of color coding of sections of instruments that are to be read qualitatively.

tions people had with each of seven different coding designs [Sabeh, Jorve, and Vanderplas, 54]. After having solicited a large number of designs initially, the investigators selected the seven shown in Figure 4-11. These were presented to 140 subjects, along with a list of seven "meanings," as follows: caution, undesirable, mixture—lean, mixture—rich, danger—upper limit, danger—lower limit, and dangerous vibration. Figure 4-11 shows the number of subjects out of 140 who selected the indicated meaning to a statistically significant level. This is another illustration of the concept of compatibility (in this case the compatibility of association with symbol "meanings") as applied to a design problem.

Sometimes what is essentially a quantitative scale is used in what is referred to as "check-reading" manner, namely, simply to determine if the value represented reflects what is a normal (satisfactory, neutral, null) condi-

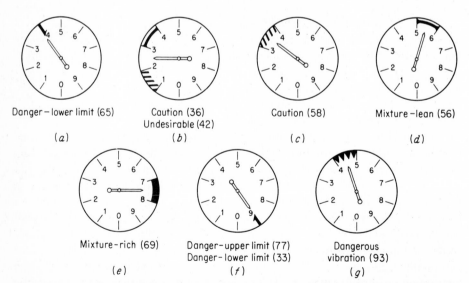

Figure 4-11 Association of coded zone markings with subjective "meaning," showing the number of individuals (out of 140) who reported significant associations. [Adapted from Sabeh, Jorve, and Vanderplas, 54.]

tion represented by a single value or a very narrow range of values, or a non-normal condition. In display design this frequently is done by having a mark along the scale that represents that value or range of values.

An interesting variation of design for check reading a qualitative instrument has been investigated by Kurke [39]. He used simulations of three variations of a quantitative instrument in which a given range of readings indicated a "danger" condition which required attention. These three variations, no indication, a red line, and a red wedge, are shown in Figure 4-12, along with mean-time scores of a group of subjects.

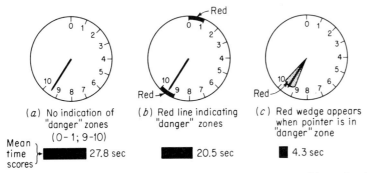

(a) No indication of "danger" zones (0–1; 9–10)

(b) Red line indicating "danger" zones

(c) Red wedge appears when pointer is in "danger" zone

Mean time scores — 27.8 sec   20.5 sec   4.3 sec

Figure 4-12 Three designs of quantitative instruments used in a check-reading situation where the "danger" condition was to be identified. Design a has no indication of the danger zone, b has a red line at the circumference, and c shows a red wedge when the pointer is in the danger zone. [From Kurke, 39, fig. 2.] (Copyright 1956 by the American Psychological Association. Reprinted by permission.)

The argument for the use of precoded displays for qualitative reading (when this is feasible) is rooted in the nature of the human perceptual and cognitive processes. To use a strictly quantitative display to determine if a given value is within one range or another range involves an additional cognitive process of allocating the value that is read to one of the possible ranges of values that represent the categories that have operational meaning. The initial perception of a precoded display immediately conveys the meaning of the display indicator.

It should be noted that quantitative displays with coded zones also can be used to reflect trends, directions, and rates of change. Further, they can also be used for quantitative reading if the scale values are included as in Figure 4-12. If in the use of quantitative data for qualitative reading it is not appropriate to precode certain zones, some conventional form of quantitative display would have to be used. The particular choice, however, would depend upon the relative importance of the qualitative versus quantitative readings, as reflected by the previously cited study by Elkin [18].

Panels of Displays for Check Reading When several or many basically quantitative displays are to be used for check reading (e.g., simply to determine if a condition is normal or not normal), the design and arrangement of the displays can facilitate the identification of any nonnormal display. With respect to arrangement of several such displays, it has been found that a systematic pattern with pointers representing the "normal" conditions usually makes it possible to identify any nonnormal (deviant) display more accurately and more rapidly than if the location of the "normal" conditions varies with different displays. For example, Dashevsky [14] found that the 12 o'clock and 9 o'clock positions of dials lent themselves equally well as the locations of the normal condition. In turn Johnsgard [30] found that dials with patterns of pointer "symmetry" (vertical or horizontal double rows of dials with the normal positions pointed toward an imaginary row between the double rows) could be read as rapidly as dials with the pointers aligned at the 9 o'clock positions. The basis for the advantage of some systematic pattern of the normal conditions of such dials is again essentially a function of our perceptual processes, in particular what is referred to as the *gestalt*, that is, the perception of the total configuration; any deviant dial "breaks up" that gestalt, thereby focusing our attention on it.

However, some patterns are more effective for this purpose than others. This was shown by Dashevsky [14], for example, in a comparison of the various patterns shown in Figure 4-13. Some of those patterns (*d*, *e*, and *f*) in-

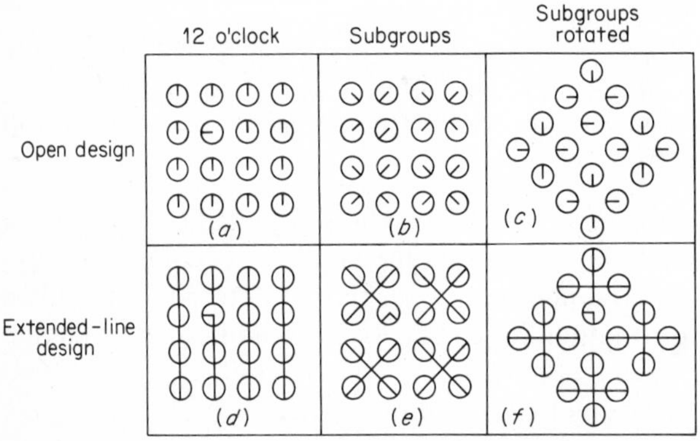

Figure 4-13 Patterns of panels of check-reading dials used in study by Dashevsky [14]. (Copyright 1964 by the American Psychological Association. Reprinted by permission.) In this study the 12 o'clock extended-line pattern *d* resulted in the lowest number of errors.

corporated lines extending from one dial to another to form continuous lines when the pointers were in their null positions. The errors resulting from this comparison are given below:

	Arrangement		
	12 o'clock	**Subgroups**	**Subgroups rotated**
Open	*a.* 53	*b.* 193	*c.* 201
Extended line	*d.* 8	*e.* 15	*f.* 41

These results show that the perception of deviant dials was more accurate with the 12 o'clock arrangement than with the subgroups, and also suggested that extending the lines between the dials enhances this perceptual detection. However, this is not the whole story. In a couple of subsequent studies by Oatman [44] a comparison was made of the detection rates of deviant dial readings with extended pointers (*c* and *d* of Figure 4-14) and of short pointers (*a* and *b*)

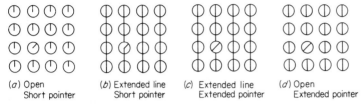

(*a*) Open   (*b*) Extended line   (*c*) Extended line   (*d*) Open
   Short pointer   Short pointer   Extended pointer   Extended pointer

Figure 4-14 Patterns of panels of check-reading dials used in a study by Oatman [44]. In this study extended-pointer designs (such as *c* and *d*) resulted in fewer errors than those without extended pointers.

in combination with the extended line between the dials (*b* and *c*) and of open patterns (*a* and *d*). On the basis of these studies he concluded that any factors that make the deviant *dial* more conspicuous (e.g., length of pointer) are apparently more significant in reducing check-reading errors than factors which make the display *pattern* simpler (e.g., extended line between the dials). However, it is reasonable to assume that the extension of lines between dials would not hinder the perceptual detection process, and possibly would help by providing a visual field that, in combination with an extended pointer, would be broken up by a deviant pointer.

STATUS INDICATORS

In a sense, some "qualitative" information approximates an indication of the "status" of a system or a component, such as the use of some displays for check reading to determine if a condition is normal or not normal, or the qualitative reading of an automobile thermometer to determine if the condition is hot, normal, or cold. However, what are more strictly status indicators reflect separate, discrete conditions, such as on and off, or (in the case of traffic lights) stop, caution, and go. In such instances the most straightforward type of display is a signal light, although other indicators can also be used. It might be added that if a quantitative instrument is to be used *only* for check-reading purposes, one could use a status indicator instead of the quantitative scale.

SIGNAL AND WARNING LIGHTS

Flashing or steady-state lights are used for various purposes, including the following: as indications of warning (as on highways); as identification of aircraft at night; as navigation aids and beacons; and to attract attention, such as to certain locations on an instrument panel. There apparently has been little research relating to such signals, but we can infer some general principles from our knowledge of human sensory and perceptual processes that might be helpful.

Detectability of Signal and Warning Lights

There are of course various factors that influence the detectability of lights. Certain such factors are discussed below.

Size, Luminance, and Exposure Time   The absolute threshold for the detection of a flash of light depends in part on a combination of size, luminance, and exposure time. Drawing in part upon some previous research by Graham and Margaria [23], Teichner and Krebs [62] have depicted the minimum sizes of lights (in terms of visual angle of the diameter in minutes of arc) that can be detected 50 percent of the time under various combinations of exposure times (in seconds) and luminance (in millilamberts), these relationships being shown in somewhat simplified form in Figure 4-15. From this it is clear that the luminance threshold decreases linearly as a function of exposure time up to a particular value, but that the exposure time at which the luminance threshold levels off decreases systematically with target size (actually the "area" of the light as implied by its diameter). Thus, the very minimal, lower-bound detectability of lights (50 percent accuracy) is seen to be a function of size, luminance, and exposure time. (Operational values should well exceed those in Figure 4-15.)

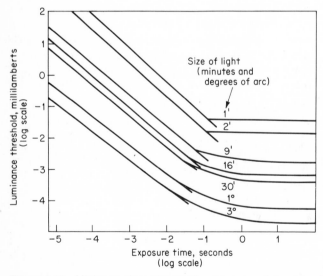

Figure 4-15   Minimum sizes of lights (in minutes and degrees of arc) that can be detected 50 percent of the time under varying combinations of exposure time and luminance. [Adapted from Teichner and Krebs, 62, as based in part on data from Graham and Margaria, 23.]

Color of Lights Another factor that is related to the effectiveness of signal lights is color. Using response time as an indication of the effectiveness of four different colors, Reynolds et al. [52] report the following order (from fastest to slowest): red, green, yellow, and white. However, the background color and ambient illumination can interact to influence the ability of people to detect and respond to lights of different colors. In general the researchers found that if a signal has good brightness contrast against a dark background, and if the absolute level of brightness of the signal is high, the color of the signal is of minimal importance in attracting attention. But with low signal-to-background brightness contrast, a red signal has a marked advantage, followed by green, yellow, and white in that order.

Flash Rate of Lights In the case of flashing lights, the flash rate should be *well* below that at which a flashing light appears as a steady light (the flicker-fusion frequency), which is approximately 30 times per second. In this regard, rates of about 3 to 10 per second (with duration of at least 0.05 s) have been recommended for attracting attention [Woodson and Conover, 68, pp. 2–26], and Markowitz [40] makes the point that the range of 60–120 flashes per minute (1 to 2/s), as now used on highways and in flyways, appears to be compatible with human discrimination capabilities and available hardware constraints.

Background of Lights As might be expected, signal lights cannot be discriminated well when other background lights are somewhat similar. (Traffic lights in areas with neon signs and Christmas tree lights represent very serious deviations from this principle.) And still another background characteristic relates to the steady versus flashing state of any background lights. In an interesting investigation of these, Crawford [13] used both steady and flashing signal lights against backgrounds of *irrelevant* lights (what we might call *noise*), these being all steady, all flashing, or some admixture of steady and flashing lights. Very briefly, his results indicated that the average time to identify the signal lights was minimal when the background-noise lights were all steady (this was especially so when the signal light was itself flashing); that the advantage of a flashing signal light (contrasted with a steady light) was completely lost if even one background-noise light was flashing; and that steady signals were more effective (could be identified more quickly) than flashing signals if the proportion of the noise lights that were flashing was any greater than 1 out of 10. In other words, flashing lights against other flashing lights really make life difficult for the viewer.

REPRESENTATIONAL DISPLAYS

Representational displays—both static and dynamic—tend to fall into two classes: (1) those that are essentially pictorial (intended to *reproduce* an object or scene, as on a TV scope or in an aerial photograph); and (2) those that are illustrative or symbolic (such as maps or aircraft-position displays). In either case the intent is to convey a visual impression that requires little or no interpretation. Since there are many varieties of such displays, we will here discuss only a few particular types.

Aircraft-Position Displays

The problem of representing the position and movement of aircraft has haunted designers—and pilots—for years. One major facet of this problem relates to the basic movement relationships to be depicted by the display, there being two such relationships, as shown in Figure 4-16 and described as follows:

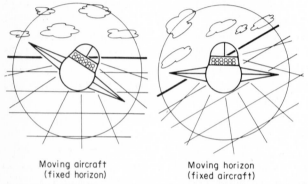

Moving aircraft   Moving horizon
(fixed horizon)   (fixed aircraft)

Figure 4-16 Illustration of the two basic movement relationships for depicting aircraft attitude, namely, the moving aircraft (outside-in) and the moving horizon (inside-out). [Adapted from Johnson and Roscoe, 33.]

• *Moving aircraft*: The earth (specifically, the horizon) is fixed, with the aircraft moving in relation to it (moving-aircraft or outside-in display).
• *Moving horizon*: The aircraft is fixed, with the horizon moving in relation to it (moving-horizon or inside-out display). (Most aircraft displays are of this type.)

This is basically a problem of visual perception, specifically with respect to what is referred to as the "figure and ground phenomenon." Is it more compatible with people's perceptions to envision the aircraft relative to a fixed earth? Or vice versa? As Johnson and Roscoe [31] lament, the experimental findings relating to this are flimsy and range from "suspect to inconclusive"; and they point out that there had been no conclusive flight experiments to nail down this issue prior to the 1970s.

Between 1970 and 1974, Roscoe and Williges and their students at the University of Illinois [74–77] completed a series of simulation and flight experiments to compare moving aircraft and moving horizon displays and a hybrid "frequency-separated" presentation in which both symbols move, the horizon in its "normal" manner (since most aircraft-position displays are of this type), and the airplane in immediate and direct response to aileron control inputs. The purpose of this arrangement was to provide control-compatible predictive indications of the direction of flight attitude changes on the moving airplane symbol while retaining the conventional moving horizon familiar to all experienced pilots.

The investigators conclude from these experiments [74]: "Non-pilots and pilots of little experience readily learn to use it (the frequency-separated attitude display) and show little tendency toward control reversals to which inexperienced pilots are subject with the conventional moving horizon. Highly experienced pilots readily adapt to it as a moving horizon display to which only a roll-rate prediction has been added, to assist them in maneuvering the airplane." For all pilot groups tested, measured performances with the frequency-separated display on a variety of flight tasks were as good as, or reliably better than, those with either the moving airplane or moving horizon displays, both in the airplane and in a simulator when operated with appropriate cockpit motion. The following results illustrate the performance of experienced pilots on a representative task.

Type of display	Tracking errors in degrees	
	In aircraft	On simulator
Moving aircraft	5.2	6.4
Moving horizon	4.6	5.1
Frequency separated	3.0	4.9

Although the moving horizon display was somewhat better than the moving aircraft display in this study, this was undoubtedly due to the fact that the subjects were pilots with at least some experience with the moving horizon type of display. Aside from the effects of previous experience, however, there is reasonable evidence to indicate that the moving part of a display normally should be displayed against a fixed scale or coordinate system (see the *principle of the moving part*, below).

In addition to the question of whether the aircraft, the horizon, or both, should move, steering commands can be presented in relation to aircraft attitude to allow either pursuit or compensatory tracking. (A pursuit display shows the movement of both the target and the pursuing aircraft against common reference coordinates, whereas a compensatory display shows only the difference, or error, in their relative positions.) In the Illinois experiments, pursuit and compensatory versions of each type of attitude presentation were compared, and, for all tasks in which performance differences occurred, the pursuit tracking displays produced the superior performance.

A more sophisticated type of aircraft-position display is what is called a *contact analog*. It provides, on a TV tube, an electronically generated representation of the terrain with the aircraft superimposed in proper position [Balding and Süskind, 3]. Figure 4-17 is an illustration of this type of display, showing an aircraft in three different relations to its predetermined *altitude-hold*

(a) On selected altitude   (b) Below selected altitude   (c) Above selected altitude

Figure 4-17 Illustration of a contact analog type of aircraft-position display, the Kaiser Flight-Path System. The display can present an electronically produced representation of the aircraft with its altitude and direction related to the landscape and its flight path. This particular illustration of the display shows the aircraft in an intended altitude-hold flight path (its "highway in the sky"). [Photographs courtesy of Kaiser Aerospace and Electronics, Palo Alto, Calif.]

flight path. This type of display can be interpreted as another example of the principle of compatibility, since it is consistent with our common experiences in viewing objects in three-dimensional space.

Principles of Aircraft-Position Displays   Although the evidence regarding certain aspects of aircraft-position displays is still not definitive, Roscoe [53] has teased out of the relevant research a few principles of display that he believes have substantial validity. Certain of these have been touched on before, in one context or another, but they will nevertheless be reiterated here.

1   *The principle of display integration:* The notion of display integration requires that *related* information be presented in a common display system which allows the relationships to be perceived directly (as illustrated in Figure 4-6, design *a*). This principle does *not* apply to the haphazard combining of unrelated information in a common display.

2   *The principle of pictorial realism:* This principle relates to the presentation of graphic relationships in such a manner that the encoded symbols can be readily identified with what they represent; in effect, the symbols are an analog of that which they represent.

3   *The principle of the moving part:* In the use of aircraft displays this principle is in conformity with the outside-in (i.e., moving-aircraft) display rather than with the inside-out (i.e., moving-horizon) display. In more general terms, it seems preferable for the image of the *moving part* (i.e., an aircraft or symbol representing any other moving object) to be displayed against a fixed scale or coordinate system.

4   *The principle of pursuit tracking:* In pursuit tracking the index of *desired* performance (sometimes referred to as the target) and the index of *actual* performance (sometimes referred to as the target) move over the display against a common scale or coordinate system. Generally this scheme results in better performance than does compensatory tracking, in which the index of either the desired or actual performance is fixed, with the moving index showing only the *error*, or difference.

5   *The principle of frequency separation:* This principle relates to the relative speed of movement of display indications; when "high-frequency"

(suddenly changing or rapidly alternating) information is displayed, the moving element must respond in the expected direction (i.e., compatibility of movement is especially critical); but when low-frequency information is displayed, the direction of movement is not as crucial.

    6 *The principle of optimum scaling:* This principle deals with the physical relationship (really the ratio) of the physical dimensions of that to be represented (i.e., features of the surface of the earth) to the dimensions on the display that represent such features, such as the number of millimeters or inches on the display that represent a mile on the earth. This problem gets all intertwined with precision requirements, but taking these into account, some optimum relationship is possible.

    Although the above principles were crystallized because of their relevance to the problems of aircraft flight and navigation displays, they probably are equally valid for numerous other reasonably corresponding display problems.

Cathode-Ray-Tube (CRT) Displays

The natures of the images presented on cathode-ray tubes (CRT's) are of course a function of the purpose of the display, but they include direct representation of scenes (as in TV), "blips" that represent objects (as in radar and aircraft-control-tower displays), graphic representations (as in various types of test and medical equipment), and generated alphanumeric and symbolic characters. Since there are so many human factors aspects of CRT's, we will here mention only one facet, namely, the problem of resolution, which is essentially associated with the number of raster scan lines—or simply, scan lines. These are the continuous narrow strips of the picture area of varying brightness formed by one horizontal sweep of the scanning spot of a CRT. (These are usually measured in terms of no./in or no./mm, or the number for the "form" presented. Most American TV broadcasts use 525 scan lines; the no./in would of course depend upon the size of a TV screen.)

    As would be expected, the ability of people to recognize images depends in part on the number of scan lines, but some evidence to demonstrate this comes from a study by Wong and Yacoumelos [67], as depicted in Figure 4-18, part (*a*). This shows the percent-correct identifications of four types of symbols in scanning maps presented on a TV screen with 5, 7, or 9 scan lines/mm. The greatest increase in accuracy was from 5 to 7 lines, but there was some further increase from 7 to 9. In this regard, Gould [22] recommends about 10 raster lines per character height for accurate detection of individual characters, with perhaps fewer for detection of words. Part (*b*) of Figure 4-18 shows the differential identification of the four types of images, namely: (1) point symbols (schools, churches, bridges, etc.); (2) alphanumeric; (3) line symbols (highways, railroads, boundaries, etc.); and (4) area symbols (vegetation, urban area, and water). Thus, we see that the type of image influences the recognition of specific items. (Further reference will be made later to alphanumeric characters as presented on CRT's.)

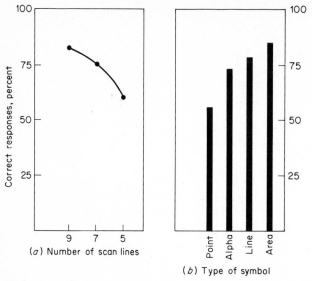

Figure 4-18 (a) Percent correct identification of various types of images on maps presented on TV with different numbers of scan lines/mm; (b) percent correct identification of four types of images (point sources, alphanumeric, lines, and areas). [From Wong and Yacoumelos, 67, figs. 7 and 9.]

Complex Configurations

Some representational displays present complex configurations of such content as land areas, traffic routes, and wiring or piping diagrams. In the development of such displays the dominant guideline is that of simplicity. Obviously, the application of this principle needs to be within the constraints imposed by the operational requirements for "fidelity" of the configuration. The argument for simplicity arises from the fact that the perceptual processes of "searching" for relevant features take longer (and are subject to higher error rates) if an image is cluttered up with what may be irrelevant material. Within the constraints mentioned above there are two possible directions of simplification—one consists simply of removal of extraneous detail; the other, of presenting a schematic representation. The latter approach is illustrated by the use of strip maps and by the representation of the London subway system as shown in Figure 4-19. With most display design problems, one should first ask (and answer) the questions: What information does the user need? and, How can that information best be presented? Obviously the simplification of complex configurations should be guided by the answers to these questions.

Graphic Representations

The format of some of the graphic representations (bar charts, pie charts, line charts, etc.) that find their way into newspapers and other publications leads one to hope that there *must* be better ways of presenting the information that the graphs presumably are intended to convey. Although there has been relatively little research to date regarding the design of graphic representations, one

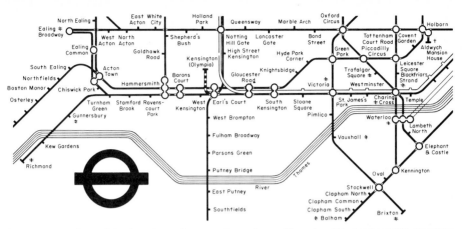

Figure 4-19 Part of the London subway system given with a simplified schematic representation. The schematic form, used in the control center, is much easier for people to use. [Photograph courtesy of London Transport.]

investigation will be summarized to demonstrate that research in this area may actually have some practical application. This study dealt with a comparison of three formats for depicting trend data, as illustrated in Figure 4-20*a*, *b*, and *c*, a

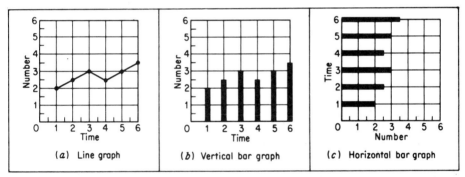

Figure 4-20 Illustrations of formats of multiple-trend charts that were compared on time and accuracy of reading; some examples had 12 or 18 points instead of the 6 shown (line graphs generally were superior). See text for discussion. [Adapted from Schutz, 55.]

line format, a vertical-bar format, and a horizontal-bar format [Schutz, 55]. For each format, there were variations in the numbers of points depicted (6, 12, or 18) and in the number of missing values. The subjects were required to estimate the trend of the data and were scored on the time required to make such estimates and on the accuracy of their estimates. On both criteria the line graph proved to be preferable, as indicated by the following mean scores:

Format	Mean relative time	Mean accuracy score
Line	6.81	1.72
Vertical bar	7.36	1.64
Horizontal bar	8.91	1.40

ALPHANUMERIC AND RELATED DISPLAYS

The effectiveness of communications that involve alphanumeric and symbolic characters depends upon various factors, including typography, content, selection of words, and writing style. Certain aspects of such communications will be discussed to illustrate their effects on the reception of the information presented.[2] Ironically, even in discussions of alphanumeric information there is a fair quota of confusion in the use of words. For our purposes we will adopt the following definitions:

- *Visibility:* The quality of a character or symbol that makes it separately visible from its surroundings. (This is essentially the same as the term *discriminability* as used in Chapter 3.)
- *Legibility:* The attribute of alphanumeric characters that makes it possible for each one to be identifiable from others. (This depends on such features as strokewidth, form of characters, contrast, and illumination.)
- *Readability:* A quality that makes possible the recognition of the information content of material when represented by alphanumeric characters in meaningful groupings, such as words, sentences, or continuous text. (This depends more on the spacing of characters and groups of characters, on their combination into sentences or other forms, on the spacing between lines, and on margins, than on the specific features of the individual characters.)

Typography

The typography of alphanumeric material refers to the various features of the characters and their arrangement.

Strokewidth   The strokewidth of alphanumeric characters is usually expressed as the ratio of the thickness of the stroke to the height of the letters or numerals. Some examples of strokewidth-to-height ratios are shown in Figure 4-21. Investigations of the legibility of strokewidth still leave some loose ends, although there are certain reasonably stable implications that have emerged from a few studies. A pair of rather basic studies are those by Berger [4, 5]. In these studies, white numbers on a black background and black numbers on a white background were used under daylight conditions outdoors, with the strokewidth-to-height ratios ranging from a very thin 1:40 to a heavier 1:5. The criterion used was the average distance at which the subjects could read the numerals; these averages are shown in Table 4-1. In the case of the white numerals, the optimum legibility (i.e., the greatest reading distance) occurred with the 1:13.3 ratio numerals, but there was a fairly wide range of ratios that gave reasonably comparable results. With black letters on a white background, however, the optimum ratio was distinctly lower, around 1:8, with a range from about 1:5.8 to 1:10 giving reasonably comparable

---

[2]For an excellent survey of legibility and related aspects of alphanumeric characters and related symbols, the reader is referred to Cornog and Rose [12].

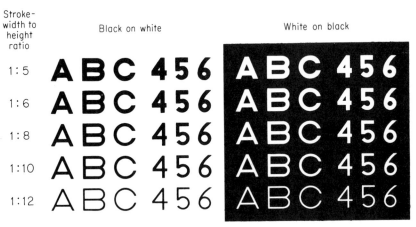

Figure 4-21   Illustrations of strokewidth-to-height ratios of letters and numerals.

results; higher ratios were poorer. This difference in optimum strokewidth of white and of black characters has been confirmed in other investigations, and is attributable to a phenomenon of irradiation, in which white features appear to spread into adjacent black areas, but not the converse. The phenomenon is especially accentuated with highly illuminated display features (in the studies by Berger, for example, under floodlighting conditions, the optimum strokewidth of white luminous numerals on black was 1:40!). Dark adaptation of individuals also tends to accentuate the effect. Because of this effect, white characters on black should have thinner strokewidths than black on white. Incidentally, when dark adaptation is required, the characters preferably should be white on black, whereas when dark adaptation is not required, black on white is preferable.

On the basis of the above and of other studies, it is possible to set forth some generalizations regarding the strokewidth-to-height ratios of alphanumeric characters as follows:

Black on white   1:6 to 1:8
White on black   1:8 to 1:10

Table 4-1   Average Distances in Meters at Which Numerals of Different Strokewidth-to-Height Ratios Can Be Read

Color of numerals	Strokewidth-to-height ratio*							
	1:40	1:20	1:13.3	1:10	1:8	1:6.6	1:5.8	1:5
White	33.9	35.8	36.5	35.5	34.7	33.4	31.4	29.4
Black	25.2	28.0	31.1	32.7	33.5	33.1	32.1	29.9

*The numerals were 42 by 80 mm in size.

Source: Adapted from C. Berger, I. Stroke-width, form and horizontal spacing of numerals as determinants of the threshold of recognition, *Journal of Applied Psychology*, 1944, vol. 28, pp. 208–231, Table 1.

Width-Height Ratios   The relationship between the width and height of alphanumeric characters usually is described as the width-height ratio (expressed as a ratio such as 4:5, or as a percent such as 80 percent). In the case of capital letters, the experimental evidence suggests that the ratio be about 1:1, although this can be reduced to about 3:5 without serious loss in legibility. In the case of numerals the rather standard recommendation is that of a ratio of about 3:5. These ratios are illustrated in Figure 4-22.

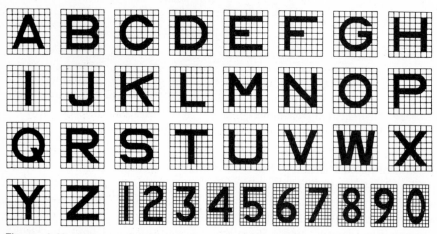

Figure 4-22 Letter and numeral font of United States Military Specification No. MIL-M-18012B (July 20, 1964); also referred to as NAMEL (Navy Aeronautical Medical Equipment Laboratory) or as AMEL. The letters as shown have a width-height ratio of 1:1 (except for I, J, L, and W). These ratios can be reduced to about 2:3 without any appreciable reduction in legibility. The numerals have a width-height ratio of 3:5 (except 1 and 4).

Font of Alphanumeric Characters   Actually, most conventional fonts of alphanumeric characters (and many of the offbeat styles) can be read with reasonable adequacy under normal conditions where size, contrast, illumination, and time permit. There are, however, significant differences in the legibility and readability of different type fonts when viewing conditions are adverse, where time is important, or where accuracy is important. In this connection, the font of capital letters and numerals shown in Figure 4-22 (United States Military Specification No. MIL-M-18012B) has been rather widely tested and found to be generally satisfactory. Although specifically designated for aircrew station displays, the characters, of course, have a wide range of applicability. Another set of characters that are rather widely used by the military services is that shown in Figure 4-23 [MIL Standard MS 33558 (ASG)]. These are sometimes

ABCDEFGHIJKLM
NOPQRSTUVWXYZ
0123456789

Figure 4-23   United States Military Standard letters and numerals MIL Standard MS 33558 (ASG) (Dec. 17, 1957). The basic strokewidth-to-height ratio is 1:8, and the width is about 70 percent of the height. These are sometimes referred to as AND (Air Force-Navy Drawing 10400).

referred to as AND (Air Force–Navy Drawing 10400). The numerals of these two sets (Figures 4-22 and 4-23) were included with a third font, that of Berger [4, 5], in a comparative study of legibility [Atkinson et al., 2] with the following average numbers of errors:

Font	Average errors (for two conditions)	
	Daylight	Transillumination
NAMEL (Figure 4-22)	5.5	11.4
Berger	8.1	13.3
AND (Figure 4-23)	9.7	14.7

Size of Alphanumeric Characters   The ability of people to make visual discriminations (such as of alphanumeric characters) depends on such factors as size, contrast, illumination, and exposure time. A systematic procedure has been proposed by Peters and Adams [45] for determining the size of alphanumeric characters that takes into account certain such factors (illumination, viewing conditions, viewing distance, and importance of reading accuracy). This procedure is based on the following formula:

$$H \text{ (height of letter, in)} = 0.0022D + K_1 + K_2$$

where D = viewing distance
     K_1 = correction factor for illumination and viewing condition
     K_2 = correction for importance (for important items such as emergency labels, $K_2 = .075$; for all other conditions, $K_2 = .0$)

This formula has been applied to various viewing distances, in combination with the other variables, to derive the heights of letters and numerals for those conditions as given in Table 4-2. The lower bounds of these values for a

Table 4-2   **Table of Heights *H* of Letters and Numerals Recommended for Labels and Markings on Panels, for Varying Distance and Conditions, Derived from Formula *H* (in) = 0.0022*D* + *K₁** + *K₂***

Viewing distance, in	0.0022D value	Nonimportant markings, $K_2 = .0$			Important markings, $K_2 = .075$		
		$K_1 = .06$	$K_1 = .16$	$K_1 = .26$	$K_1 = .06$	$K_1 = .16$	$K_1 = .26$
14	0.0308	0.09	0.19	0.29	0.17	0.27	0.37
28	0.0616	0.12	0.22	0.32	0.20	0.30	0.40
42	0.0926	0.15	0.25	0.35	0.23	0.33	0.43
56	0.1232	0.18	0.28	0.38	0.25	0.35	0.45

*Applicability of K_1 values:

   $K_1 = .06$ (above 1.0 fc, favorable reading conditions)
   $K_1 = .16$ (above 1.0 fc, unfavorable reading conditions)
   $K_1 = .16$ (below 1.0 fc, favorable reading conditions)
   $K_1 = .26$ (below 1.0 fc, unfavorable reading conditions)

Source: Based on formula of Peters and Adams [45]; see text.

reading distance of 28 in (for K_1 values of 0.06) are 0.12 in for nonimportant and 0.20 in for important markings. These correspond well with the minimum values proposed by Grether and Baker [26, p. 107] of 0.10 in for noncritical and 0.20 in for critical information or adverse reading conditions. For greater viewing distances, of course, the sizes of characters need to be increased.

One additional point should be made about the size of characters, as demonstrated by Poulton [50]. Some styles of letters have longer "ascenders" and "descenders" than others (i.e., the tip of letters such as b, and the tails of letters such as y). He found that legibility is largely influenced by what he calls "x height," the height of the main body of the letters, and is not influenced much by the length of the ascenders or descenders, nor by the body size (the total height of the block of metal upon which the letter is cast, referred to as 6-pt., 10-pt., etc.). Further, an "x height" of lower-case letters of about 1.2 mm is very close to the lower bounds of legibility. In a separate study involving the scanning of lists of foods for specific items Poulton [49] found that housewives did significantly better with a 6-pt. "Invers" type (which has an x height of about 1.2 mm) than with a smaller typeface, and indicates that this typeface is as small as food manufacturers should use on food containers if the print is to be reasonably legible to housewives of all ages (and probably to non-housewives as well!).

Readability The readability of printed or typed text, and its comprehension, is a function of a wide assortment of factors such as type style (font), type form (capital, lowercase, boldface, italics, etc.), size, contrast, leading (spacing) between lines, length of lines, and margins. This is obviously a very broad spectrum of variables, and it is not in our province here to pull together and synthesize the research in this area. However, to illustrate such research, let us summarize the results of one particular study [Poulton, 48]. In the experiment, a comparison was made of the comprehension of material printed in the four types and formats shown in Table 4-3. The subjects (275 scientists) were tested on their comprehension of the subject matter. Condition A was considerably easier to comprehend than the others, probably owing largely to style of type (compared with C, which was comparable otherwise in point size, number of columns, etc.) and perhaps in part to point size and number of columns (compared with B, the same style but in smaller point size and with two columns instead of one). Such a study is, of course, far from conclusive, but at least it suggests some of the variables that may influence readability and comprehension.

As another example of factors that influence readability, Tinker [63] had some subjects read regular type such as this (actually roman type), *had other subjects read italicized type such as this*, AND HAD OTHER SUBJECTS READ CAPITALIZED TYPE SUCH AS THIS. The regular-type group read significantly faster than those who read material in all capital letters. These results undoubtedly are due in large part to familiarity of people with the conventional upper- and lowercase type in continuous text.

Table 4-3 Effects of Printing Types and Formats on Reading Comprehension

Condition	Style of type			
	A 7 Modern Extended No. 1	**B** 7 Modern Extended No. 1	**C** 101 Imprint	**D** 327 Times New Roman
Point size	11	9	11	9
Letter height, in, total	0.15	0.13	0.15	0.13
Letter height, lower case x	0.07	0.05	0.07	0.06
Leading between lines, in	0.03	0.01	0.02	0.01
Line length	5.2	2.8	5.0	2.8
Columns per page	1	2	1	2
Comprehension scores, %	63	56	58	58

A 7 Modern Extended No. 1: point size 11; 0.03-in leading between lines, 5.2-in line length; 1 column per page.

B 7 Modern Extended No. 1: point size 9; 0.01-in leading between lines; 2.8-in line length; 2 columns per page.

C 101 Imprint: point size 11; 0.02-in leading between lines; 5.0-in line length; 1 column per page.

D 327 Times New Roman: point size 9; 0.01-in leading between lines; 2.8-in line length; 2 columns per page.

Source: From Poulton [48].

In the use of words as labels (such as on instrument panels, for identification), however, the shoe is on the other foot; words in all capital letters in this type of situation generally are more readable than those in lowercase or mixed type.

The design and layout of *Applied Ergonomics* was developed to take advantage of the results of research relating to legibility and readability of printed material [Poulton, Warren, and Bond, 51]. Some of the authors' design decisions (based also in part on practical and aesthetic grounds) are as follows (for what interest they may provide in reflecting the practical application of what the researchers referred to as ergonomic principles to journal design): two columns of 8.8 mm (3.3 in) with from 5.0 to 7.5 mm (0.2 to 0.3 in) separation; 13-mm (0.5-in) margin; unjustified lines; Press Roman type, 10-pt.; 1-pt. leading between lines; and titles and subheadings in upper- and lowercase.

CRT and Other Illuminated Symbols

Under some circumstances alphanumeric and symbolic characters are generated by some form of illumination, as on CRT's, by the use of electroluminescent (EL) lights, by the miniaturization of incandescent lights, etc. Depending on the particular technique used, the configurations reproduced can be conventional characters or those produced by generating dots, lines, or segmented lines. Various investigators have studied the adequacy of different forms of such characters. For example, Vartabedian [64] compared upper-

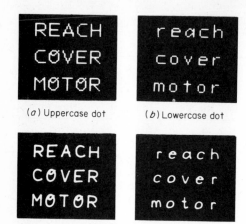

(a) Uppercase dot   (b) Lowercase dot

(c) Uppercase line   (d) Lowercase line

Figure 4-24 Examples of words used in cathode-ray-tube (CRT) displays for comparison of four types of generated letters. [From Vartabedian, 64, fig. 1.]

case and lowercase letters formed with dots and those formed with lines, as shown in Figure 4-24. Displays consisting of 27 words were searched by the subjects for specific words given by the experimenter. The following conclusions were reported: (1) displays with uppercase letters were searched faster than those with lowercase letters; (2) search time was about the same for the dot and stroke designs, and for each of the three sizes used (0.12, 0.14, and 0.16 in); and (3) the subjects preferred uppercase letters and those formed from dots rather than lines.

In a subsequent study [Vartabedian, 65] it was found that readability of a 7×9 dot pattern of characters was about the same as for conventional stroke symbols (formed with a Leroy lettering set) and that there was even a slight edge in favor of the dot patterns (in 12 out of 21 subjects). Thus, the types of characters that can be generated most economically with CRT's are at least as readable as conventional characters and as acceptable for the users.

The advantage of uppercase letters reflected by the results of the study by Vartabedian [64] is consistent with the use of uppercase letters in other contexts where the task is more one of searching or identification than one of reading. (The conclusion regarding size is slightly at odds with that of Giddings [20], who found that symbols of 0.187 in were better than those of 0.125 in). In another investigation Plath [47] compared the legibility of conventional numerals (the NAMEL numerals shown in Figure 4-22) with two forms of segmented numerals, slanted and vertical, as shown in Figure 4-25. The errors in legibility under time-constrained conditions were least for the NAMEL numerals, as shown in that figure. Such results argue against the use of segmented figures where accuracy is critical and when time is severely limited. However, as indicated earlier, if there are no time limitations in viewing alphanumeric symbols and if they are of reasonable size and presented under adequate illumination, virtually any form can be read. Alphanumeric design becomes critical under adverse conditions.

Although the discriminability of symbols used on CRT displays is

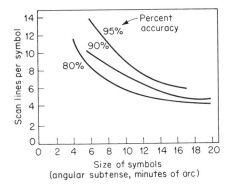

Type of numeral	Examples	Number of errors
NAMEL	**3 4 5**	187
Slanted segmented	⌐𝟣𝟧	391
Vertical segmented	⌐𝟦𝟧	388

Figure 4-25 Forms of numerals investigated by Plath, along with errors made in their identification under time-controlled conditions. [Adapted from Plath, 47, fig. 1 and table 1.]

influenced by their design and size, it is also influenced by the number of scan lines of the CRT (which were referred to before). Using 16 symbols (squares, stars, triangles, half-circles, pentagons, etc.), Hemingway and Erickson [27] first explored their legibility when presented in various sizes and with various numbers of scan lines, and then synthesized their results with those of other investigations (albeit those in which alphanumeric characters were used) and ended up with the pattern shown in Figure 4-26. This shows the combinations

Figure 4-26 Number of scan lines and size of alphanumeric and geometric symbols required for each of three levels of identification accuracy. [From Hemingway and Erickson, 27, fig. 10.]

of scan lines and angular subtense (i.e., the size in visual angle) for each of three levels of accuracy. It is clear from this that there is a trade-off between number of scan lines and size of symbols, but the results still indicate that for a high accuracy it is desirable to provide for 8 or 10 or more scan lines unless the symbols are quite large.

Readability of Groupings of Letters and Numerals

We are frequently required to store in long-term memory sequences of several numerals and/or letters such as our own identification numbers, license numbers, and account numbers, or to store in short-term memory such sequences as telephone numbers and numbers to be keypunched. Such long-term or short-term storage usually is facilitated when we slice up the entire sequence into groups. In the case of numerals, Klemmer [37], in synthesizing

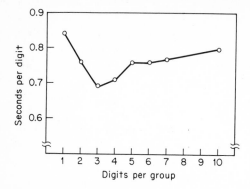

Figure 4-27 Average time per digit to key numbers of 18 to 21 digits as a function of grouping of the digits. [From Klemmer, 37, fig. 1.]

the results of various studies, points out that the typical "bunching" is in terms of groups of three or four. This pattern is reflected by results of his own study, as shown in Figure 4-27. This particular study dealt with the presentation of 18 to 21 digits in various groupings, in a keying task using a push-button telephone. In the keying of seven-digit numbers, presented with various groupings, the average keying times are those shown below:

Grouping of digits	Average keying time, s
376 5934	6.7
37 65934	7.3
3765934	7.6
3 765934	8.2

Here, again, is the impression of the advantage of grouping in threes and fours.

The proclivity of people to deal with alphanumeric symbols in "groups" is also reflected by the results of a study by Karmeier et al. [36] that dealt with 16 combinations of five or six letters and/or numerals that might be used on license plates. The criterion used was the percentage of correct identifications of the characters. These percentages are given below for a few of the 16 combinations, these results being from a sample of police officers:

Form	Percent correct	Form	Percent correct
12 345	99	123 456	75
AB 123	93	AB 1234	70
123 A4	83	ABC 123	57

The implication of such indications about the way in which people tend to "group" numerals and letters is of course that of presenting them (whenever appropriate) in the groupings that are most natural or compatible for people.

Saying What Is Meant

Your experiences in everyday reading and writing will confirm the fact that recorded verbal material does not always convey reliably the meaning that is intended. The unambiguous, understandable use of language is very pertinent to various tangents of human factors engineering, including the preparation of training materials, job aids, instructions, directions, and labels. Chapanis [8] ran across one horrible example of a notice beside an elevator that read as shown in the first box of Figure 4-28. That figure also indicates what the sign

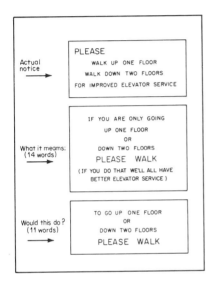

Figure 4-28 Examples of an elevator sign and of suggested revisions. [Adapted from Chapanis, 8.]

really meant, and presents an improved version that is both clearer and shorter. When the material to be presented involves complex relationships, it may be desirable to forgo flowing prose in favor of some more straightforward presentation, such as the use of a "logic tree," as suggested by Wright [69].

VISUAL CODES, SYMBOLS, AND SIGNS

Our present civilization abounds with a wide assortment of visual codes, symbols, and signs that are intended to convey meaning to us, their use being part of almost all phases of human activity, as: travel, business, medicine, the sciences, religion, engineering, and recreation. Actually the use of graphic symbols as a means of communication goes back into the early history of man; it became part of the folklore of most cultures. Dreyfus, who developed a data bank of 20,000 graphic symbols used all over the world [15], considered such symbols as being of one of three types: (1) representational (fairly accurate, simplified pictures of objects, such as a skull and crossbones to represent danger, or of actions, such as a man on a bicycle representing a bicycle path); (2) abstract (symbols that reduce essential elements of a message to graphic terms, retaining only faint resemblance to the original concept, as the signs of the zodiac); and (3) arbitrary (symbols that have been invented and that then need to be learned, such as the triangle "yield" traffic sign).

In the intentional use of visual codes, symbols, and signs for specific purposes, we need to keep in mind the several guidelines for coding as set forth in the previous chapter and, in particular, the requirements for detectability, discriminability, and meaningfulness. Meaningfulness, of course, can be based on symbolic representation (what Dreyfus refers to as representational and abstract symbols), or on learned association (the need to "learn" the meaning of what Dreyfus refers to as arbitrary symbols). Where possible, visual coding systems should capitalize on the use of symbolic representations or whatever learned associations people already have acquired. In discussing visual codes, symbols, and signs, let us first summarize the results of a sample of illustrative research investigations.

Unidimensional Visual Codes

Some of the studies of individual visual codes (quaintly called *alphabets*) have involved a comparison of the effectiveness of different types of codes for use in relation to various types of "tasks." One such study is that of Hitt [28], who used the five different codes shown in Figure 4-29: numerals, letters, geometric

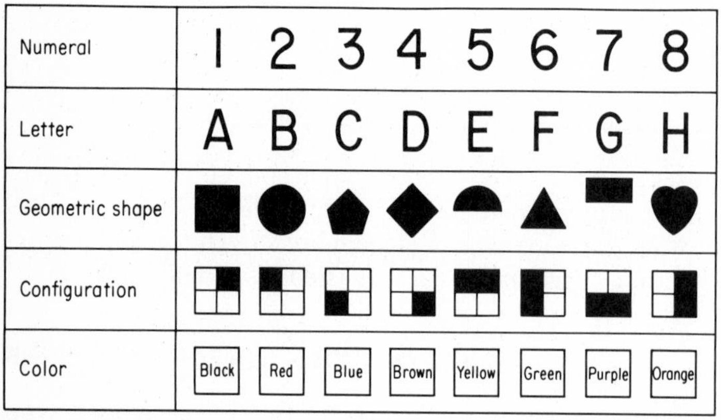

Figure 4-29 Illustration of code symbols used in comparing coding of targets. [From Hitt, 28.]

shapes, configurations, and colors. The code symbols were used in maplike displays with eight columns and five rows. Eight symbols of each class were used to represent eight types of buildings, facilities, industries, etc. Each display included only one type of code, but under different conditions the number of different symbols of the type was varied (2, 4, and 8) and the density (the number of different symbols in a cell) was varied (1, 2, and 3).

The subjects performed five different tasks with these displays; *identification* (e.g., identify type of industry in one cell); *location* (e.g., locate cell that includes only one steel mill); *counting* (e.g., count number of aluminum plants in row C); *comparison* (e.g., compare number of petroleum refineries in one cell with that in another); and *verification* (e.g., the plant in a given cell is a steel mill—true or false). Their chore was to go down a list of questions of these types and record their answers. A summary of the major results is given in Figure 4-30, which shows, for the different codes and tasks, the number of correct responses per minute.

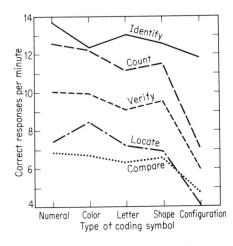

Figure 4-30 Relationship between coding method and performance in five tasks comparing targets on a display. [Adapted from Hitt, 28.]

In general terms it can be seen that the numerical and color codes were best for most of the tasks and that the configuration code was consistently the last, with the letter and shape codes falling in between, in some instances comparing reasonably well with the numerical and color codes.

Let us look at one more study of unidimensional visual codes, this one by Smith and Thomas [61]. They used four codes, namely, color, military symbols, geometric forms, and aircraft shapes, as shown in Figure 4-31, each code having five symbols. For each of the three shape codes, displays were prepared

	C-54	C-47	F-100	F-102	B-52
Aircraft shapes					
Geometric forms	Triangle	Diamond	Semicircle	Circle	Star
Military symbols	Radar	Gun	Aircraft	Missile	Ship
Colors (Munsell notation)	Green (2.5 G 5/8)	Blue (5 BG 4/5)	White (5 Y 8/4)	Red (5R 4/9)	Yellow (10 YR 6/10)

Figure 4-31 Four sets of codes used in a study by Smith and Thomas [61] (copyright 1964 by the American Psychological Association and reproduced by permission). The notations under the color labels are the Munsell color matches [72] of the colors used.

with 20, 60, or 100 symbols of the type in question, these being randomly allocated to any of 400 positions in a 20 by 20 imaginary matrix. In various parts of the study three sets of displays were used (each set having separate displays for each of the three shape codes), as follows:

1 Sets with shape symbols colored randomly
2 Sets with shape symbols all the same color (but with different displays for each color)
3 Sets in which each of the five symbols of a shape class was coded a unique color, with different displays of each symbol-color combination

The task of the subjects was to count the number of items of a predesignated *target* class, such as *red, gun, circle,* or *B*-52, depending upon the set of displays used in the particular phase of the study. Both time and errors were recorded, the results being shown in Figure 4-32. As would be expected, time

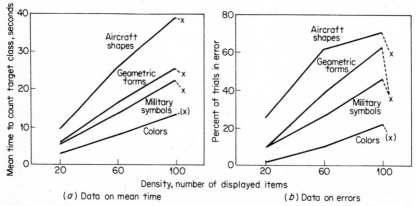

(*a*) Data on mean time   (*b*) Data on errors

Figure 4-32 Mean time *a* and errors *b* in counting items of four classes of codes as a function of display density [Smith and Thomas, 61] (copyright 1964 by the American Psychological Association and reprinted by permission). The ×'s indicate comparison data for displays of 100 items with color (or shape) held constant.

and errors increased with density (the number of items in a display); but more importantly, it was obvious that time and errors differed for the various types of codes, with color generally being the best.

Apart from the specific results of these two studies, it is obvious from such studies that the use of various types of visual coding systems can result in differential effectiveness in their use by people. But we should hasten to add the point that the utility of a given code system can depend upon the context in which it is used—in particular, the nature of the task being performed.

Realizing that various factors (including good judgment) must enter into the selection of visual codes for specific purposes, a comparison such as that given in Table 4-4 can serve as at least partial guidance. In particular, that table indicates the approximate number of levels of each of various visual codes that can be discriminated, along with some sideline comments about certain of the methods.

Table 4-4 Summary of Certain Visual Coding Methods
(Numbers refer to number of levels which can be discriminated on an absolute basis under optimum conditions.)

Alphanumeric	Single numerals, 10; single letters, 26; combinations, unlimited. Good; especially useful for identification; uses little space if there is good contrast. Certain items easily confused with each other.
Color (of surfaces)	Hues, 9; hue, saturation, and brightness combinations, 24 or more. Preferable limit 9. Particularly good for searching and counting tasks; poorer for identification tasks. Affected by some lights; problem with color-defective individuals.†‡
Color (of lights)	10. Preferable limit 3. Limited space required. Good for qualitative reading.◖
Geometric shapes	15 or more. Preferable limit 5. Generally useful coding system, particularly in symbolic representation; good for CRT's. Shapes used together need to be discriminable; some sets of shapes more difficult to discriminate than others.◖
Angle of inclination	24. Preferable limit 12. Generally satisfactory for special purposes such as indicating direction, angle, or position on round instruments like clocks, CRT's, etc.§
Size of forms (such as squares)	5 or 6. Preferable limit 3. Takes considerable space. Use only when specifically appropriate.
Visual number	6. Preferable limit 4. Use only when specifically appropriate, such as to represent numbers of items. Takes considerable space; may be confused with other symbols.
Brightness of lights	3–4. Preferable limit 2. Use only when specifically appropriate. Weaker signals may be masked.◖
Flash rate of lights	Preferable limit 2. Limited applicability if receiver needs to differentiate different flash rates. Flashing lights, however, have possible use in combination with controlled time intervals (as with lighthouse signals and naval communications) or to attract attention to specific areas.

†Feallock et. al. [19].
‡Jones [34].
§Muller et. al. [41].
◖Grether and Baker [26].

Multidimensional Visual Codes

As one example of a multidimensional visual code we will review briefly a study by Anderson and Fitts [1] in which a redundant-combination color-numeric code (black numerals on color patches) was compared with two single-coding dimensions, one a color code (nine color patches) and the other a numeric code (nine black numerals). By varying the number of items of each of these three classes, it was possible to vary the amount of stimulus information to be presented in different "messages" from 9.51 to 25.36 bits. Without going into details, each message (i.e., specific colors, numerals, or combinations) was presented to the subjects for 0.1 s, and the subjects recorded what they thought the message was. The results of the study showed that the total information transmitted to the subjects (actually received by them) generally was higher for the redundant color-numeric code than for either the color or the numeric code.

Substantial confirmation of this comes from a study by Smith [60] in which specific alphanumeric items in a field of 20, 60, or 100 such items were to be searched or counted. In some displays the items to be searched or counted were color-coded, and in others they were not. The percents of *reduction* in mean times when the alphanumeric items were color-coded (as opposed to being black and white) are summarized below, these results showing reductions for both tasks, but particularly for the search task.

	Reduction in mean time, %
Search task	45 to 70
Counting task	63 to 70

By and large, it appears that redundant visual coding systems tend to result in improved performance by people, but we should hasten to add that this is not a universal pattern, as reported by Kanarick and Petersen [35]. (Their particular study dealt with a "keeping-track" task in which the stimuli were redundant combinations of letters and colored backgrounds.) Their study reinforced the point that the nature of the task at hand can affect the utility of a given coding system. But although a redundant coding system might not, in a specific instance, result in an improved performance, such a coding system would not be expected to bring about any degradation in performance (as contrasted with the use of a single code).

Color Coding

Since color is a fairly common visual code, we will discuss it somewhat further. An important question relating to color deals with the number of distinct colors persons with normal color vision can differentiate on an absolute basis. It has generally been presumed that the number was relatively moderate; for example, Jones [34] indicated that the normal observer could identify about nine surface colors, varying primarily in hue. However, it appears that, with training, people can learn to make upwards of a couple dozen such discriminations when combinations of hue, saturation, and lightness are prepared in a nonredundant manner [Feallock et al., 20]. On the basis of this study 24 colors were so chosen that only two of those within the set were confused by subjects with others in the set under four experimental lighting conditions.[3]

When relatively untrained people are to use color codes, however, the better part of wisdom would argue for the use of smaller numbers. In this regard Conover and Kraft [11] present four sets of colors for coding purposes when

[3]The identifications of these colors in the Federal Standard on Colors [73] follow, with an (×) after those that were identifiable under all lighting conditions, are 32648 (×), 31433, 30206 (×), 30219, 30257, 30111, 31136 (×), 31158, 32169, 32246, 32356, 33538 (×), 33434, 33695, 34552 (×), 34558, 34325 (×), 34258 (×), 34127, 34108 (×), 35189, 35231 (×), 35109, and 37144 (×).

absolute recognition is required (barring the use of color-blind people). These sets include, respectively, eight, seven, six, and five colors.[4]

Color is clearly a very useful coding dimension in some contexts. This was illustrated, for example, by the results of a study by Shontz et al. [56] in which subjects were to locate (i.e., to search for) specific types of 28 checkpoints (e.g., airfields, bridges, etc.) on aeronautical charts. The three experimental conditions used were:

1 Printed labels of checkpoints, plain chart
2 Color dots ($^3/_{16}$ in) to represent checkpoints, plain chart
3 Color dots ($^3/_{16}$ in) to represent checkpoints, gray (achromatic) chart

The results, given in Figure 4-33, show the cumulative proportion of

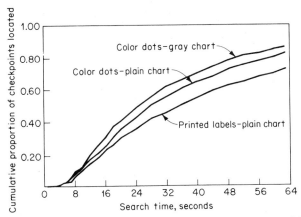

Figure 4-33 Search time for various checkpoints (e.g., airfields, bridges, etc.) on aeronautical charts with checkpoints identified by color codes and by printed labels. [Adapted from Shontz et al., 56, fig. 2.]

checkpoints located over the experimental period. The figure indicates a systematic difference in favor of the color-coded conditions (2 and 3). Essentially similar, but perhaps more accentuated, results had previously been reported by Smith [60], in an experiment involving the searching for, and counting of, alphanumeric symbols in a two-dimensional display, in which a comparison was made when the symbols were, and were not, color-coded. And Konz and Koe [38] report a modest reduction in errors (10 to 15 percent) in a card-filing task in which the letters of the cards were color-coded, as contrasted with not being color-coded.

But lest we jump to the conclusion that color coding is a universally superi-

[4]The Munsell notations for the colors in these four sets are given below:

8-color: 1R; 9R; 1Y; 7GY; 9G; 5B; 1P; 3RP
7-color: 5R; 3YR; 5Y; 1G; 7BG; 7PB; 3RP
6-color: 1R; 3YR; 9Y; 5G; 5B; 9P
5-color: 1R; 7YR; 7GY; 1B; 5P

or mode of visual coding, we should hark back to the study by Kanarick and Peterson in which it was found that color did not improve performance on a "keeping-track" task (which depended very much on remembering previously displayed information). In reflecting about the use of color codes, it appears that they have their greatest utility in circumstances in which there is some type of searching, scanning, or locating task, presumably because they "catch the eye" more rapidly than most other visual codes.

Road Signs

Road signs represent a natural situation for use of symbolic codes, such as the shape of signs and the shape of configurations on them. Such signs, of course, have to be visible at appropriate distances and under many conditions and discriminable from other signs, and any alphanumeric characters on them must be readable. However, there are two other characteristics that it would be desirable for road signs to have. First, insofar as possible they should be visually suggestive of that which they are intended to symbolize, in order to minimize the "recoding" of symbols. And second, they should generally be standardized across those geographical boundaries that are commonly traversed. The international road signs generally conform to these principles. Certain examples are shown in Figure 4-34. The international sign system has

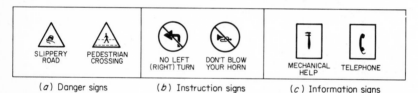

(*a*) Danger signs   (*b*) Instruction signs   (*c*) Information signs

Figure 4-34 Examples of a few International Road Signs. These signs conform to two useful design principles, namely, they are directly symbolic of their meaning, and they are standardized across countries.

much higher marks on these principles than do the sign systems of the states in the United States, although the United States is moving in the direction of standardization.

Actual Use of Road Signs   Even the best-designed signs in the world, however, serve no purpose if they are not used. And there is (unfortunately) evidence that road signs frequently are not used by drivers. On the basis of a field study in Sweden, Johansson and Backlund [29] come to the dismal conclusion that " . . . the road sign system to a high degree does not achieve its purpose." Their study was carried out by placing any one of six signs (shown in Figure 4-35) on a main highway each in such a position that drivers would have a clear view of it about 400 meters away. Beyond the sign, around a slight rise and curve, was a police barrier, out of sight from the position where the sign could be viewed. The drivers were stopped by police at the barrier and asked certain questions, including: What was the last road sign you passed? The percentages

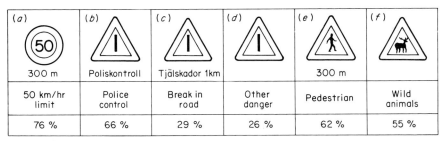

Figure 4-35 Road signs used in a field study on a main highway in Sweden, along with the percent of drivers who recalled having seen each sign about a minute after passing it. [Adapted from Johansson and Backlund, 29.]

of 2525 drivers who responded correctly to this question are given in Figure 4-35, these values ranging from 26 percent (sign *d*) to 76 percent (sign *a*). (These values can be considered as reflecting the human "reliability" associated with the signs.) Interestingly enough, there is marked difference in the recall of the various signs despite the fact that they were very similar from the point of view of perceptual impressiveness. These differences suggest some sort of differential "selectivity" in giving attention to the signs, perhaps in terms of their potential relevance to the drivers. As an addend, it should be noted that a comparison of the responses of four different groups of drivers showed that those drivers who knew in advance of the experiment did much better (as one would expect) than those drivers who had known nothing about it.

The fact that many drivers presumably pass relevant road signs by with nary a glance is of course very disturbing. Since there is, at the moment, no obvious approach to the modification of this pattern of behavior, this problem must be retained on the human factors agenda as a piece of unfinished business.

Perceptual Principles of Symbolic Designs

In discussing the use of coding symbols, Easterby [16,17] makes the point that the effective use of such displays is predicated on perceptual processes, and postulates a number of principles that are rooted in perceptual research that generally would enhance the use of such displays. Certain of these principles will be summarized briefly and illustrated. The illustrations are in Figure 4-36. Although these particular examples are specifically applicable to machine displays [Easterby, 17], the basic principles would be applicable in other contexts.

Figure/Ground Clear and stable figure-to-ground articulation is essential, as illustrated in (*a*) of Figure 4-36.

Figure Boundaries A contrast boundary (essentially a solid shape) is preferable to a line boundary, as shown in (*b*) of Figure 4-36. With different elements of a display to be depicted, the following practice should be followed:

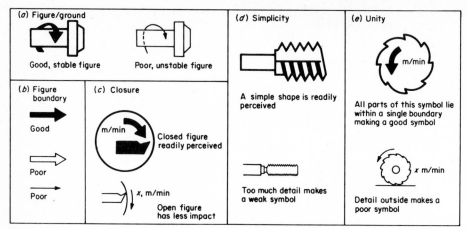

Figure 4-36 Examples of certain perceptual principles relevant to the design of visual code symbols. These particular examples relate to codes used with machines. [Adapted from Easterby, 17.]

symbol dynamic—solid; moving or active part—outline; stationary or active part—solid.

Closure A closed figure as illustrated in (c) of Figure 4-36 enhances the perceptual process and should be used unless there is reason for the outline to be discontinuous.

Simplicity The symbols should be as simple as possible, consistent with the inclusion of features that are necessary, as illustrated in (d) of Figure 4-36.

Unity Symbols should be as unified as possible. For example, when solid and outline figures occur together, the solid figure should be within the line outline figure, as shown in (e) of Figure 4-36.

GENERAL GUIDELINES IN DESIGNING VISUAL DISPLAYS

In this chapter we have discussed and illustrated a variety of visual displays. The sample of displays covered, however, is only suggestive of the wide range of visual displays that are in actual use. In the selection or design of visual displays for certain specific purposes, the basic type of display to use is sometimes virtually dictated by the nature of the information to be presented and the use to which it is to be put. In other circumstances, however, options may be available.

Although it is not possible to provide specific guidelines to follow in resolving every design problem, there are certain general guidelines and principles that can be followed in many situations. A few are given below, along with some words of caution where circumstances sometimes will justify deviations from these. These guidelines deal largely with some of the more conventionally used visual displays.

Quantitative Scales

• Digital or open-window is preferable if values remain long enough to read.
• Fixed-scale, moving-pointer designs are usually preferable to moving-scale, fixed-pointer designs
• For long scales, moving scale with tape on spools behind panel or a counter plus circular scale has practical advantage over fixed scale.
• For values subject to continuous change, display all (or most) of range used (as with circular or horizontal scale).
• If two or more items of *related* information are to be presented, consider integrated display.
• Smallest scale unit to be read should be represented on scale by about 0.05 in or more.
• Preferably use marker for each scale unit, unless scale has to be very small.
• Use conventional progression system of 1, 2, 3, 4, etc., unless there is reason to do otherwise, with major markers at 0, 10, 20, etc.

Qualitative Scales

• Preferably use fixed scale with moving pointer (to show trends).
• For groups, use circular scales, and arrange null positions systematically for ease of visual scanning, as at 9 o'clock or 12 o'clock positions.
• Preferably use extended pointers, and possibly extended lines between scales.

Status Indicators

• If basic data represent discrete, independent categories, or if basically quantitative data are always used in terms of such categories, use display that represents each.

Signal and Warning Lights

• Minimum size used must be consistent with luminance and exposure time.
• With low signal-to-background contrast, red light is more visible.
• Flash rate of flashing lights of 1 to 10 per second presumably can be detected by people.

Representational Displays

• A moving element (such as an aircraft) should be depicted against a fixed background (as the horizon).
• Graphic displays that depict trends are read better if they are formed with lines than with bars.

- Pursuit displays usually are easier for people to use than compensatory displays.
- Cathode-ray-tube (CRT) displays are most effective when there are seven to nine or more scan lines per mm.
- In the design of displays of complex configurations (such as traffic routes and wiring diagrams), avoid unnecessary detail and use schematic representation if consistent with uses.

Alphanumeric Displays

- The typography of alphanumeric characters (design, size, contrast, etc.) is especially critical under adverse viewing conditions.
- Alphanumeric characters should be presented in groups of three or four for optimum short-term memory.
- Capital letters and numerals used in visual displays are read most accurately (*a*) when the ratio of strokewidth to height is about 1:6 to 1:8 for black on white and somewhat higher (up to 1:10) for white on black, and (*b*) when the width is at least two-thirds the height.

Symbolic Displays

- Symbolic displays should be designed on the basis of the following perceptual principles: figure/ground; figure boundaries; closure; simplicity; and unity.

REFERENCES

1   Anderson, N. S., and P. M. Fitts: Amount of information gained during brief exposures of numerals and colors, *Journal of Experimental Psychology*, 1958, vol. 56, pp. 362–369.
2   Atkinson, W. H., L. M. Crumley, and M. P. Willis: *A study of the requirements for letters, numbers, and markings to be used on trans-illuminated aircraft control panels. Part 5. The comparative legibility of three fonts for numerals*, Naval Air Material Center, Aeronautical Medical Equipment Laboratory, Report TED NAMEL–609, part 5, June 13, 1952.
3   Balding, G. H., and C. Süskind: Generation of artificial electronic displays, with application to integrated flight instrumentation, *IRE Transactions on Aeronautical and Navigational Electronics*, September, 1960, vol. ANE–7, no. 3, pp. 92–98.
4   Berger, C.: I, Stroke-width, form and horizontal spacing of numerals as determinants of the threshold of recognition, *Journal of Applied Psychology*, 1944, vol. 28, pp. 208–231.
5   Berger, C.: II, Stroke-width, form and horizontal spacing of numerals as determinants of the threshold of recognition, *Journal of Applied Psychology*, 1944, vol. 28, pp. 336–346.
6   Burg, A.: Visual acuity as measured by static and dynamic tests: A comparative evaluation, *Journal of Applied Psychology*, 1966, vol. 50, no. 6, pp. 460–466.
7   Burg, A., and S. F. Hulbert: Dynamic visual acuity and other measures of vision, *Perceptual and motor skills*, 1959, vol. 9, p. 334.

8 Chapanis, A.: Words, words, words, *Human Factors*, February, 1967, vol. 7, no. 1, pp. 1–17.

9 Christensen, J. M.: *Quantitative instrument reading as a function of dial design, exposure time, preparatory fixation and practice*, USAF AML Report AF 52/116, 1952.

10 Cohen, E., and R. L. Follert: Accuracy of interpolation between scale graduations, *Human Factors*, 1970, vol. 12, no. 5, pp. 481–483.

11 Conover, D. W., and C. L. Kraft: *The use of color in coding displays*, USAF, WADC, TR 55–471, October, 1958.

12 Cornog, D. Y., and F. C. Rose: *Legibility of alphanumeric characters and other symbols: II. A reference handbook*, National Bureau of Standards, Miscellaneous 262–2, Superintendent of Documents, Washington, D. C., February, 1967.

13 Crawford, A.: The perception of light signals: the effect of mixing flashing and steady irrelevant lights, *Ergonomics*, 1963, vol. 6, pp. 287–294.

14 Dashevsky, S. G.: Check-reading accuracy as a function of pointer alignment, patterning, and viewing angle, *Journal of Applied Psychology*, 1964, vol. 48, pp. 344–347.

15 Dreyfus, H.: *Symbol sourcebook*, McGraw-Hill Book Company, New York, 1972.

16 Easterby, R. S.: Perceptual organization in static displays for man/machine systems, *Ergonomics*, 1967, vol. 10, pp. 195–205.

17 Easterby, R. S.: The perception of symbols for machine displays, *Ergonomics*, 1970, vol. 13, no. 1, pp. 149–158.

18 Elkin, E. H.: *Effect of scale shape, exposure time and display complexity on scale reading efficiency*, USAF, WADC, TR 58–472, February, 1959.

19 Feallock, J. B., J. F. Southard, M. Kobayashi, and W. C. Howell.: Absolute judgments of colors in the Federal Standards System, *Journal of Applied Psychology*, 1966, vol. 50, pp. 266–272.

20 Giddings, B. J.: Alpha-numerics for raster displays, *Ergonomics*, 1972, vol. 15, no. 1, pp. 65–72.

21 Goodson, J. E., and J. W. Miller: *Dynamic visual acuity in an applied setting*, USN School of Aviation Medicine, Pensacola, Fla., Report 16, Project NM 17 01 99 Subtask 2, May 25, 1959.

22 Gould, J. D.: Visual factors in the design of computer-controlled CRT displays, *Human Factors*, 1968, vol. 10, no. 4, pp. 359–376.

23 Graham, C. H., and R. Margaria: Area and intensity-time relation in the peripheral retina, *American Journal of Physiology*, 1935, vol. 113, pp. 299–305.

24 Graham, N. E.: The speed and accuracy of reading horizontal, vertical, and circular scales, *Journal of Applied Psychology*, 1956, vol. 40, pp. 228–232.

25 Grether, W. F.: Instrument reading: I. The design of long-scale indicators for speed and accuracy of quantitative readings, *Journal of Applied Psychology*, 1949, vol. 33, pp. 363–372.

26 Grether, W. F., and C. A. Baker: "Visual presentation of information," in H. A. Van Cott and R. G. Kinkade (eds.), *Human engineering guide to equipment design*, rev. ed., chap. 3, U. S. Government Printing Office, Washington, D. C., 1972.

27 Hemingway, J. C., and R. A. Erickson: Relative effects of raster scan lines and image subtense on symbol legibility on television, *Human Factors*, 1969, vol. 11, no. 4, pp. 331–338.

28　Hitt, W. D.: An evaluation of five different coding methods, *Human Factors*, July, 1961, vol. 3, no. 2, pp. 120–130.

29　Johansson, G., and F. Backland: Drivers and road signs, *Ergonomics*, 1970, vol. 13, no. 6, pp. 749–759.

30　Johnsgard, K. W.: Check-reading as a function of pointer symmetry and uniform alignment, *Journal of Applied Psychology*, 1953, vol. 37, pp. 407–411.

31　Johnson, S. L., and S. N. Roscoe: What moves, the airplane or the world?, *Human Factors*, 1972, vol. 14, no. 2, pp. 107–129.

32　Johnson, S. L., R. C. Williges, and S. N. Roscoe: *An approach to motion relations for flight director displays*, Aviation Research Laboratory, Institute of Aviation, University of Illinois, Savoy, Ill., Technical Report ARL-71/20/ONR-71-3/AFOSR-71-6, October, 1971.

33　Jones, J. C., A. J. Ward, and P. W. Haywood: *Reading dials at short distances*, AEI Engineering (Associated Electrical Industries Ltd., London), Jan.–Feb., 1965, vol. 5, no. 1, pp. 28–32.

34　Jones, M. R.: Color coding, *Human Factors*, 1962, vol. 4, pp. 355–365.

35　Kanarick, A. F., and R. C. Petersen: Redundant color coding and keeping-track performance, *Human Factors*, 1971, vol. 13, no. 2, pp. 183–188.

36　Karmeier, D. F., C. G. Herrington, and J. E. Baerwald: "A comprehensive analysis of motor vehicle license plates," in H. O. Orland (ed.), *Highway Research Board, Proceedings of the thirty-ninth annual meeting*, NRC Publication 773, 1960, vol. 39, pp. 416–440, National Academy of Sciences, Washington, D. C.

37　Klemmer, E. T.: Grouping of printed digits for manual entry, *Human Factors*, 1969, vol. 11, no. 4, pp. 397–400.

38　Konz, S. A., and B. A. Koe: The effect of color coding on performance of an alphabetic filing task, *Human Factors*, 1969, vol. 11, no. 3, pp. 207–212.

39　Kurke, M. I.: Evaluation of a display incorporating quantitative and check-reading characteristics, *Journal of Applied Psychology*, 1956, vol. 40, pp. 233–236.

40　Markowitz, J.: Optimal flash rate and duty cycle for flashing visual indicators, *Human Factors*, 1971, vol. 13, no. 5, pp. 427–433.

41　Muller, P. F., Jr., R. C. Sidorsky, A. J. Slivinske, E. A. Alluisi, and P. M. Fitts: *The symbolic coding of information on cathode ray tubes and similar displays*, USAF, WADC, TR 55-375, October, 1955.

42　Murrell, K. F. H.: *Human performance in industry*, Reinhold Publishing Corporation, New York, 1965.

43　Murrell, K. F. H., W. D. Laurie, and C. McCarthy: The relationship between dial size, reading distance and reading accuracy, *Ergonomics*, 1958, pp. 182–190.

44　Oatman, L. C.: Check-reading accuracy using an extended-pointer dial display, *Journal of Engineering Psychology*, 1964, vol. 3, pp. 123–131.

45　Peters, G. A., and B. B. Adams: These three criteria for readable panel markings, *Product Engineering*, May 25, 1959, vol. 30, no. 21, pp. 55–57.

46　Petersen, H. E., and D. J. Dugas: The relative importance of contrast and motion in visual detection, *Human Factors*, 1972, vol. 14, no. 3, pp. 207–216.

47　Plath, D. W.: The readability of segmented and conventional numerals, *Human Factors*, 1970, vol. 12, no. 5, pp. 493–497.

48　Poulton, E. C.: *Effects of printing types and formats on the comprehension of scientific journals*, Applied Psychology Research Unit, Cambridge, England, Report APU 346, 1959.

49　Poulton, E. C.: Skimming lists of food ingredients printed in different sizes, *Journal of Applied Psychology*, 1969, vol. 53, no. 1, pp. 55–58.

50   Poulton, E. C.: Size, style, and vertical spacing in the legibility of small typefaces, *Journal of Applied Psychology*, 1972, vol. 56, no. 2, pp. 156–161.

51   Poulton, E. C., T. R. Warren, and J. Bond: Ergonomics in journal design, *Applied Ergonomics*, 1970, vol. 1, no. 4, pp. 207–209.

52   Reynolds, R. E., R. M. White, Jr., and R. L. Hilgendorf: Detection and recognition of colored signal lights, *Human Factors*, 1972, vol. 14, no. 3, pp. 227–236.

53   Roscoe, S. N.: Airborne displays for flight and navigation, *Human Factors*, 1968, vol. 10, no. 4, pp. 321–332.

54   Sabeh, R., W. R. Jorve, and J. M. Vanderplas: *Shape coding of aircraft instrument zone markings*, USAF, WADC, Technical Note 57–260, March, 1958.

55   Schutz, H. G.: An evaluation of formats for graphic trend displays—Experiment II, *Human Factors*, 1961, vol. 3, pp. 99–107.

56   Shontz, W. D., G. A. Trumm, and L. G. Williams: Color coding for information location, *Human Factors*, 1971, vol. 13, no. 3, pp. 237–246.

57   Simon, C. W., and S. N. Roscoe: *Altimetry studies: II. A comparison of integrated versus separated, linear versus circulatar, and spatial versus numerical displays*, Hughes Aircraft Company, Culver City, Calif., Technical Memorandum 435, May, 1956.

58   Sinclair, H. J.: Digital versus conventional clocks—a review, *Applied Ergonomics*, September, 1971, pp. 178–181.

59   Sleight, R. B.: The effect of instrument dial shape on legibility, *Journal of Applied Psychology*, 1948, vol. 32, pp. 170–188.

60   Smith, S. L.: *Display color coding for visual separability*, MITRE Corp., Bedford, Mass., MITRE Report MTS-10, August, 1963.

61   Smith, S. L., and D. W. Thomas: Color versus shape coding in information displays, *Journal of Applied Psychology*, 1964, vol. 48, pp. 137–146.

62   Teichner, W. H., and M. J. Krebs: Estimating the detectability of target luminances, *Human Factors,* 1972, vol. 14, no. 6, pp. 511–519.

63   Tinker, M. A.: Prolonged reading tasks in visual research, *Journal of Applied Psychology*, 1955, vol. 39, pp. 444–446.

64   Vartabedian, A. G.: The effects of letter size, case, and generation method on CRT display search time, *Human Factors*, 1971, vol. 13, no. 4, pp. 363–368.

65   Vartabedian, A. G.: Developing a graphic set for developing cathode-ray-tube display using a 7×9 dot matrix, *Applied Ergonomics*, 1973, vol. 4, no. 1, pp. 11–16.

66   Vernon, M. D.: *Scale and dial reading*, Cambridge University, England, Medical Research Council, Unit in Applied Psychology, Report APU 49, June, 1946.

67   Wong, K. W., and N. G. Yacoumelos: Identification of cartographic symbols from TV displays, *Human Factors*, 1973, vol. 15, no. 1, pp. 21–31.

68   Woodson, W. E., and D. W. Conover: *Human engineering guide for equipment designers*, 2d ed., University of California Press, Berkeley, 1964.

69   Wright, P.: Writing to be understood: why use sentences? *Applied Ergonomics*, 1971, vol. 2, no. 4, pp. 207–209.

70   Zeff, C.: Comparison of conventional and digital time displays, *Ergonomics*, 1965, vol. 8, no. 3, pp. 339–345.

71   Applied ergonomics handbook: pt 1, chap. 3, "Displays," *Applied Ergonomics*, 1970, vol. 1, no. 2, pp. 86–94.

72   *Munsell book of color*, Munsell Color Co., Baltimore, 1959.

73   New Federal Standard on Colors, *Journal of the Optical Society of America*, 1957, vol. 47, pp. 330–334.

74 Beringer, D. B., R. C. Williges, and S. N. Roscoe: The transition of experienced pilots to a frequency-separated aircraft attitude display, *Human Factors*, 1975, vol. 17, no. 4, in press.

75 Ince, R., R. C. Williges, and S. N. Roscoe: Aircraft simulator motion and the order of merit of flight attitude and steering guidance displays, *Human Factors*, 1975, vol. 17, no. 4, in press.

76 Jacobs, R. S., R. C. Williges, and S. N. Roscoe: Simulator motion as a factor in a flight-director display evaluation, *Human Factors*, 1973, vol. 15, pp. 569–582.

77 Roscoe, S. N., and R. C. Williges: Motion relationships in aircraft attitude and guidance displays: A flight experiment, *Human Factors*, 1975, vol. 17, no. 4, in press.

Chapter 5

Auditory and Tactual Displays

We all depend upon the auditory and tactual senses in many mundane aspects of life, including the sensing of conventional auditory signals (such as horns, bells, and buzzers) and—in the case of blind persons—the use of Braille printing. With regard to sensory modalities generally, it is becoming increasingly possible to convert stimuli that are intrinsically associated with one modality into stimuli associated with another modality. Such technological developments may then result in increased use of the auditory and tactual senses in at least certain special circumstances, such as in using buzzers to warn blind people of physical objects ahead.

In this chapter we will discuss the use of the auditory and tactual senses as means of communication, except that the topic of speech will be deferred until the next chapter.

HEARING

In discussing the hearing process, let us first describe the physical stimuli to which the ear is sensitive, namely, sound vibrations.

The Nature and Measurement of Sound

Sound is originated by vibrations from some source. While such vibrations can be transmitted through various media, our primary concern is with those transmitted through the atmosphere to the ear. Two primary attributes of sound are *frequency* and *intensity* (or amplitude).

Frequency of Sound Waves   The frequency of sound waves can be visualized if we think of a simple sound-generating source such as a tuning fork. When it is struck, the tuning fork is caused to vibrate at its "natural" frequency. In so doing it causes the air particles to be moved back and forth. This alternation creates corresponding increases and decreases in the air pressure. The number of alternations per second is the frequency of the sound expressed in hertz (Hz) or cycles per second (c/s). The frequency of a physical sound gives rise to the human sensation of pitch.

The vibrations of a simple sound-generating source such as a tuning fork form *sinusoidal*, or *sine*, waves that can be represented as the projection of the movement of a point around a circle that is revolving at a constant rate, as shown in Figure 5-1. As point P revolves around its center O, its vertical ampli-

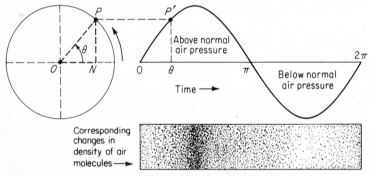

Figure 5-1   Reproduction of sinusoidal, or sine, wave. The magnitude of the alternating changes in air pressure caused by a sound-generating source with a given frequency can be represented by a sine wave. A sine wave is represented as the projection of a point on the circumference of a circle as that point rotates about its center at a constant speed. The lower part of the figure depicts the changes in the density of molecules of the air caused by the vibrating source.

tude, as a function of time, will be that represented by the sine wave. The height of the wave above the midline at any given point in time represents the amount of above-normal air pressure at that point in time, the crest, of course, being the maximum. Positions below the midline, in turn, represent the reduction in air pressure below normal. On the musical scale, middle C has a frequency of 256 Hz. Any given octave has double the frequency of the one below it. In general terms, the human ear is sensitive to frequencies in the range from about 20 to 20,000 Hz, although there are marked differences among individuals.

Intensity of Sound   Since sounds are pressure waves that vary above and below normal air pressure, the measurement of the amplitude or intensity of sound is related to such variations in air pressure. The various absolute measures of air pressure (or acoustic power) include such measures as watts, newtons, micronewtons, and microbars per unit area. Because the range of such values for common sounds is so tremendous, however, it is the usual practice to use a logarithmic measure of sound-pressure level (SPL) in characterizing sounds, such a measure expressing a ratio between two sound pressures. The *bel* (named after Alexander Graham Bell) is the basic ratio used for this purpose. The number of bels is the logarithm (to the base 10) of the ratio of the two intensities. Actually the most convenient and most common measure of sound intensity is the *decibel* (dB). A decibel is $1/_{10}$ of a bel;[1] like the bel, it expresses a ratio. Still another step in the conventional standardization of intensity measurement is that of using, for the lower of the two sounds, a standard reference level that represents zero decibels. The most common reference level is 20 micronewtons per square meter (abbreviated 20 μN/m^2). "Micro" stands for a factor of 1/1 million with an abbreviation of the Greek letter μ (mu). The μN stands for 0.000001 newton; m^2 refers to square meters. This can also be expressed as 0.0002 dyn/cm^2 or 0.0002 μbar.[2] Figure 5-2 shows the decibel scale with examples of several sounds that fall at varying positions along the scale. That figure also shows the power ratios with increasing decibel levels; an increase of 10 dB reflects a tenfold increase in relative sound power.

Complex Sounds   There are very few sounds that are pure tones. Even tones from musical instruments are not pure, but rather consist of a fundamental frequency in combination with certain others (especially harmonic frequencies that are multiples of the fundamental). Most complex sounds, however, are nonharmonic. Complex sounds can be depicted in two ways. One of these is a wave form which is the composite of the wave forms of the individual component sounds; such a wave form is illustrated in Figure 5-3. The other method of depicting complex sounds is by the use of a sound spectrum that shows the intensity of various frequency bands, as illustrated in Figure 5-4. The four curves illustrate spectral analyses of the noise of a rope-closing machine, using analyzers of varying bandwidths, namely, an octave, a half octave, a third of an octave, and thirty-fifth of an octave. The narrower the bandwidth, the greater

[1]The difference in the intensities of two sounds, given in decibels, is expressed as follows:

$$\text{Number of decibels} = 20 \log \frac{P_1}{P_2}$$

in which P_1 and P_2 represent the pressures of the two sounds.

[2]It should be noted that the American National Standards Institute (ANSI)—formerly the United States of American Standards Institute (USASI) and the American Standards Association—has established a standard to which sound-level meters should conform. This standard requires that three alternate *frequency-response* characteristics be provided in such instruments, these consisting of *weighting* networks (designated *A*, *B*, and *C*) which selectively discriminate against low and high frequencies in accordance with certain equal-loudness contours (to be discussed later). Recommended practice provides for using all three weightings, but, in any event designating any single one that is used in presenting data, such as "the *A*-weighted sound level is 45 dB," "sound level (A) = 45 dB," "SLA = 45 dB," or "45 dB(A)."

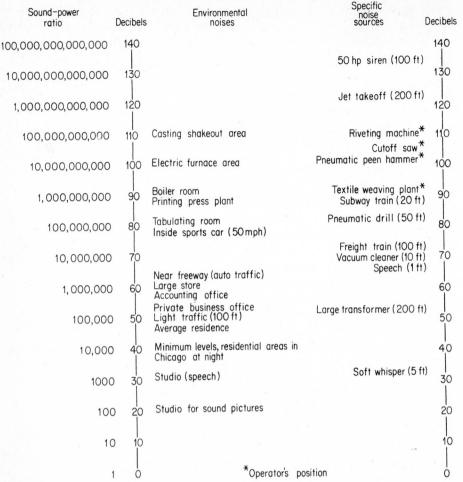

Sound–power ratio	Decibels	Environmental noises	Specific noise sources	Decibels
100,000,000,000,000	140			140
			50 hp siren (100 ft)	
10,000,000,000,000	130			130
			Jet takeoff (200 ft)	
1,000,000,000,000	120			120
100,000,000,000	110	Casting shakeout area	Riveting machine*	110
			Cutoff saw*	
10,000,000,000	100	Electric furnace area	Pneumatic peen hammer*	100
1,000,000,000	90	Boiler room Printing press plant	Textile weaving plant* Subway train (20 ft)	90
100,000,000	80	Tabulating room Inside sports car (50 mph)	Pneumatic drill (50 ft)	80
10,000,000	70		Freight train (100 ft) Vacuum cleaner (10 ft) Speech (1 ft)	70
1,000,000	60	Near freeway (auto traffic) Large store Accounting office		60
100,000	50	Private business office Light traffic (100 ft) Average residence	Large transformer (200 ft)	50
10,000	40	Minimum levels, residential areas in Chicago at night		40
1000	30	Studio (speech)	Soft whisper (5 ft)	30
100	20	Studio for sound pictures		20
10	10			10
1	0	*Operator's position		0

Figure 5-2 Decibel levels (dB) and sound power ratios for various sounds. Decibel levels are A-weighted sound levels measured with a sound-level meter. (See footnote 2 for discussion of A-weighting.) [Examples from Peterson and Gross, 24, fig. 2-1, p. 4.]

the detail of the spectrum and the lower the level of each bandwidth. There have been various practices in the division of the sound spectrum into octaves, but the current preferred practice as set forth by the American National Standards Institute (ANSI) is that of dividing the audible range into 10 bands, with the *center* frequencies being 31.5, 63, 125, 250, 500, 1000, 2000, 4000, 8000, and 16,000 Hz [40]. (Many existing sets of sound and hearing data are presented with the previous practice of defining octaves in terms of the *ends* of the class intervals instead of their *centers*.)

The Anatomy of the Ear

The ear has three primary anatomical divisions, namely, the outer ear, the middle ear, and the inner ear. These are shown schematically in Figure 5-5.

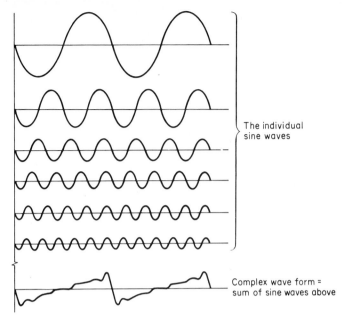

Figure 5-3 Waveform of a complex sound formed from six individual sine waves. [Adapted from·Lindsay and Norman, 18, fig. 6-6.]

The Outer Ear   The outer ear consists of the external part (called the *pinna* or *concha*), the auditory canal (the *meatus*), which is a tube about an inch long that leads inward from the external part, and the eardrum (the *tympanic membrane*) at the end of the auditory canal.

The Middle Ear   The middle ear is separated from the outer ear by the membranous eardrum. The middle ear includes a chain of three small bones called *ossicles*, the hammer, the anvil, and the stirrup (also called the *malleus*, the *incus*, and the *stapes*). These three ossicles, by their interconnections, transmit vibrations from the eardrum to the oval window of the inner ear. The stirrup acts something like a piston on the oval window, its action transmitting the changes in sound pressure to the fluid of the inner ear, on the other side of

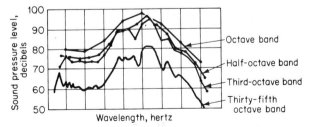

Figure 5-4   Spectral analyses of noise of a rope-closing machine, using analyzers of varying bandwidths. The narrower the bandwidth, the greater the detail and the lower the level of any single bandwidth. [Adapted from *Industrial noise manual*, 37, p. 25.]

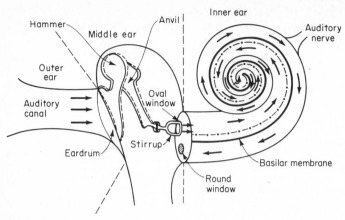

Figure 5-5 Schematic drawing of the ear, showing auditory canal through which sound waves travel to tympanic membrane, which in turn vibrates the ossicles of the middle ear. This vibration is transmitted through the membrane of the oval window to the cochlea, where the vibrations are transmitted by liquid through membranes to sensitive hair cells, which send nerve impulses to the brain.

the oval-window membrane. In this process, the extremely minute pressure changes on the eardrum are amplified about 22 times by the time they reach the fluid of the inner ear [Békésy, 1].

The Inner Ear The inner ear or cochlea is a spiral-shaped affair that resembles a snail. If uncoiled it would be about 30 mm long, with its widest section (near the oval window) being about 5 or 6 mm. The inner ear is filled with a fluid. The stirrup of the middle ear acts on this fluid like a piston, driving it back and forth in the rhythm of the sound pressure. These movements of the fluid force into vibration a thin membrane called the "basilar membrane," which in turn transmits the vibrations to the organ of Corti, which contains hair cells and nerve endings that are sensitive to very slight changes in pressure. The neural impulses picked up by these nerve endings are transmitted to the brain via the auditory nerve.

The Conversion of Sound Waves into Sensations

While the mechanical processes that are involved in the ear have been known for some time, the procedures by which sound vibrations are "heard" and differentiated are still not entirely known. In this regard, the numerous theories of hearing fall generally into two classes. The *place* (or resonance) theories are postulated on the notion that the fibers at various positions (or places) along the basilar membrane act, as Geldard puts it [10, p. 243], like harp or piano strings. Since these fibers vary in length, they are differentially sensitive to different frequencies and thus give rise to sensations of pitch. In turn, overtones stimulate a series of fibers spaced down the membrane. On the other hand, the *frequency* (or telephone) theories are based on the tenet that the basilar membrane vibrates as a whole, thus functioning much like the diaphragm of a telephone or microphone. The vibrations transmitted by the stirrup to the inner ear are also present in the movements of the basilar membrane and hair cells; this complete pattern of stimulation is reproduced in the response of the auditory nerve, thus imposing upon the brain the chore of dissecting and thus interpreting for us the neural impulses.

As Geldard comments [10, p. 257], it is not now in the cards to be able to accept one basic theory to the exclusion of any others as explaining all auditory phenomena. Actually, Geldard indicates that the current state of knowledge would prejudice one toward putting reliance in a place theory as related to high tones and in a frequency theory as related to low tones, one principle thus giving way to the other in the middle range of frequencies. As with theories in various areas of life, this notion of the two theories being complementary to each other needs still to be regarded as tentative pending further support or rejection from additional research.

Loudness and Related Sensations

The sensation of loudness is in large part a function of intensity, but it is also influenced in part by sound frequency. Over the years research has been carried out regarding various subjective attributes of sound, such as "loudness," "noisiness," and "perceived magnitude." As pointed out by Stevens [33], however, human judgments made about these indicate that these attributes may be considered essentially synonymous. For our purposes here we will discuss three measures that are relevant to the basic attribute of loudness, these being the *phon*, the *sone*, and *perceived level* (PLdB).

Loudness Level in Phons   Some years ago, Fletcher and Munson [7] developed what have become known as *equal-loudness* contours. They obtained judgments of subjects with regard to sounds of different combinations of frequencies and intensities to find out which ones were judged to have equal loudness. More recently Robinson and Dadson [28] in Great Britain have developed somewhat more refined equal-loudness curves, these being shown in Figure 5-6. Each contour shows the decibel intensities of different frequencies that were judged to be equal in loudness to that of a 1000-Hz tone of the specified decibel level of intensity. To illustrate the curves, one can see that a tone of 50 Hz of about 62 dB is judged equal in loudness to a 1000-Hz tone of only 40 dB.

The unit *phon* (of German origin) is used to indicate the loudness level of sounds. Thus, any point along a given contour in Figure 5-6 represents sounds of the same number of phons. The loudness level in phons, then, is numerically equal to the decibel level of a tone of 1000 Hz, which is judged equivalent in loudness.

Loudness in Sones   The phon tells us only about subjective *equality* of various sounds, but it tells us nothing of the *relative subjective loudness* of different sounds. For such comparative purposes we need still another yardstick.

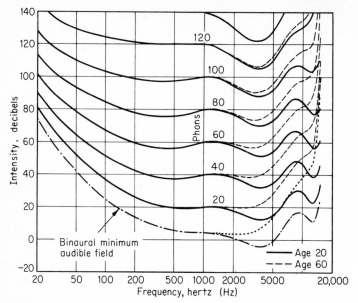

Figure 5-6 Equal-loudness curves of pure tones. Each curve represents intensity levels of various frequencies that are judged to be equally loud. The lowest curve shows the minimum intensities of various frequencies that typically can be heard. [From Robinson and Dadson, 28. Crown copyright reserved. Courtesy National Physical Laboratory, Teddington, Middlesex, England.]

Fletcher and Munson [7] developed such a scale—a ratio scale of loudness. Stevens [30], in turn, labeled the scale, using the term *sone*. In developing this scale (as with the phon) a reference sound was used. One sone is defined as the loudness of a 1000-Hz tone of 40 dB. A sound that is judged to be twice as loud as the reference sound has a loudness of 2 sones, a sound that is judged to be three times as loud as the reference sound has a loudness of 3 sones, etc. In turn, a sound that is judged to be half as loud has a loudness of $1/2$ sone. To provide some basis for relating sones to our own experiences, the following examples are given [Bonvallet, 38, p. 43]:

Noise source	Decibels	Loudness, sones
Residential inside, quiet	42	1
Household ventilating fan	56	7
Automobile, 50 ft	68	14
"Quiet" factory area	76	54
18-in automatic lathe	89	127
Punch press, 3 ft	103	350
Nail-making machine, 6 ft	111	800
Pneumatic riveter, 4 ft	128	3000

A procedure for estimating the loudness of complex sounds has been developed and revised by Stevens [31, 32]. This procedure uses a table of data for deriving a loudness index for each octave band. Excerpts from this table are presented in Table 5-1 [Peterson and Gross, 24]. First, the measurement of the band level is taken, in decibels, of each octave. Given these values, one proceeds as follows:

1 From the table (Table 5-1), find the proper loudness index for each band level (S).
2 Add all the loudness indexes (ΣS).
3 Multiply this sum by 0.3.
4 Add this product to 0.7 of the index that has the largest index (S_{max}). The total loudness in sones then is ($0.3 \, \Sigma S + 0.7 S_{max}$).
5 This total loudness (sones) can be converted to loudness level (phons) by using the two columns at the right of Table 5-1.

Perceived Level of Noise (PLdB): Mark VII   More recently Stevens [33] has developed a procedure for deriving a measure of the perceived level of noise (PLdB), "Mark VII," which is related to the sone (which he called "Mark VI"). It is based on a standard reference sound of $^1/_3$ octave band centered at 3150 Hz, which has certain advantages over the 1000-Hz tone that serves as the reference sound for the sone. (Since a 32-dB sound of 3150 Hz is about equivalent in loudness level to a 40-dB sound of 1000 Hz—which is 1 sone—the PLdB measure is about 8 dB lower than the corresponding values based on the Mark VI procedures.) The use of the 3150-Hz reference sound presumably brings about an improvement in accuracy of measurement. Further, the resulting value (PLdB) is expressed in dB, which facilitates a comparison with the dB readings of a sound-level meter.[3] In addition, the PLdB measure can be extended to lower values than the sone can. The procedures involved in deriving the PLdB involve the use of certain tables, which will not be repeated here. [See Stevens, 33, or Peterson and Gross, 24, pp. 28–33.]

Masking

"Masking" is a condition in which one component of the sound environment reduces the sensitivity of the ear to another component. Operationally defined, masking is the amount by which the threshold of audibility of a sound (the masked sound) is raised by the presence of another (masking) sound. In studying the effects of masking, an experimenter typically measures the absolute threshold (the minimum audible level) of a sound (the sound to be masked) when presented by itself and then measures its threshold in the presence of the masking sound. The difference is attributed to the masking effect.

The effects of masking vary with the type of masking sound and with the masked sound itself—whether pure tones, complex sound, white noise, speech,

---

[3]This comparison is enhanced with the use of an "*E*"-weighted—standing for "ear weighting"—sound-level meter. Such a weighting "network" of a sound-level meter—as well as the *A*, *B*, and *C* weightings—are intended to "adjust" the sound level readings for variations in the sensitivity of the ear to different frequencies.

Table 5-1 Excerpts from Table for Use in Calculating Loudness (Sones) of Complex Sounds

Band level, dB	Band loudness index (S) for octave bands — Midpoint of octave band									Loudness, sones	Loudness level, phons
	31.5	63	125	250	500	1000	2000	4000	8000		
20						0.18	0.30	0.45	0.61	0.25	20
30				0.16	0.49	0.67	0.87	1.10	1.35	0.50	30
40		0.07	0.37	0.77	1.18	1.44	1.75	2.11	2.53	1.00	40
50	0.26	0.62	1.13	1.82	2.24	2.68	3.2	3.8	4.6	2.00	50
60	0.94	1.56	2.44	3.4	4.1	4.9	5.8	7.0	8.3	4.00	60
70	2.11	3.2	5.0	6.2	7.4	8.8	10.5	12.6	15.3	8.00	70
80	4.3	6.7	9.3	11.1	13.5	16.4	20.0	24.7	30.5	16.0	80
90	8.8	13.6	17.5	21.4	26.5	32.9	41	52	66	32.0	90
100	18.7	28.5	35.3	44	56	71	90	113	139	64.0	100
110	44	61	77	97	121	149	184	226	278	128	110
120	105	130	160	197	242	298	367			256	120

Source: Peterson and Gross [24, Table 3-1, pp. 25, 26].

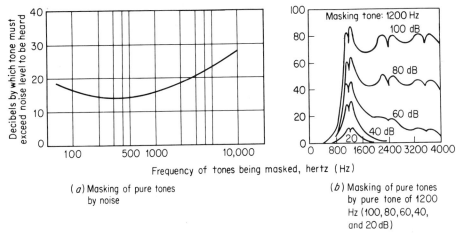

(a) Masking of pure tones by noise

(b) Masking of pure tones by pure tone of 1200 Hz (100, 80, 60, 40, and 20 dB)

Figure 5-7 The effects of masking pure tones a by noise and b by other tones. The curves show the number of decibels by which tones of various frequencies are masked by noise or other tones. [a Adapted from W. A. Munson in C. M. Harris, 11, pp. 5–16; b from Wegel and Lane, 36.]

etc. Two examples of masking effects of pure tones are shown in Figure 5-7. In the masking effects of noise (a) we can see that the threshold of pure tones is increased by 15 to nearly 30 dB, the increase being greatest in the case of the tones of higher frequencies. In the masking effects of pure tones on other pure tones (b) we can see that the masking effects are much more complex, with the effects being generally greatest in the frequencies around that of the masking tone and its harmonic overtones. The nature of masking effects depends very much upon the nature of the two sounds in question, as discussed by Geldard [33, pp. 215–220], and Deatherage [6]. Although these complex effects will not be discussed here, the effects of masking in our everyday lives are of considerable consequence, as, for example, the noise of office machines drowning out conversation.

AUDITORY DISPLAYS

The nature of the auditory sensory modality offers certain unique advantages for presenting information as contrasted with the visual modality, which has its own advantages. One set of comparisons of these two senses was given in Table 3-1, Chapter 3. On the basis of such comparisons and of other cues, it is possible to crystallize certain types of circumstances in which auditory displays usually would be preferable to visual displays. Some of these circumstances are given below:

- When the origin of the signal is itself a sound
- When the message is simple and short
- When the message will not be referred to later
- When the message deals with events in time
- When sending warnings or when the message calls for immediate action

- When presenting continuously changing information of some type, such as aircraft, radio range, or flight-path information
- When the visual system is overburdened
- When speech channels are fully employed (in which case auditory signals should be clearly detectable from the speech, such as tones)
- When illumination limits use of vision
- When the receiver moves from one place to another

Obviously the application of the above set of guidelines should be tempered with judgment rather than being followed rigidly. And there are of course other circumstances in which auditory displays would be preferable. Of the above guidelines particular mention should be made of the desirability of restricting auditory messages to those that are short and simple (except in the case of speech), since people do not do well in short-term storage of complex messages.

In a sense there are three types of human functions involved in the reception of auditory signals, these depending on the nature of the signal in question, as follows: (1) *detection* (determining if a given signal is or is not present, such as a warning signal); (2) *relative discrimination* (differentiating between two or more signals when presented close together); and (3) *absolute identification* (identifying a particular signal of some class, when only the one is presented). Relative discrimination and absolute identification can be made on the basis of any of several stimulus dimensions, such as intensity, frequency, duration, and direction (the difference in intensity of signals transmitted to the two ears).

Detection of Signals

The detection of auditory signals can be viewed in the frame of reference of signal detection theory (as discussed in Chapter 3), even though the signals occur in "peaceful" surroundings or in environments that are permeated by some ambient noise. As indicated in the discussion of signal detection theory, if at all possible the frequency distribution (at points in time) of the signal plus noise (SN) should not overlap the distribution of the noise (N) itself. When there is some overlap, the signal cannot invariably be detected from the noise, this confusion in part being a function of masking. When the signal (the masked sound) occurs in the presence of noise (i.e., the masking sound), the threshold of detectability of the signal is elevated—and it is this "elevated" threshold that should be exceeded by the signal if it is to be detected accurately.

In reasonably quiet surroundings (when one doesn't have to worry about clacking typewriters, whining machines, or screeching tires), a sound of about 40 to 50 dB above absolute threshold normally would be sufficiently above the ambient background noise to be detected. However, such detectability would vary somewhat with the frequency of the signal [Pollack, 25] and its duration. With respect to duration, the ear does not respond instantaneously to sound. For pure tones it takes about 200 to 300 ms to "build up" [Munson, 21], and about 140 ms to decay [Stevens and Davis, 34], although wide-band sounds build up and decay more rapidly. Because of these lags, signals of less than about 200 to 500 ms in duration do not sound as loud as those of longer duration. Thus, auditory signals (especially pure tones) should be at least 300 ms in duration; if they have to be shorter than that, their intensity should be increased to compensate for reduced audibility.

In rather noisy conditions, the signal intensity has to be set at a level far enough above that of the noise to insure detectability. In this regard Deatherage [6] has suggested a rule of thumb for specifying the optimum signal level, as follows: that the signal intensity (at the entrance of the ear) be about midway between the masked threshold of the signal in the presence of noise and 110 dB.

The Use of Filters   Under some circumstances it is possible to enhance the detectability of signals by filtering out some of the noise. This is most feasible when the predominant frequencies of the noise are different from those of the signal. In such an instance some of the noise can be filtered out and the intensity of the remaining sound raised (signal plus the nonfiltered noise). This has the effect of increasing the signal-to-noise ratio, thus making the signal more audible.

Relative Discrimination of Auditory Signals

The relative discrimination of signals on the basis of intensity and frequency (which are the most commonly used dimensions) depends in part on interactions between these two dimensions.

Discrimination of Intensity Differences   An impression of human abilities to discriminate intensity differences is given in Figure 5-8 [Deatherage, 6].

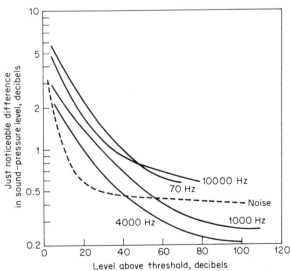

Figure 5-8 The just-noticeable differences (JND's) in sound intensity for pure tones of selected frequencies and for wide-band noise. [From Deatherage, 6, p. 147, as based on data from Riesz, 27, and Miller, 20.]

This reflects the just-noticeable differences (JND's) for certain pure tones and for wide-band noise of various sound-pressure levels. It is clear that the smallest differences can be detected with signals of higher intensities, such as at least 60 dB above the absolute threshold. The JND's of signals above 60 dB are minimal for intermediate frequencies (1000 and 7000 Hz) signals, these differences of course having implications if the auditory signals are to be discriminated by intensity.

Discrimination of Frequency Differences   Some indication of the ability of people to tell the difference between pure tones of different frequencies is reflected by the results of an early study by Shower and Biddulph [29] as shown in Figure 5-9. This figure shows the JND's for pure tones of various

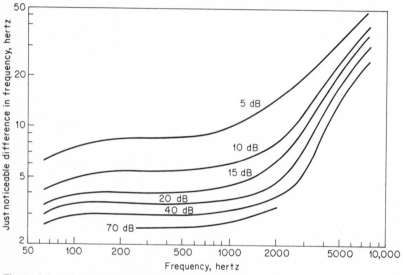

Figure 5-9   Just-noticeable differences (JND's) for pure tones at various levels above threshold. [From Shower and Biddulph, 29.]

frequencies at levels above threshold. The JND's are smaller for frequencies below about 1000 Hz (especially for high intensities) but increase rather sharply for frequencies above that. Thus, if signals are to be discriminated on the basis of frequency, it usually would be desirable to use those of lower frequencies. This practice may run into a snag, however, if there is much ambient noise (which usually tends to consist of lower frequencies) that might mask the signals. A possible compromise is that of using signals in the 500 to 1000-Hz range [Deatherage, 6]. It is also obvious from that figure that the JND's are smaller for signals of high intensity than for low intensity, thus suggesting that signals probably should be at least 30 dB above absolute threshold.

Absolute Identification of Auditory Signals

The JND for a given stimulus dimension reflects the minimum difference that can be discriminated on a relative basis. But in many circumstances it is necessary to make absolute identification of an individual stimulus (such as the frequency of a single tone) presented by itself. The number of "levels" along a continuum that can be so identified usually is quite small, as mentioned in the discussion of information theory in Chapter 3. Table 5-2 gives for each of cer-

Table 5-2 Summary of Certain Auditory Coding Methods

Intensity (pure tones)	30 to 5 levels. Intensity changes easier to detect than single intensities. For pure tones restrict to 1000 to 4000 Hz, but preferably use wide-band noise.
Frequency	Variously estimated at from 4 to 7 levels. Preferably use fewer and space widely apart. Set intensity at least 30 dB above threshold.
Sound duration	Use clear-cut differences, preferably 2 or 3.
Sound direction	Difference in intensity between two ears should be distinct; particularly relevant for directional information (i.e., right versus left).
Intensity and frequency	9 levels.

Source: Deatherage [6] and VanCott and Warrick [35].

tain auditory dimensions some indication of the number of levels that can be so identified.

Multidimensional Coding If the amount of information to be transmitted by auditory codes is substantial (meaning that discriminations among numerous signals need to be made), it is possible to use a multidimensional code system. Referring back to an earlier mention of the study by Pollack and Ficks [26], it will be recalled that various combinations of several parameters (dimensions) were used, such as direction (right ear versus left ear), frequency, intensity, repetition rate, on-and-off time fraction, and duration. Such a system obviously imposes the requirement for training the receiver regarding the "meaning" of each unique combination. In using such multidimensional codes, however, it is generally better to use more dimensions with fewer "steps" or "levels" of each (such as eight dimensions with two steps of each dimension) than to use fewer dimensions and more levels of each (such as six dimensions with five steps of each). (Both of these systems "transmitted" approximately the same amount of information.) It is on the basis of the many different facets (dimensions) of sounds that we can identify the voices of individuals, a dripping water faucet, or a squeaking hinge.

Principles of Auditory Display

As in other areas of human factors engineering, most guidelines and principles have to be accepted with a few grains of salt, since specific circumstances, trade-off values, etc., may argue for their violation. With such reservations in mind, a few guidelines for the use of auditory displays are given below. These generally stem from research findings and experience. Some of them are drawn in part from Mudd [20] and Licklider [17].

I   General principles

     A   *Compatibility:* Where feasible, the selection of signal dimensions and their encoding should exploit learned or natural relationships of the users, such as high frequencies being associated with up or high, and wailing signals with emergency.

     B   *Approximation:* Two-stage signals should be considered when complex information is to be presented, these stages to consist of:

         1   Attention-demanding signal: to attract attention and identify a general category of information.

         2   Designation signal: to follow the attention-demanding signal and designate the precise information within the general class indicated above.

     C   *Dissociability:* Auditory signals should be easily discernible from any ongoing audio input (be it either meaningful input or noise). For example, a warning bell would not stand out if the operator happened to be a Swiss bell ringer. If a person is to listen concurrently to two or more channels, the frequencies of the channels should be different if it is possible to make them so.

     D   *Parsimony:* Input signal to the operator should not provide more information than is necessary for him.

     E   *Invariance:* The same signal should designate the same information at all times.

II   Principles of presentation

     F   *Avoid extremes of auditory dimensions.*

     G   *Establish intensity relative to ambient noise level.*

     H   *Use interrupted or variable signals:* Where feasible, avoid steady-state signals and, rather, use interrupted or variable signals. This will tend to minimize perceptual adaptation.

III   Principles of installation of auditory displays

     I   *Test signals to be used:* Such tests should be made with a representative sample of the potential user population, to be sure the signals can be detected by them.

     J   *Avoid conflict with previously used signals:* Any newly installed signals should not be contradictory in meaning to any somewhat similar signals used in existing or earlier systems.

     K   *Facilitate changeover from previous display:* Where auditory signals are to replace some other mode of presentation (such as visual), preferably continue both modes for a while, to help people become accustomed to new auditory signals.

Auditory Displays for Specific Purposes

For illustrative purposes, a few types of auditory displays for specific purposes will be discussed below.

Warning and Alarm Signals   The unique features of the auditory system lend auditory displays to special use for signalling warnings and alarms. For this purpose the various types of available devices have their individual characteristics and corresponding advantages and limitations. A summary of such characteristics and features is given in Table 5-3 [from Deatherage, 6, p. 126].

Table 5-3 The Characteristics and Features of Certain Types of Audio Alarms

Alarm	Intensity	Frequency	Attention-getting ability	Noise-penetration ability
Diaphone (foghorn)	Very high	Very low	Good	Poor in low-frequency noise
Horn	High	Low to high	Good	Good
Whistle	High	Low to high	Good if intermittent	Good if frequency is properly chosen
Siren	High	Low to high	Very good if pitch rises and falls	Very good with rising and falling frequency
Bell	Medium	Medium to high	Good	Good in low-frequency noise
Buzzer	Low to medium	Low to medium	Good	Fair if spectrum is suited to background noise
Chimes and gong	Low to medium	Low to medium	Fair	Fair if spectrum is suited to background noise
Oscillator	Low to high	Medium to high	Good if intermittent	Good if frequency is properly chosen

Source: Deatherage [6, Table 4-2].

In the selection or design of warning and alarm signals, the following design recommendations have been proposed by Deatherage [6, pp. 125, 126] and by Mudd [21], these being in slightly modified form.

• Use frequencies between 200 and 5000 Hz, and preferably between 500 and 3000 Hz, because the ear is most sensitive to this middle range.
• Use frequencies below 1000 Hz when signals have to travel long distances (over 1000 ft), because high frequencies do not travel as far.
• Use frequencies below 500 Hz when signals have to "bend around" major obstacles or pass through partitions.
• Use a modulated signal (1 to 8 beeps per second or warbling sounds varying from 1 to 3 times per second), since it is different enough from normal sounds to demand attention.
• Use signals with frequencies different from those that dominate any background noise, to minimize masking.
• Preferably use high-intensity sudden-onset signals to alert receiver. Where earphones are used, consider *dichotic* presentation (alternating signal from one ear to the other).
• If different warning signals are used to represent different conditions, each should be discriminable from the others.
• Where feasible, use a separate communication system for warnings, such as loudspeakers, horns, or other devices not used for other purposes.

Radio Range Signals An example of a situation that requires discrimination between sounds is the reception of radio range signals in aircraft, in which the A (dot-dash) or the N (dash-dot) is heard if the pilot is to the left or to the right, respectively, of the center beam. The signal of one beam is on when that of the other is off. When the beams are of equal strength, the pilot hears a continuous signal meaning that he is on the center beam. Under adverse noise conditions the difference between the A and N signals may not be properly identified, and the pilot may think he is to the right of the beam when actually he is to the left, or vice versa.

Some of the aspects of this particular problem have been investigated under simulated signal and noise conditions that were somewhat like those encountered in an airplane [Flynn et al., 8]. Without discussing the details of this investigation, the results probably have important implications for various auditory displays. In particular the signal-to-noise ratio was found to be a much more critical factor in effective auditory discriminations than the intensity of the signal itself is.

THE CUTANEOUS SENSES

In everyday life people depend upon their cutaneous (or somesthetic or skin) senses much more than they realize. There is, however, an ongoing question as to how many cutaneous senses there are, this confusion existing in part because of the basis on which the senses can be classified. As Geldard points out [10, pp. 258, 259], we can classify them *qualitatively* (on the basis of their observed similarity—that is, the sensations generated); in terms of the *stimulus* (i.e., the form of energy that triggers the sensation, as thermal, mechanical, chemical, or electrical); or *anatomically* (in accordance with the nature of the sense organs or tissues involved). With respect to the anatomical structures involved, it is still not clear how many distinct types of nerve endings there are, but for convenience Geldard considers the skin as housing three more or less separate systems of sensitivity, one for pressure reception, one for pain, and one responsive to temperature changes. Although it has been suggested that any particular sensation is triggered by the stimulation of a specific corresponding type of nerve ending or receptor, this notion seems not to be warranted. Rather, it is now generally believed that the various receptors are specialized in their functions through the operation of what Geldard refers to as the principle of *patterned response*. Some of the cutaneous receptors are responsive to more than one form of energy (such as mechanical pressure and thermal changes) or to certain ranges of energy. Through complex interactions among the various types of nerve endings as they are stimulated by various forms and amounts of energy, we experience a wide "variety" of specific sensations that we endow with such labels as "touch," "contact," "tickle," and "pressure" [Geldard, 10].

TACTUAL DISPLAYS

Although we depend upon our cutaneous senses very much in everyday living, these senses have been used only to a limited degree as the basis for the "intentional" transmission of information to people by the use of tactual displays. The primary uses of the cutaneous senses for tactual displays to date have included the use of Braille print for the blind and the use of shape-coded control devices. In recent years, however, there has been a modest flurry of interest relating to the cutaneous senses for transmitting either static or dynamic information in certain special circumstances.

Braille Print

Braille print for the blind consists of raised "dots" formed by the use of all the possible combinations of 6 dots numbered and arranged thus:

```
1 . . 4
2 . . 5
3 . . 6
```

A particular combination of these represents each letter, numeral, or common word. The critical features of these dots are position, distance between dots, and dimension (diameter and height), all of which have to be discriminable to the touch.

Tactual Coding of Control Devices

Another use of the tactual sense is with respect to the design of control knobs and related devices. Although these are not "displays" in the conventional sense of the term, the need to correctly identify such devices may be viewed within the general framework of displays. The coding of such devices for tactual identification includes their shape, texture, and size.

Discrimination of Shape-coded Controls   The discrimination of shape-coded controls is essentially one of tactual sensitivity. The procedure generally used in the selection of controls that are not confused with each other is illustrated by the study by Jenkins [15] in which he had 25 controls mounted on a rotating lazy Susan. Each subject, blindfolded, was presented with one knob which he touched for 1 s. The experimenter then rotated the turntable to a predesignated point from which the subject went from knob to knob, feeling each in turn, until he found the one he thought was the one he had previously touched. It was then possible to determine which knobs were confused with which other knobs. While the statistical results will not be presented, it can be said that two sets of eight knobs were identified, such that the knobs within each group were rarely confused with each other. These two sets of knobs are shown in Figure 5-10.

Figure 5-10   Two sets of knobs for levers that are distinguishable by touch alone. The shapes in each set are rarely confused with each other. [From Jenkins, 16.]

Following essentially the same tack as that mentioned above, the United States Air Force has developed 15 knob designs which are not often confused with each other. These designs are of three different types, each type being designed to serve a particular purpose [Hunt, 15]:

  • *Class A: Multiple rotation.* These knobs are for use on controls (1) which require twirling or spinning, (2) for which the adjustment range is one full turn or more, and (3) for which the knob position is not a critical item of information in the control operation.
  • *Class B: Fractional rotation.* These knobs are for use on controls (1) which do not require spinning or twirling, (2) for which the adjustment range usually is *less* than one full turn, and (3) for which the knob position is not a critical item of information in the control operation.

- *Class C: Detent positioning.* These knobs are for use on discrete setting controls.

The 15 knobs in these three classes are shown in Figure 5-11.[4] In connection

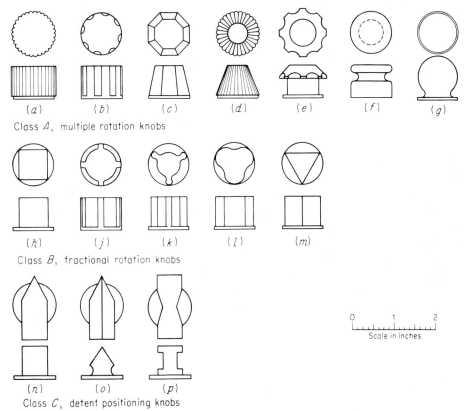

(a)   (b)   (c)   (d)   (e)   (f)   (g)

Class *A*, multiple rotation knobs

(h)   (j)   (k)   (l)   (m)

Class *B*, fractional rotation knobs

(n)   (o)   (p)

0   1   2
Scale in inches

Class *C*, detent positioning knobs

Figure 5-11 Knob designs of three classes that are seldom confused by touch. [Adapted from Hunt, 15.]

with sizes of knobs in these three classes, Hunt suggests that they be not more than 4 in in their maximum dimension and not less than $^1/_2$ in (except for class C, for which he suggests a $^3/_4$- in miminum). In height they should not be less than $^1/_2$ in but need not be more than 1 in.

If in addition to being individually discriminable by touch, the controls have shapes that are symbolically associated with their use, the learning of their use usually is simplified. In this connection, the United States Air Force has developed a series of 10 knobs that have been standardized for aircraft cockpits. These standard knob shapes, besides being distinguishable from each

---

[4]A few of these knobs may be confused with each other, and such combinations should not be used together if identification is critical. These combinations were *ab, co, cd#, do#, eg#, kp, ln, lo, np,* and *op#.* Those with a number sign (#) were confused only with gloves on and were not confused without gloves.

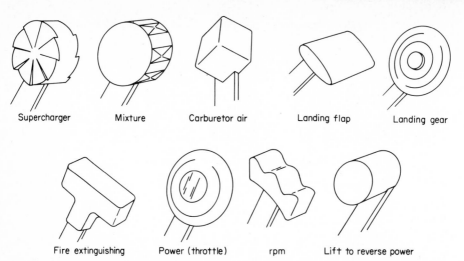

Figure 5-12 Standardized shape-coded knobs for United States Air Force aircraft. A number of these have symbolic associations with their functions, such as a wheel representing the landing-gear control. [*Personnel Subsystems*, 39, chap. 2, sec. 2D18, p. 3.]

other by touch, include some that also have symbolic meaning. In Figure 5-12, which includes these shapes, it will be seen, for example, that the landing-gear knob is like a landing wheel, the flap control is shaped like a wing, and the fire-extinguishing control resembles the handle on some fire extinguishers.

   Texture Coding of Controls   In addition to shape, control devices can be varied in their surface texture. This characteristic was studied (along with certain other variables) in a series of experiments with flat cylindrical knobs such as those shown in Figure 5-13 [Bradley, 3]. In one phase of the study, knobs of

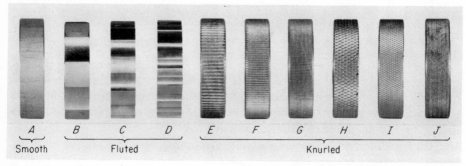

Figure 5-13 Illustration of some of the knob designs used in study of tactual discrimination of surface textures. Smooth: *A*; fluted: *B* (6 troughs), *C* (9), *D* (18); and knurled: *E* (full rectangular), *F* (half rectangular), *G* (quarter rectangular), *H* (full diamond), *I* (half diamond), and *J* (quarter diamond). [From Bradley, 3.]

this type of 2-in diameter were used, and subjects were presented with individual knobs through a curtained aperture and were asked to identify the particular design they felt. The results are shown as a "confusion matrix" in Figure 5-14, this indicating the number of times each knob was identified correctly and in-

Knob that was felt

	Smooth	Fluted			Knurled					
	A	B	C	D	E	F	G	H	I	J
A	45									
B		42	6							
C		3	33	1						
D			6	44						
E					29	11	1	4	1	
F					8	8	7	8	5	6
G					1	7	27		15	9
H					6	7		27	1	2
I					1	7	6	5	2	18
J						5	4	1	21	10

(Row label at left: Response (the knob that the "felt" knob was "identified" as))

Figure 5-14 "Confusion matrix," showing results of study in which knobs with different surface textures were presented through an aperture to 45 subjects. (The knobs are those shown in Figure 5-13.) The numbers in any given column are the numbers of times each knob was "identified" as the one that was felt by hand, when the one actually felt was the one identified at the top of the column. The numbers of *correct* identifications (out of 45) are those in the cells along the diagonal. [Adapted from Bradley, 3.]

correctly (and in such cases the knobs with which the one was confused). The smooth knob was not confused with any other, and vice versa; the three fluted designs were confused with each other, but not with other types; and the knurled designs were confused with each other, but not with other designs. It should be added that with gloved hands and with smaller-sized knobs (in a later phase of the study) there was some cross-confusion among classes, but this was generally minimal. The investigator proposes that three surface characteristics can thus be used with reasonably accurate discrimination, namely, smooth, fluted, and knurled.

Size Coding of Controls Size coding of controls is not as useful for coding purposes as shape, but there may be some instances where it is appropriate. When such coding is used, the different sizes used should of course be such that they are discriminable one from the others. Part of the study by Bradley reported above [3] dealt with the discriminability of cylindrical knobs of varying diameters and thickness. It was found that knobs that differ by $1/2$ in in diameter and by $3/8$ in in thickness can be identified by touch very accurately, but that smaller differences between them sometimes result in confusion of knobs with each other. Incidentally, Bradley proposes that a combination of three surface textures (smooth, fluted, and knurled), three diameters ($3/4$, $1 1/4$, and $1 3/4$ in), and two thicknesses ($3/8$ and $3/4$ in) could be used in all combinations to provide 18 tactually identifiable knobs.

Aside from the use of size coding for individual control devices, size coding is part and parcel of ganged control knobs, where two or more knobs are mounted on concentric shafts with various sizes of knobs super-imposed on each other like the layers of a wedding cake. When this type of design is dictated by engineering considerations, the differences in the sizes of superimposed knobs need to be great enough to make them clearly distinguishable, as illustrated in Figure 9-6 in Chapter 9 [Bradley, 4].

Dynamic Tactual Displays

The more exciting possibilities in the use of tactual displays are in transmission of dynamic information. Gilmer [in Hawkes, 12, pp. 76–84] suggests the possible use of tactual communications for transmitting information of the following classes: quantitative information; coordinates in space; direction; rates; language; attention demanding; and vigilance. Because of the inherent advantages of the visual and auditory senses for communications it would not be expected that dynamic tactual displays would become widespread in use, but that they might rather be used in special circumstances. Considering the effects on the skin of various forms of energy, the adaptation of the skin to stimuli, and other factors, it is probable that mechanical and electrical stimulation offer the greatest promise.

Mechanical Stimuli   There are two general approaches to the use of mechanical stimuli in tactual communications. The first of these consists of the use of some type of vibrator affixed to the surface of the skin. The coding of the vibrations can be based on such physical parameters as location of vibrators, frequency of vibrations, intensity, and duration. In this connection Geldard [9] developed a vibratory language of 5 chest locations, 3 levels of intensity, and 3 durations, which provided 45 unique patterns ($5 \times 3 \times 3$), which in turn were used to code 26 letters, 10 numerals, and 4 frequently used words. Well-trained subjects could receive as many as 67 words per minute, well above the most demanding military requirements for Morse code reception of 24 words per minute. Thus, it appears that the combinations of vibratory parameters would permit the transmission of alphanumeric information, where such a method would be useful. Other less complicated vibratory signals likewise appear to be potentially useful for special purposes.

The second general approach to the use of mechanical stimuli is by the transmission to the skin of amplified speech sounds transmitted by a single vibrator [Myers, 23]. In the various phases of the experiment, 16 selected words and numbers and 17 basic speech sounds (phonemes such as e, g, th, s, z, i, and ng) were recorded on tape. The frequencies of these recordings were then reduced by rerecordings to one-half (for the words and numerals) or one-eighth (for the phonemes). These reduced-frequency recordings were then amplified and transmitted to the skin via a vibrator. To indicate the possible utility of their procedure, two subjects were able to receive the phonemes with a median accuracy of 91 percent after eight practice sessions.

Electrical Stimuli   A basic problem in the possible use of electrical stimuli is that they are, as Gilmar [in Hawkes, 12, pp. 76–84] suggests, between the boundaries of "pain" and "painless pulses." These boundaries take in several parameters, such as intensity, polarity, duration, and interval, and also electrode type, size, and spacing. A few of these parameters have been explored. For example, Gilmer suggests that it probably would be possible to use three levels of intensity, six loci (three on each palm), and two *duration-depth* dimensions in an electropulse communication system, which would allow for 36 discrete symbols such as 26 letters and 10 digits. Additional locations would, of course, increase the range of possibilities. Although considerable progress has been made in the development of hardware components for electrocutaneous communication systems [Hennessy, 13], such systems are not as yet at the stage of practical use—although that time is coming.

As an example of the use of electrical stimulation in an experimental system, Hofmann and Heimstra [14] used electrodes applied to the right and left sides of the neck to transmit directional information to subjects in a tracking task. The magnitude of the error was represented by the intensity of the stimulation. The study also involved the use of visual and auditory displays. Although the three types of displays resulted in some differences in the results as reflected by four different criteria used, the average percents of "efficiency" across all four measures were as follows:

Vision: 56%
Audition: 67%
Electrocutaneous: 64%

The fact that the electrocutaneous display resulted in nearly as efficient tracking performance as auditory signals offers reasonable promise of the potential use of such displays in at least certain tracking tasks.

Optical-to-Tactile Image Conversion

Because of the cost of Braille printing there are only limited amounts of Braille reading material available to blind persons. The combination of research relating to the sensitivity of the skin and technological developments seems now to offer promise for converting optical images into tactile vibrations that can be interpreted by blind persons. The Optacon (for *optical-to-tactile converter*), developed at the Stanford University Electronics Laboratories, consists of essentially two components. One of these is an optical pickup "camera" and a "slave" vibrating tactile stimulator that reproduces images that are scanned by the camera [Linvill, 19; Bliss et al., 2]. A photograph of the Optacon is shown in Figure 5-15. The 144 phototransistors of the camera in a 24 by 6 array serve to control the activation of 144 corresponding vibrating pins that "reproduce" the visual image. The vibrating pins can be sensed by one fingertip. The particular features of the vibrating pins and of their vibrations were based on exten-

Figure 5-15 Photograph of Optacon for use in converting visual images into tactile vibrations. [Photograph courtesy of Stanford Electronics Laboratories. The Optacon is produced by Telesensory Systems, Inc.]

sive previous psychophysical research relating to the sensitivity of the skin to vibrations. (For what interest they may be, the specifications of the vibrating pins were: size of pin, 10 mils; space between 6 rows, 100 mils; space between 24 pins in each row, 50 mils; frequency of vibrations, 12.5 to 250 Hz; depth of skin indentation, 2.6 mils [Bliss et al., 2].)

Learning to "read" with the Optacon takes time. For example, one subject took about 160 hours of practice and training to achieve a reading rate of about 50 words per minute. This rate is well below that of most readers of Braille, some of whom achieve a rate of 200 words per minute. At the same time, such a device does offer the possibility for blind persons of being able to read conventional printing at at least a moderate rate.

In much the same manner as printed material can be converted into tactile form, so objects in the environment can be reproduced in the form of tactile stimulation. A recent development of this type is a Tactile Vision Substitution System (TVSS) developed at the Smith-Kettlewell Institute of Visual Sciences and the Department of Visual Sciences of the University of the Pacific [Collins and Bachy-y-Rita, 5]. This system, illustrated in Figure 5-16, senses objects with a pair of "seeing eyeglasses" and transmits the images via a flexible-fiber optic bundle to a lightweight television camera, which in turn converts the visual image into a corresponding pattern of electronic impulses. Although such a system is still in the developmental stage, the back-up research evidence suggests that with adequate training the brain can integrate the "information"

Figure 5-16 A portable electronic image projection seeing-aid system for the blind (the Tactile Vision Substitution System [TVSS]). The image sensed by the "seeing" eyeglasses is transmitted by fiber optics to a small TV camera that converts the image into corresponding electronic pulses by an array of small electrodes shown here in the left hand. When is use, the components shown here are worn around the waist, the array of electrodes being in contact with the skin. [Photograph courtest of Smith-Kettlewell Institute of Visual Sciences, University of the Pacific.]

so transmitted into the total sensory "pool" of information, and thus utilize it in the analysis of the individual's environment.

Discussion

Relative to the other senses, the cutaneous senses seem generally suitable for transmitting only a limited number of discrete stimuli. Although we would then not expect tactual displays to become commonplace, there are at least two types of circumstances for which future developments of such displays would seem to be appropriate. In the first place, such displays offer some promise of being useful in certain special circumstances in which the visual and auditory senses are overburdened; in such instances tactual displays might be used, such as for warning purposes. And in the second place, they have definite potential as aids to blind persons.

REFERENCES

1 Békésy, G. von: The ear, *Scientific American*, August, 1957.
2 Bliss, J. C., M. H. Katcher, C. H. Rogers, and R. P. Shepard: Optical-to-tactile image conversion for the blind, *IEEE Transactions on Man-Machine Systems*, March, 1970, vol. MMS-11, no. 1, pp. 58–65.
3 Bradley, J. V.: Tactual coding of cylindrical knobs, *Human Factors*, 1967, vol. 9, no. 5, pp. 483–496.
4 Bradley, J. V.: Desirable dimensions for concentric controls, *Human Factors*, 1969, vol. 11, no. 3, pp. 213–226.
5 Collins, C. C., and P. Bachy-y-Rita: Transmission of pictorial information through the skin, *Advances in biological and medical physics*, 1973, vol. 14, pp. 285–315.
6 Deatherage, B. H.: Auditory and other sensory forms of information presentation, in H. P. Van Cott and R. G. Kinkade (eds.), *Human engineering guide to equipment design*, U.S. Government Printing Office, Washington, D.C., 1972, chap. 4.
7 Fletcher, H., and W. A. Munson: Loudness, its definition, measurement, and calculation, *Journal of the Acoustical Society of America*, 1933, vol. 5, pp. 82–108.
8 Flynn, J. P., S. J. Goffard, I. P. Truscott, and T. W. Forbes: *Auditory factors in the discrimination of radio range signals: collected informal reports*, Harvard University, Psycho-acoustic Laboratory, OSRD Report 6292, Dec. 31, 1945.
9 Geldard, F. A.: Adventures in tactile literacy, *American Psychologist*, 1957, vol. 12, pp. 115–124.
10 Geldard, F. A.: *The human senses*, 2d ed., John Wiley & Sons, Inc., New York, 1972.
11 Harris, C. M. (ed.): *Handbook of noise control*, McGraw-Hill Book Company, New York, 1957.
12 Hawkes, G. R. (ed.): *Symposium on cutaneous sensitivity*, USA Medical Research Laboratory, Ft. Knox, Ky., Report 424, Dec. 22, 1960.
13 Hennessy, J. R.: Cutaneous sensitivity communications, *Human Factors*, 1966, vol. 8, no. 5, pp. 463–470.
14 Hofmann, M. A., and N. W. Heimstra: Tracking performance with visual, auditory, or electrocutaneous displays, *Human Factors*, 1972, vol. 14, no. 2, pp. 131–138.
15 Hunt, D. P.: *The coding of aircraft controls*, USAF, WADC, Technical Report 53–221, August, 1953.
16 Jenkins, W. O.: "The tactual discrimination of shapes for coding aircraft-type controls," in P. M. Fitts (ed.), *Psychological research on equipment design*, Army Air Force, Aviation Psychology Program, Research Report 19, 1947.
17 Licklider, J. C. R.: *Audio warning signals for Air Force weapon systems*, USAF, WADD, TR 60–814, March, 1961.
18 Lindsay, P. H., and D. A. Norman: *Human information processing*, Academic Press, New York, 1972.
19 Linvill, J. G.: *Research and development of tactile facsimile reading aid for the blind (the Optacon)*, Stanford Electronics Laboratories, Stanford University, Stanford, Calif., March, 1973.
20 Miller, G. A.: Sensitivity to changes in the intensity of white noise and its relation

to masking and loudness, *Journal of the Acoustical Society of America*, 1947, vol. 19, pp. 609–619.

21 Mudd, S. A.: *The scaling and experimental investigation of four dimensions of pure tone and their use in an audio-visual monitoring problem*, unpublished Ph.D. thesis, Purdue University, Lafayette, Ind., 1961.

22 Munson, W. A.: The growth of auditory sensitivity, *Journal of the Acoustical Society of America*, 1947, vol. 19, p. 584.

23 Myers, R. D.: *A study in the development of a tactual communication system*, Symposium on the Air Force Human Engineering, Personnel, and Training Research, 1960, NAS, NRC, Publication 783, pp. 238–243.

24 Peterson, A. P. G., and E. E. Gross, Jr.: *Handbook of noise measurement*, 7th ed., General Radio Co., New Concord, Mass., 1972.

25 Pollack, I.: Comfortable listening levels for pure tones in quiet and noise, *Journal of the Acoustical Society of America*, 1952, vol. 24, p. 158.

26 Pollack, I., and L. Ficks: Information of elementary multidimensional auditory displays, *Journal of the Acoustical Society of America*, 1954, vol. 26, pp. 155–158.

27 Riesz, R. R.: Differential intensity sensitivity of the ear for pure tones, *Physiological Review*, 1928, vol. 31, pp. 867–875.

28 Robinson, D. W., and R. S. Dadson: Threshold of hearing and equal-loudness relations for pure tones, and the loudness function, *Journal of the Acoustical Society of America*, 1957, vol. 29, no. 12, pp. 1284–1288.

29 Shower, E. G., and R. Biddulph: Differential pitch sensitivity of the ear, *Journal of the Acoustical Society of America*, 1931, vol. 3, pp. 275–287.

30 Stevens, S. S.: A scale for the measurement of a psychological magnitude: loudness, *Psychological Review*, 1936, vol. 43, pp. 405–416.

31 Stevens, S. S.: Calculation of the loudness of complex noise, *Journal of the Acoustical Society of America*, 1956, vol. 28, pp. 807–832.

32 Stevens, S. S.: Procedure for calculating loudness: Mark VI, *Journal of the Acoustical Society of America*, 1961, vol. 33, no. 11, pp. 1577–1585.

33 Stevens, S. S.: Perceived level of noise by Mark VII and decibels (E), *Journal of the Acoustical Society of America*, 1972, vol. 51, no. 2, pt. 2, pp. 575–601.

34 Stevens, S. S., and H. Davis: *Hearing, its psychology and physiology*, John Wiley & Sons, Inc., New York, 1938.

35 Van Cott, H. P., and M. J. Warrick: Man as a system component, in H. P. Van Cott, and R. G. Kinkade (eds.), *Human engineering guide to equipment design*, rev. ed., Superintendent of Documents, Washington, D.C., 1972, Chap. 2.

36 Wegel, R. L., and C. E. Lane: The auditory masking of one pure tone by another and its probable relation to the dynamics of the inner ear, *Physiological Review*, 1924, vol. 23, pp. 266–285.

37 *Industrial noise manual*, 2d ed., American Industrial Hygiene Association, Detroit, 1966.

38 *Noise*, lectures presented at the Inservice Training Course on the Acoustical Spectrum, Feb. 5–8, 1952, sponsored by the University of Michigan, School of Public Health and Institute of Industrial Health, University of Michigan Press, Ann Arbor, Mich.

39 *Personnel subsystems*, USAF, AFSC design handbook, Series 1–0, General, AFSC DH 1–3, 1st ed., Jan. 1, 1969, Headquarters, AFSC.

40 *Standard specification for octave, half-octave, and third-octave band filter sets*, American National Standards Institute (ANSI) S1.11, 1966.

Chapter 6

Speech Communication

Each language is a system of coding, there being parallel written and spoken forms for transmitting meaning via the visual and auditory senses. Our interest in this chapter is primarily with speech, in which various speech sounds are used as "codes" to transmit the information associated with the codes in question.

HUMAN FACTORS ASPECTS OF SPEECH

In a human factors framework, speech is both an "output" (from a talker) and an "input" (to a receiver). Realizing this dual aspect, however, it seems most appropriate to discuss it here in connection with other "inputs."

In many circumstances in which speech is used there is no particular problem in its transmission and reception. However, speech transmission can be affected adversely by noise, or by the communication system (telephone, intercommunication system, radio, etc.), and it is in such circumstances that speech and its transmission take on human factors implications. Such concern is especially relevant when the communications are particularly critical, such as in airport control tower operations. In the process of designing a speech communication system it should be a first order of business to establish the criteria or standards that the system should meet in terms of the reception of speech, such as the relative importance of intelligibility (as in an airport control tower system), or perhaps naturalness or quality (as in a home telephone system—where it is important to be able to recognize voices).

With any such criteria or standards nailed down, the designer can consider the implications of the various "components" of the speech communication system as they might influence the fulfillment of these criteria, these components including the following: the message to be transmitted; the talker; the transmission system itself (i.e., telephone, radio, intercommunication, etc.), including whatever noise may be present; and the receiver. Before discussing these, however, we should mention briefly certain characteristics of speech and the measurement of speech intelligibility.

CHARACTERISTICS OF SPEECH

The speech of any given language consists of speech sounds called *phonemes*. These phonemes are generated by the *articulators* (the lips, tongue, teeth, and palate) as these interact to interrupt or constrict the breath stream. In the formation of possible phonemes by the processes of articulation, there are five types of articulation [Miller, 10, p. 17]:

- *Plosives*, or *stops*, produced by completely stopping the passage of air, as *p* in *pop*
- *Fricatives*, or *spirants*, produced by forming a narrow slit or groove for air passage, as *th* in *their*
- *Laterals*, formed by closing the middle line of the mouth, and leaving an air passage around one or both sides, as *l* in *let*
- *Trills*, caused by rapid vibration of an articulator, as the trilled *r* in certain European languages
- *Vowels*, produced by an unobstructed air passage

Each basic type of articulation has several possible variations that can be produced by changes in the positions of the articulators, which gives a wide range of possible phonemes. The English language includes about 16 different vowel sounds, the specific number depending on whether certain closely related sounds are counted as the same or different; and it includes about 22 consonant sounds, making a total of about 38 phonemes. Other languages employ some phonemes that we do not use, and vice versa.

As any speech phoneme or other sound is generated it produces variations in air pressure (the "speech wave") that can be represented in various ways. One such representation is the waveform, which shows the variations in air pressure over time. Another representation is the spectrum, which depicts the combinations of frequency and intensity (speech power) of the speech sound. The combinations of many individual speech sounds in turn can be represented by an "overall" or "continuous" spectrum which shows the combinations of frequencies and intensities of the many individual sounds; further, it is possible to derive an estimate of the overall intensity level across all frequencies.

Intensity of Speech

The average intensity, or speech power, of individual phonemes varies tremendously, with the vowels generally having much greater speech power than the consonants. For example, the *o* as pronounced in *talk* has roughly 680 times the speech power of *th* as pronounced in *then* [Fletcher, 5]. This is a difference of about 28 dB.

The overall intensity of speech, of course, varies from one person to another and also for the same individual. Fletcher [5, pp. 76–77], for example, shows that when one talks almost as softly as possible, speech has a decibel level of about 46, and when one talks almost as loudly as possible, it has a decibel level of about 86. The results of a survey by Fletcher [5, p. 77] of the telephone-speech levels of a good-sized sample of people showed a range from about 50 to about 75 dB, with a mean of about 66 dB. There was, however, a heavy concentration of cases from about 60 to 69 dB, with about 40 percent falling within 3 dB above or below the mean.

$$\frac{0.525}{1.0} = \frac{3}{x}$$

$$\therefore \sigma = 5.71 \text{ dB}$$

Frequency of Speech

As indicated above, each phoneme has its own unique spectrum of several frequencies, although there are, of course, differences in these among people, and each individual can (and does) shift his spectrum up and down the register depending in part on the circumstances, such as talking in a quiet conversation, when the pitch and loudness are usually relatively low, or talking to a group, when the pitch and loudness typically are higher, or screeching at the children to stay out of the wet varnish on the floor (when the pitch and loudness usually are near their peak). Across the board, men have a somewhat lower spectrum than women, as shown in Figure 6-1. That figure shows the average "overall"

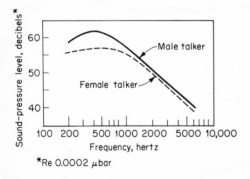

Figure 6-1 The octave-band spectrum of speech of samples of males and females. [Adapted from Kryter, 7, fig. 5-2, p. 164.]

level of speech of samples of males and females. There are greater differences between the two curves in the lower frequencies than in the higher frequencies; the overall level for females is about 3 dB less than for males.

Quality of Speech

The quality of speech is really the mixture of pitch and loudness. In a sense every phoneme has its own distinguishing quality, and the speech of individuals also has its own quality that makes it possible to identify the voices of individuals.

SPEECH INTELLIGIBILITY

In many communication situations the criterion of voice communication is its intelligibility to a receiver. For evaluating speech communication under different conditions (e.g., under noise, with different communication systems, and at various distances), and for research, we need some measure of the intelligibility of speech.

Tests of Speech Intelligibility

The most straightforward method of measuring speech intelligibility is by the use of some test. Such tests usually are by the transmission of speech material to a receiver, who is then asked to repeat what he hears, or to make some other response. The material usually is given by a trained "talker" using a microphone, but may be presented to the receivers by use of recorded voices. The types of test material include the following [Kryter, 7, p. 168]:

   1 Nonsense syllable tests. Examples: *deeg* and *zak*. Basis of scoring: percents of accuracy of responses of the entire syllables, of consonants, and of vowels.
   2 Phonetically balanced (PB) word lists. Examples: *clove* and *rib*: the 20 lists from which words are selected are proportionately representative of various types of speech sounds in everyday speech. Basis of scoring: percent accuracy.
   3 Modified rhyme tests (MRT). Examples: *rang* and *pit*. Basis of scoring: percent accuracy of recognition of one of the phonemes in each test word, either the initial or the final consonant.
   4 Sentence tests (two types). Example of type 1: What day comes before Wednesday? Basis of scoring: determining if the response indicates that the receiver understood the question. Example of type 2: Deal the cards from the top, you bully. Basis of scoring: percent of "key" words recorded correctly.

Speech intelligibility tests actually can serve various purposes, including: speech research; evaluating speech communication systems; evaluating the effects of noise and other conditions upon intelligibility; and measuring the speech effectiveness and hearing abilities of individuals.

The Articulation Index (AI)

Speech intelligibility tests frequently are not practical for use in evaluating communication systems or the effects of noise because of the time and trouble involved in their administration. In such circumstances the articulation index (AI) can serve as the basis for estimating the intelligibility of speech. Although there are several methods of deriving an articulation index, they all involve about the same basic approach. The procedures of such approach are summarized below, in particular, the one-third-octave-band method [Kryter, 7; ANSI, 15]. (Other methods provide for the use of data based on 20 bands or octave bands.)

1 For each one-third-octave band, plot on a worksheet, such as in Figure 6-2, the band levels of the speech peaks reaching the listener's ear. The specific procedures for deriving such band levels are given by Kryter [7], but

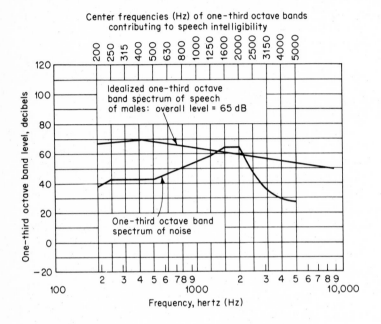

1. Band	2. Speech peaks minus noise, dB	3. Weight	4. Column 2 x 3
200	30	0.0004	0.0120
250	26	0.0010	0.0260
315	27	0.0010	0.0270
400	28	0.0014	0.0392
500	26	0.0014	0.0364
630	22	0.0020	0.0440
800	16	0.0020	0.0320
1000	8	0.0024	0.0192
1250	3	0.0030	0.0090
1600	0	0.0037	0.0000
2000	0	0.0038	0.0000
2500	12	0.0034	0.0408
3150	22	0.0034	0.0758
4000	26	0.0024	0.0624
5000	25	0.0020	0.0500
		AI =	0.4738

Figure 6-2 Example of the calculation of an articulation index (AI) by the one-third-octave-band method. In any given situation the difference (in dB) between the level of speech and the level of noise is determined for each band. These differences are multiplied by their weights. The sum of these is the AI. [Adapted from Kryter, 7.]

an "idealized" spectrum for males is presented in Figure 6-2 as the basis for illustrating the derivation of the articulation index.

   2 Plot on the worksheet the band levels of steady-state noise reaching the ear of the listener. An example of such a spectrum is presented in Figure 6-2.

   3 Determine at the center frequency of each of the bands on the worksheet the difference in decibels between the level of speech and that of the noise. When the noise level exceeds that of speech, assign a zero value. When the speech level exceeds that of noise by more than 30 dB, assign a difference of 30 dB. Record this value in column 2 of the table given as part of Figure 6-2.

   4 Multiply the value for each band derived by step 3 above by the weight for that band as given in column 3 of the table given with Figure 6-2, and enter that value in column 4.

   5 Add the values in column 4. This sum is the AI.

Intelligibility Score

An intelligibility score is simply the percentage of spoken material that can be understood. The AI is not itself an index of intelligibility, but it can be converted into an estimate of intelligibility of various types of speech material, such as shown in Figure 6-3. We will refer back to this figure later. The relationships shown in that figure are approximations in that they depend upon such factors as speaking skills of the speakers and the hearing skills of the lis-

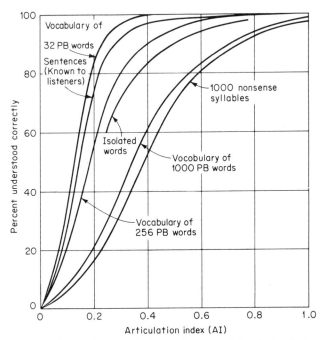

Figure 6-3 Relationship between the articulation index (AI) and the intelligibility of various types of speech-test materials. [Adapted from Kryter, 7, and from French and Steinberg, 6.]

teners. The figure shows that intelligibility depends very much on the nature of the speech material. However, it has been suggested that if the AI of a speech communication system is less than about 0.3, the system would be inadequate for transmission of conventional speech material [Beranek, 1].

Speech Interference Level

The speech interference level (SIL) is another index that is used in estimating the effects of noise on speech intelligibility, and has been used by engineers as a gross basis for comparing the relative effectiveness of speech transmission under different environments of reception. For any given situation, it is actually the simple numerical average of the decibel level of noise in three octave bands, namely, those with centers at 500, 1000, and 2000 Hz. Thus, if the decibel levels of noise in the three octave bands are 70, 80, and 75 dB, respectively, the SIL would be their average, 75 dB. The SIL is useful as a rough index for estimating the effects of noise on speech intelligibility, especially if the noise spectrum is relatively flat. It loses some of its value if the noise has intense low-frequency components, has an irregular spectrum, or consists primarily of pure tones.

Noise Criteria Curves

A further elaboration of the SIL theme resulted in the development of a set of *noise criteria* (NC) curves [Beranek, 2] as shown in Figure 6-4. The derivation of these will not be discussed, but in general they take into account the SIL and loudness level in phons (as discussed in Chapter 5); each NC curve has a loudness level in phons that is 22 units greater than the SIL in dB, which is expressed by the NC number of the curve. In use, individual NC curves are recommended as permissible noise levels in the eight octave bands; no single octave band should exceed the NC curve used as the standard. Each recommended NC value (such as given later in Table 6-1) is intended to permit adequate speech communications required for the room or space in question and to minimize annoyance in that area.

COMPONENTS OF SPEECH COMMUNICATIONS

If we need to do something about improving the intelligibility of speech communication in some system, we need to do so in terms of the individual components involved, such as the message itself, the talker, the transmission system, and the receiver.

The Message

Under adverse communication conditions such as noise, some speech messages or message units are more susceptible to degradation than others. Under

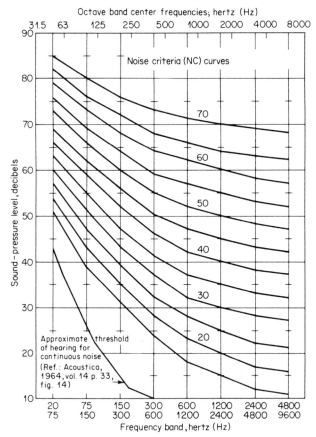

Figure 6-4 Noise criteria (NC) curves. For any specified standard, such as an NC of 30, no frequency band of the noise should exceed the NC curve in question. Each NC curve has a loudness level in phons that is 22 units greater than the SIL in decibels that is expressed by the NC number of the curve. [Adapted from Beranek, 2.]

such conditions, then, it behooves one to construct messages in such a way as to increase the probability of their getting through.

The Vocabulary Used   If we picked a word randomly from the dictionary, the probability of your guessing the correct one would be infinitesimally small. (P. S. Would you have guessed *pedantic*?) But if we arbitrarily restrict our language to two words (for example, *pedantic* and *jostle*), your probability of guessing the right one would be 50–50. Under extremely noisy conditions, when it is difficult to make out the speech sounds, the total number of possible words that *might* be used has a marked influence on the correct recognition of the words—the smaller the possible vocabulary, the greater the probability of recognition. This general principle has been confirmed in experiments such as the one by Miller, Heise, and Lichten [11]. In this experiment subjects were presented with words from vocabularies of various sizes (2, 4, 8, 16, 32, and

256 words, and unselected monosyllables). These were presented under noise conditions with signal-to-noise ratios ranging from −18 to 9 dB. The results, shown in Figure 6-5, show clearly that the percent of words correctly recognized was very distinctly related to the size of the vocabulary that was used.

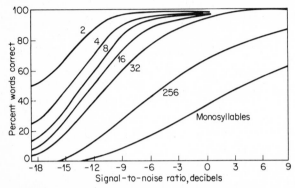

Figure 6-5   Intelligibility of words from vocabularies of different sizes under varying noise conditions. The numbers 2, 4, 8, etc., refer to the number of words in the vocabulary used. [From Miller, Heise, and Lichten, 11. Copyright 1951 by the American Psychological Association and reproduced by permission.]

In certain types of operational situations (such as control-tower operations and voice radio) the vocabularies that are used are in fact very restricted; actual experience probably had demonstrated the principle that speech reception is better when there is a limited assortment of possible words or other components of the message.

The Context of the Message   Closely related to the size of the vocabulary used is the context of the message or message components. This is essentially a problem of expectancy. If one were to hear the expression "a rolling ——— gathers no moss," but failed to distinguish the third word, we could readily supply it in the context of the expression. But it would be hard for us to fill in this blank: "On Wednesday he ———." The effect of context on intelligibility was illustrated in Figure 6-3. In that figure it can be seen that, for any given AI (as derived by the method discussed earlier in this chapter), sentences are more intelligible than isolated words, and isolated words, in turn, are more intelligible than separate syllables. Thus, components in the context of a meaningful conceptual message stand a better chance of being picked up against a noisy background than do those that have no such contextual backdrop.

Phonetic Aspects of Message Components   Within limits it is possible to use in speech communications those speech sounds which have high levels of speech power, and which therefore stand a better chance of "getting through" adverse conditions than sounds which have low levels of speech power. Recognition of this is reflected in the use of alphabetical equivalents (such as the now familiar *Roger*) by certain military organizations and in international aviation, in which a particular letter is represented by a phonetically distinct word.

The Talker

As we all know, the intelligibility of speech depends in part on the character of the talker's speaking voice. Although, in common parlance, we refer to such features of speech as enunciation, research has made it possible to trace down certain specific features of speech that affect its intelligibility. In a study by Bilger, Hanley, and Steer [4], for example, it was found that the speech of "superior" talkers (as contrasted with less intelligible talkers) had a longer "syllable duration," and that they spoke with greater intensity, utilized more of the total time with speech sounds (and less with pauses) and varied their speech more in terms of fundamental vocal frequencies.

The differences in intelligibility of talkers generally are due to the structure of their articulators (the organs that are used in generating speech) and the speech habits people have learned. Although neither of these factors can be modified very much, it has been found that appropriate speech training usually can bring about moderate improvements in the intelligibility of talkers.

The Transmission System and Environment

Speech transmission systems (such as telephones and radios) can produce a variety of forms of distortion, such as frequency distortion, filtering, amplitude distortion, and modifications of the time scale. If intelligibility (and not fidelity) is important in a system, it should be noted that certain types of distortion (especially amplitude distortion) still can allow intelligibility. Since high-fidelity systems are very expensive, it is useful to know what effects various forms of distortion have on intelligibility to be able (it is hoped) to make better decisions about the design or selection of communication equipment. For illustrative purposes, we shall discuss the effects of a couple types of distortion, namely, filtering and amplitude distortion, along with the effects of certain aspects of the environment such as noise.

Effects of Filtering on Speech   The filtering of speech consists basically in blocking out certain frequencies and permitting only the remaining frequencies to be transmitted. Filtering may be the fortuitous, unintentional consequence, or in some instances the intentional consequence, of the design of a component. Most filters eliminate frequencies *above* some level (a *low-pass* filter) or frequencies *below* some level (a *high-pass* filter). Typically, however, the cutoff, even if intentional, usually is not precisely at a specific frequency, but rather tapers off over a range of adjacent frequencies. A given filtering affects the intelligibility of certain phonemes more than others. Fletcher [5, pp. 418–419], for example, points out that long *e* is recognized correctly about 98 percent of the time if the frequencies either above or below 1700 Hz are eliminated. But while *s* is affected only slightly by eliminating frequencies below 1500, its intelligibility is practically destroyed by eliminating frequencies above 4000. Most short vowels have important sound components below 1000 Hz, and 20 percent error in their recognition occurs when frequencies below that level are eliminated; on the other hand, the elimination of frequencies above 2000 has little effect on their intelligibility.

However, the intelligibility of normal speech does not depend entirely upon intelligibility of each and every speech sound. For example, filters that eliminate all frequencies above, or all frequencies below, 2000 Hz will *in the quiet* still transmit speech quite intelligibly, although the speech does not sound natural [Rosenblith and Stevens, 13, p. 112]. The distortion effects on speech of filtering out certain frequencies are summarized in Figure 6-6. This shows, for high-pass filters and for low-pass filters with various cutoffs, the percentage of intelligibility of the speech that is transmitted. It can be seen, for example, that the filtering out of frequencies above 4000 Hz or below about 600 Hz has relatively little effect on intelligibility. But look at the effect of filtering out frequencies above 1000 Hz or below 3000 Hz! Such data as those given in Figure 6-6 can provide the designer of communications equipment with some

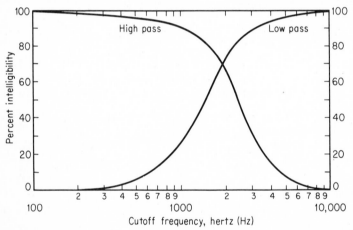

Figure 6-6 Effect on intelligibility of elimination of frequencies by the use of filters. A low-pass filter permits frequencies below a given cutoff to pass through the communication system, and eliminates frequencies above that cutoff. A high-pass filter, in turn, permits frequencies above the cutoff to pass and eliminates those below. [Adapted from French and Steinberg, 6.]

guidelines to follow when trying to decide how much filtering can be tolerated in the system.

Effects of Amplitude Distortion on Speech   Amplitude distortion has been defined as the deformation which results when a signal passes through a nonlinear circuit [Licklider, 8]. One form of such distortion is *peak clipping*, in which the peaks of the sound waves are clipped off and only the center part of the waves left. Although peak clipping is produced by electronic circuits for experiments, some communication equipment has insufficient amplitude-handling capability to pass the peaks of the speech waves and at the same time provide adequate intensity for the lower-intensity speech components, thus reducing intelligibility. Since peak clipping impairs the quality of speech and music, it is not used in regular broadcasting, but it is sometimes used in military and commercial communication equipment. In such cases premodulation clippers are built into the transmitters, thus reducing the peaks, but the available power is then used for transmitting the remainder of the speech waves. Center clipping, on the other hand, eliminates the amplitudes below a given value and leaves the peaks of the waves. Clipping is more of an experimental procedure than one used extensively in practice. The amount of clipping can be controlled through electronic circuits. Figure 6-7 illustrates the speech waves that would result from both forms of clipping.

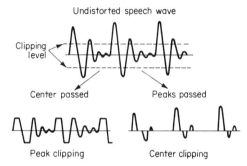

Figure 6-7 Undistorted speech wave and the speech waves that would result from peak clipping and center clipping. [From Miller, 10, p. 72.]

The effects of these two forms of clipping on speech intelligibility are very different, as shown in Figure 6-8. It can be seen that peak clipping does not

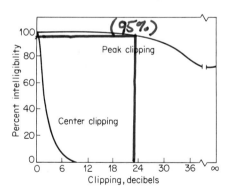

Figure 6-8 The effects on speech intelligibility of various amounts of peak clipping and of center clipping. [From Licklider and Miller, 9.]

cause major degradation of intelligibility even when the amount of clipping (in decibels) is reasonably high. On the other hand, even a small amount of center clipping results in rather thorough garbling of the message. The reason for this difference in the effects of peak and center clipping is the difference in phonetic characteristics of the vowels (which generally have more speech power) and of the consonants (which generally have lower speech power). Thus, when the peaks are lopped off, we reduce the power of the vowels, which are less critical in intelligibility, and leave the consonants essentially unscathed. But when we cut out the center amplitudes, the consonants fall by the wayside, thus leaving essentially the high peaks of the vowels. Since intelligibility is relatively insensitive to peak clipping, the communications engineers can shear off the peaks and repackage the wave forms, using the available power to amplify the weaker but more important consonants.

Effects of Noise on Speech   Some noise is ever with us, but in large doses it is a potential bugaboo as far as speech communications are concerned, whether it be ambient noise (noise in the environment) or in-line noise in a communication system. The effects of noise in face-to-face communications as presented by Webster [14] are shown in Figure 6-9. This shows the level of

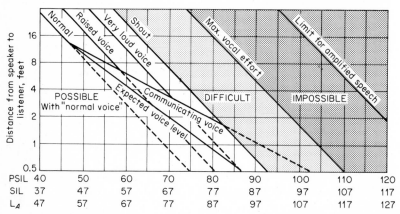

Figure 6-9   Voice level and distance between talker and listener for satisfactory face-to-face communication as limited by ambient noise level (expressed in SIL). For any given noise level (such as 57) and distance (such as 8 ft) it is possible to determine the speech level that would be required (in this case a "raised" voice). [Adapted from Webster, 14, fig. 19, p. 69.]

voice that would be required for reasonable intelligibility of speech for varying combinations of SIL and distance from the talker. For the distance and noise level combinations in the white area of the figure, a normal voice would be satisfactory, but for those combinations above the "maximum vocal effort line" it would probably be best to get out the boy scout wigwag flags or send up Indian smoke signals.

The subjective reactions of people to noise levels in private offices and in large offices (secretarial, drafting, business machine offices, etc.) were elicited by Beranek and Newman [3] by questionnaires. The results of this survey were used to develop a rating chart for office noises, as shown in Figure 6-10. The line for each group represents SIL's (base lines) that were judged to exist at certain subjective ratings (vertical scale). The dot on each curve represents the judged upper limit for intelligibility. The judged limit for private offices (nor-

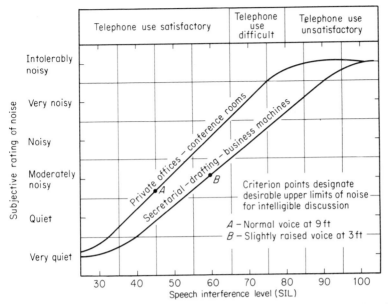

Figure 6-10 Rating chart for office noises. [Based on data from Beranek and Newman, 3, as modified by Peterson and Gross, 12, to reflect the current practice of using octave bands with centers at 500, 1000, and 2000 Hz.]

mal voice at 9 ft) was slightly above 45 dB, and for larger offices (slightly raised voice at 3 ft) was 60 dB.

The judgments with regard to telephone use are given below [Peterson and Gross, 12, p. 38]:

SIL, dB	Telephone use
Less than 60	Satisfactory
60 to 75	Difficult
Above 80	Impossible

These judgments were made for long-distance or suburban calls. For calls within a single exchange, about 5 dB can be added to each of the above levels, since there is usually better transmission within a local exchange.

Criteria for control of background noise in various communication situations have been set forth by Peterson and Gross [12] based on earlier standards by Beranek and Newman [3]. These, expressed as SIL's, are given in Table 6-1, along with the corresponding NC recommended subsequently by Beranek [2]. (Aside from certain minor differences, the systematic difference of about 5 dB is due to the fact that the SIL data have been modified to reflect the current practice of using bands with centers at 500, 1000, and 2000 Hz.)

Effects of Reverberation on Speech Reverberation is the effect of noise bouncing back and forth from the walls, ceiling, and floor of an enclosed room. As we know from experience in some rooms or auditoriums, this rever-

Table 6-1 Speech Interference Levels and Noise Criteria Recommended for Certain Types of Rooms

Type of room	Maximum permissible level (measured in vacant rooms)	
	SIL	NC
Secretarial offices, typing	60	50–55
Coliseum for sports only (amplification)	55	50
Small private office	45	30–35
Conference room for 20	35	30
Movie theater	35	30
Conference room for 50	30	20–30
Theaters for drama, 500 seats (no amplification)	30	20–25
Homes, sleeping areas	30	25–35
Assembly halls (no amplification)	30	
Schoolrooms	30	25
Concert halls (no amplification)	25	15–20

Source: SIL data from Beranek and Newman [3] as modified by Peterson and Gross [12, Table 3-5, p. 39] to reflect current practice of using octave bands with centers at 500, 1000, and 2000 Hz; NC data from Beranek [2].

beration seems to obliterate speech or important segments of it. Figure 6-11 shows approximately the reduction in intelligibility that is caused by varying degrees of reverberation (specifically, the time in seconds that it takes the noise to die down). This relationship essentially is a straight-line one.

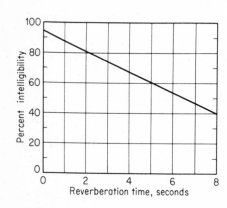

Figure 6-11 Intelligibility of speech in relation to reverberation time. The longer the reverberation of noise in a room, the lower the intelligibility of speech. [Adapted from Fletcher, 5.]

Effects of Earplugs on Speech In a sense, earplugs are part of the transmission system, since they intervene between the environment and the receiver. Although their purpose is to prevent or minimize hearing loss, one might expect that they would also reduce the intelligibility of speech. In fact, using earplugs may actually increase the intelligibility of speech under high noise levels. Under low noise levels, however, the use of earplugs may impair speech intelligibility somewhat. It is in high-noise-level situations, however, that there would be a greater likelihood that earplugs would be worn, and in such conditions they can be effective. The explanation for this is that at high noise levels a point is reached where additional intensity cannot be discriminated; at such levels the difference between the intensity of the signal (in this case speech) and of its background noise cannot be discriminated. The value of earplugs under such circumstances is to bring the levels of both the signal and the background noise down to the point where the difference between them can be discriminated.

The Receiver

The receiver is the last link in the communication chain. For receiving speech messages under noise conditions, the receiver should have normal hearing, should be trained in the types of communications to be received, should be reasonably durable in withstanding the stresses of the situation, and should be able to concentrate on one of several conflicting stimuli.

DISCUSSION

When feasible, of course, speech communications should be carried out under favorable conditions, uncluttered with noise. However, in many circumstances it is not possible to reduce noise at its source; (one cannot stop the rolling mills of a steel mill for people to communicate with others). Under these and other circumstances it is necessary to look to other elements of the total communication system, rather than to the noise source itself, for possibilities of improving the intelligibility of speech. On the engineering design side of the coin, the possibilities to consider are those of minimizing the transmission of noise if possible (through acoustical treatment and other means), improving the design of the communication equipment, and modifying the nature of the messages to be used; and on the personnel side of the coin, the possibilities are those of selection and training of talkers and receivers, where these are feasible.

REFERENCES

1   Beranek, L. L.: The design of speech communication systems, *Proceedings of the Institute of Radio Engineers*, New York, 1947, vol. 35, pp. 880–890.
2   Beranek, L. L.: Revised criteria for noise in buildings, *Noise Control*, 1957, vol. 3, no. 1, pp. 19–27.
3   Beranek, L. L., and R. B. Newman: Speech interference levels as criteria for rating background noise in offices, *Journal of the Acoustical Society of America*, 1950, vol. 22, p. 671.
4   Bilger, R. C., T. D. Hanley, and M. D. Steer: *A further investigation of the relationships between voice variables and speech intelligibility in high level noise*, TR for SDC, 104–2–26, Project 20–F–8, Contract N6ori–104, Purdue University, Lafayette, Ind. (mimeographed).
5   Fletcher, H.: *Speech and hearing in communication*, D. Van Nostrand Company, Inc., Princeton, N. J., 1953.

6 French, N. R., and J. C. Steinberg: Factors governing the intelligibility of speech sounds, *Journal of the Acoustical Society of America*, 1947, vol. 19, pp. 90–119.

7 Kryter, K. D.: Speech communication, in H. P. Van Cott and R. G. Kinkade (eds.), *Human engineering guide to equipment design*, U. S. Government Printing Office, Washington, D. C., 1972, chap. 5.

8 Licklider, J. C. R.: Effects of amplitude distortion upon intelligibility of speech,. *Journal of the Acoustical Society of America*, 1946, vol. 18, pp. 429–434.

9 Licklider, J. C. R., and G. A. Miller: The perception of speech, in S. S. Stevens (ed.), *Handbook of experimental psychology*, John Wiley & Sons, Inc., New York, 1951, chap. 26.

10 Miller, G. A.: *Language and communication*, paperback ed., McGraw-Hill Book Company, New York, 1963.

11 Miller, G. A., G. A. Heise, and W. Lichten: The intelligibility of speech as a function of the context of the test materials, *Journal of Experimental Psychology*, 1951, vol. 41, pp. 329–335.

12 Peterson, A. P. G., and E. E. Gross, Jr.: *Handbook of noise measurement*, 7th ed., General Radio Co., New Concord, Mass., 1972.

13 Rosenblith, W. A., and K. N. Stevens: *Handbook of acoustic noise control. Vol. II. Noise and man*, USAF, WADC, TR 52–204, June, 1953; Report PB 111, 274, U. S. Department of Commerce, Office of Technical Services.

14 Webster, J. C.: *Effects of noise on speech intelligibility*, The American Speech and Hearing Association, Washington, D.C., ASHA Reports 4, 1969, pp. 49–73.

15 *Methods for the calculation of the articulation index*, American National Standards Institute (ANSI), Report ANSI 535–1969.

Human Output and Control Processes

Chapter 7

Human Activities: Their Nature and Effects

Human "work" (whether on a job, playing golf, or playing bridge) ranges over a spectrum from that which is strictly mental, through that which is essentially psychomotor, to that which is dominantly physical. Any given work activity can vary in terms of its "type" and in terms of its level of "intensity." Snook and Irvine [61] postulate that high levels of work intensity increase the likelihood of exhaustion, and that low levels increase the likelihood of boredom, with intermediate levels being optimum, as illustrated in Figure 7-1. (It should be noted, however, that "boredom" is the reaction of the individual, and not an attribute of the work activity.) There is reason to concur in the notion of there being an "optimum" level of work intensity, although this might well differ for individuals; where feasible, human factors efforts should be directed toward achieving this optimum.

The immediate outputs of human activity in most job situations include the execution of physical responses or communications, which accomplish some (hopefully) desired objective. This chapter will deal primarily with the nature of human physical responses and activities, since these have implications for many aspects of human factors, such as: the design of control devices, hand tools, and other devices; the handling of materials; the physical layout and arrangement of work spaces; and work methods and procedures. Since our

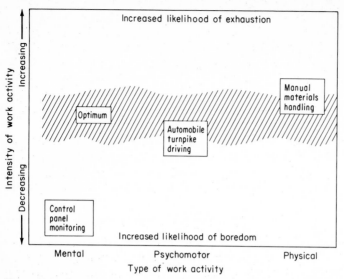

Figure 7-1 Generalized model of human work, varying in terms of the type of work and the level of intensity. The figure shows approximately where three types of work would fit into this model. For any given type there is probably some optimum that is most suitable for human involvement. [Adapted from Snook and Irvine, 61, fig. 1.]

present interest deals primarily with physical activities, we will first discuss the physical and physiological bases of such activities.

It should be pointed out, however, that some combination of physiological processes is involved in virtually every form of human activity—not only with physical acts such as operating a machine or carrying out the garbage, but also with mental work and "nonphysical" acts such as worrying about the monthly bills or taking an afternoon snooze.

BASES OF HUMAN MOTOR ACTIVITIES

The abilities of people to perform various types of motor activities depend essentially on the physical structure of the body (the skeleton), the skeletal muscles, the nervous system, and the metabolic processes.

The Skeletal Structure

The basic structure of the body consists of the skeleton, there being 206 bones that form the skeleton. Certain bone structures serve primarily the purpose of housing and protecting essential organs of the body, such as the skull (which protects the brain) and the rib cage (which protects the heart, lungs, and other internal organs). The other (skeletal) bones—those of the upper and lower extremities and the articulated bones of the spine—are concerned primarily with the execution of physical activities, and it is these that are particularly relevant to our subject. The skeletal bones are connected at body joints, there being two general types of joints that are principally used in physical activities, namely, synovial joints and cartilaginous joints. The synovial joints include hinge joints (such as the fingers and knees), pivot joints (such as the elbow, which is also a hinge joint), and ball-and-socket joints (such as the shoulder and hip). The primary examples of cartilaginous joints are those of the vertebrae of the spine, which make possible, collectively, considerable rotation and forward bending of the body.

The Skeletal Muscle System

The bones of the body are held together at their joints by ligaments. The skeletal muscles (also called "striated" or "voluntary" muscles) consist of bundles of muscle fibers that have the property of contractility; the muscle fibers serve to convert chemical energy into mechanical work. The two ends of each muscle blend into tendons, which, in turn, are connected to different skeletal bones in such a manner that when the muscles are activated, they apply some form of mechanical leverage.

Neural Control of Muscular Activity

The nerves entering a muscle are of two classes, namely, sensory nerves and motor nerves. Some of the sensory nerves are associated with the cutaneous senses (touch, heat and cold, pain, etc.). The other sensory nerves are the proprioceptors, which are distributed through the muscles, the tendons, and the covering of the bones and which provide kinesthetic feedback to aid in muscular control. The motor nerves actually control the actions of the muscles.

The execution of physical activities depends upon learning which can be viewed in a two-stage, hierarchical frame of reference. To borrow terms from computer lingo, the first stage, the *executive program*, deals with the overall purpose or plan of the act and is essentially under conscious control by the central nervous system. The second stage, the *subroutine*, deals with the control of the specific movements that are required to complete the act. Through practice the subroutines typically become so learned that they are performed automatically once the "executive plan" is put into action. The "learning" of the subroutines is based on the establishment of neural connections at the neural center of the motor nerve pathways. Thus, if at the executive program level one decides to go upstairs, the subroutines of doing so typically are executed without conscious control.

Muscle Metabolism

Metabolism is the collective chemical process of the conversion of foodstuffs into two forms, namely, mechanical work and heat. Some of the mechanical work is of course used internally in the processes of respiration and digestion. Other mechanical work is used externally, as in walking and performing physical tasks. In either case, heat is generated, usually in amounts that are excessive to body needs; this surplus heat must be dissipated by the body. The contraction of a muscle requires energy, the basic source of this energy being glycogen, which can be thought of as a large number of glucose molecules bound together to form one large molecule. The conversion of glycogen into energy consists of a chemical reaction that ends in the production of lactic acid. However, the lactic acid needs to be dissipated by being broken down into water and carbon dioxide. The first stage (the breakdown of glycogen into lactic acid) does not require oxygen, and is said to be *anaerobic*. The second stage (the breakdown of lactic acid into water and carbon dioxide) is said to be *aerobic*.

At the initiation of physical activity the muscles can utilize the glycogen which is already available. But the amount of glycogen and hence glucose available is small; so if the activity is continued, the body needs to replenish these nutrients from the blood, along with a supply of oxygen, which is required in the second stage. When an adequate supply of oxygen is provided, little or no lactic acid accumulates. If the level of activity requires more oxygen than is provided by the normal rate of blood flow through the cardiovascular system, the system adjusts itself to fulfill the increased demands, in particular by increasing the breathing rate to bring additional oxygen into the lungs, and by increasing the heart rate to pump more blood through the "pipes" of the cardiovascular system. From the heart the blood is pumped through the lungs, where it picks up a supply of oxygen, which is then carted by the blood to the muscles where the oxygen is needed. With at least moderate rates of work, the heart rate and breathing rate are normally increased to the level that provides enough oxygen to perform the physical activities over a continuing period of time. However, when the amount of oxygen delivered to the muscles fails to meet the requirements (as when the level of physical activity is high), lactic acid tends to accumulate in the blood. If the rate or duration of physical activity results in continued accumulation of lactic acid, the muscles will ultimately cease to respond. If the rate of removal of lactic acid does not keep pace with its formation, additional oxygen must be supplied after cessation of the activity to remove the remaining lactic acid. This is referred to as the *oxygen debt*. Since this debt has to be paid back, the heart rate and breathing rate do not immediately settle back to prework levels when work ceases, but rather slow down gradually until the borrowed oxygen is replaced.

Basal Metabolism   The basal metabolic rate is that which is required simply to maintain the body in an inactive state. Although it varies from individual to individual, the average for adults usually ranges from about 1500 to 1800 kcal/day [Tuttle and Schottelius, 65].[1] Considering the basal level plus the energy required for a relatively sedentary existence, Passmore [49] estimates that about 500 kcal are required for 8 hours in bed plus about 1400 kcal for

---

[1]The energy unit generally used in physiology is the kilocalorie (abbreviated kcal) or Calorie (with a capital C to distinguish it from the gram-calorie). The kilocalorie is the amount of heat required to raise the temperature of a kilogram of water from 15 to 16°C. The relation of the kilocalorie to certain other units of energy measurement is

1 kcal = 426.85 kg-m
1 kcal = 3087.4 ft-lb
1 kcal = 1000 cal = 1 C

nonworking time, adding up to a total of 1900 kcal per day; in turn, Lehmann [41] estimates the corresponding requirements (basal metabolism and leisure) at around 2300 kcal/day, and Tuttle and Schottelius [65] estimate that the typical adult who lives a fairly sedentary life utilizes about 2400 kcal/day. Thus, various estimates of the total nonworking calorie requirements range from about 1900 to about 2400 kcal/day. (The physiological costs of work will be mentioned later.)

THE CONCOMITANTS OF HUMAN ACTIVITY

Human activity that falls within the optimum level of intensity illustrated in Figure 7-1 generally can be carried out without, as they say, any "stress or strain." However, some activities and some environments may be outside of reasonable limits, and thus impose some form of stress upon the individual.

The Nature of Stress and Strain

Stress refers to any aspect of human activity or the environment acting upon the individual which results in some undesirable cost to, or reaction upon, the individual. Some of the possible sources of stress are given in Figure 7-2, these being grouped into those of a physiological and those of a psychological nature. *Strain*, in turn, is the "cost" or effect or consequence on the individual, of stress. Some of the typical measures of physiological and psychological strain are also given in Figure 7-2.

Source of stress		*Physiological*	*Psychological*
	Work	Heavy work Immobilization	Information overload Vigilance
	Environment	Atmospheric Noise/vibration Heat/cold	Danger Confinement
	Circadian	Sleep loss	Sleep loss

Measures of strain	Chemical	Electrical	Physical	Activity	Attitudes
	Blood content Urine content Oxygen consumption Oxygen deficit Oxygen recovery curves Calories	EEG (electroencephalogram) EKG (electrocardiogram) EMG (electromyograph) EOG (electrooculogram) GSR (galvanic skin response)	Blood pressure Heart rate Sinus arrhythmia Pulse volume Pulse deficit Temp. of body Respiratory rate	Work rate Errors Blink rate	Boredom Other attitudinal factors

Figure 7-2 Primary sources of stress and primary measures of strain as induced by stress. [Adapted from Singleton, 59, fig. 4.]

Measures of Physiological Functions

The physiological measures shown in Figure 7-2 can be used to evaluate the physiological activity of individuals, or that associated with some work activity or environment generally (using groups of individuals). Certain measures are indicative in their own way of the general level of physiological demands of the activity in question, whereas other measures reflect the activity of specific muscle groups, and still others tend to be more relevant as indices of mental activity.

Measures of General Physiological Demands   The physiological demands of work made upon people are frequently measured by heart rate (or some derivative thereof) and oxygen consumption. These two measures are rather closely correlated with each other across varying levels of general dynamic muscular work, but neither is a very sensitive index of static muscular work, "local" dynamic muscular work (that of specific muscles or muscle groups), or mental work [Burger, 8]. Heart rate is reasonably indicative of the effects of heat stress and emotional stress, but is also related to individual factors (constitution, physical condition, sex, etc.), and is therefore less suitable as an absolute index of the load imposed by various types of work than is oxygen consumption.

Since heart rate and oxygen consumption have their limitations as measures of physiological demands on an across-the-board basis, Burger [8] argues for the use of an index of circulatory load as something of a combined index of the load of different organic systems. An approximation of this is the product of heart rate, mean blood pressure, and stroke volume.

There are certain derivatives of oxygen consumption and heart rate that are sometimes used. The oxygen debt, for example, is the amount of oxygen that is required by the muscles after the beginning of work, over and above that which is supplied to them by the circulatory system during their activity. This debt needs to be "repaid" after the cessation of work, and is reflected in the elevated rate (i.e., above resting level) of oxygen consumption in the recovery process. A somewhat related measure based on the heart rate is what Davies [16] refers to as the *pulse deficit* (PD). The PD reflects the deficit in cardiac output in the early stages of exercise and is thus an indication of the level of anaerobic metabolism. Operationally, the PD is calculated by taking the heart rate for the fifth minute as a reference, and calculating the area bounded by the pulse/time curve for the first four minutes of exercise. However, Shephard [58] raises questions about this procedure, suggesting that the pulse rate usually continues to rise during the second five minutes of exercise, and believes that a pulse deficit measure based on the second five minutes would have more significance as a measure of the intensity of exercise.

Another measure based on heart rate is the heart-rate recovery curve as used by Brouha [5], this being a curve of the heart rate measured at certain intervals after work (such as 1, 2, and 3 minutes). In general terms, the more strenuous the work activity, the longer it takes the heart rate to settle down to its prework level.

The energy cost of work is frequently measured in terms of calories or oxygen consumption, usually expressed in units per minute or per hour.

Measures of Local Muscular Work   One of the measures of "local" muscular work is *electromyographic* (EMG) recordings. These are inked tracings of the electrical impulses that occur during work, and provide estimates of the magnitude of voluntary muscular activity. Examples of such recordings are shown in Figure 7-3 for four muscles of one subject when applying a constant figure of 60 ft-lb and of 15 ft-lb on a steel socket. There have been several problems associated with the recording, quantification, and interpretation of such recordings. In this regard Khalil [37] has developed procedures for summating the action potentials of several muscles that are monitored simultaneously. The values given in Figure 7-3 show the individual and summated val-

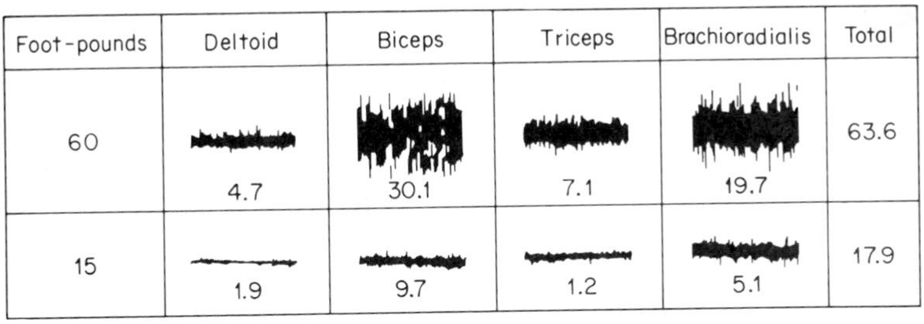

Foot-pounds	Deltoid	Biceps	Triceps	Brachioradialis	Total
60	4.7	30.1	7.1	19.7	63.6
15	1.9	9.7	1.2	5.1	17.9

Figure 7-3 Electromyograms recorded for four muscles of a subject maintaining a constant torque of 60 ft-lb and of 15 ft-lb. The sum of the four values is an index of the total amount of energy expended. [Adapted from Khalil, 37, fig. 3.]

ues for the four muscles. Khalil provides some evidence that the method is suitable for both static and dynamic muscular exertion, and suggests that it might be used for evaluating the expenditure of effort required in the performance of various types of work activities.

Measures of Mental Activity Since mental activities involve physiological processes, it is reasonable to believe that refined physiological measures could serve as indexes of the mental loads of different types of mental activities. Certain measurement techniques used to date (such as EEG and EMG), however, have not been entirely satisfactory, so further efforts are being made to derive physiological measures of mental load. One such measure is the sinus (or cardiac) arrhythmia. This is essentially a measure of the irregularity of the heart action. The problems in measurement of this are discussed by Kalsbeek [34, 35] and by Luczak and Laurig [42] and need not be described here, except to say that there are numerous possible variations in measurement. In general terms Kalsbeek [34] concludes that an increase in mental load (as measured by the number of binary choices made per minute) is reflected by a decrease in the scored irregularity of the heart-rate pattern. In one analysis that he reports, for example, the sinus arrhythmia (s.a.) scores for three levels of mental load were as follows:

Condition	Sinus arrhythmia score
At rest	13.0
Low level of mental load	1.5
High level of mental load	0.5

One cannot say today that measures of sinus arrhythmia are definitely "the answer" to the measurement of mental load, but they seem to offer some such promise.

BIOMECHANICS OF MOTION

The *biomechanics* of motion deal with the various aspects of the physical movements of the body and the body members.[2] The operation of the body members can be characterized in terms of kinematics (the science of motion). The bones, connected at their joints, in combination with their associated muscles, serve as levers.

Types of Movements of Body Members

Certain of the movements which the arms, legs, and other body members are capable of performing can be considered as basic. Some of these, with their associated jargon in biomechanics, are given below [Damon, Stoudt, and McFarland, 13]:

- *Flexion:* bending, or decreasing the angle between the parts of the body
- *Extension:* straightening, or increasing the angle between the parts of the body
- *Adduction:* moving toward the midline of the body
- *Abduction:* moving away from the midline of the body
- *Medial rotation:* turning toward the midline of the body
- *Lateral rotation:* turning away from the midline of the body
- *Pronation:* rotating the forearm so that the palm faces downward
- *Supination:* rotating the forearm so that the palm faces upward

Essentially, the above describe the movements of body members in terms of the functioning of the muscles (e.g., flexion and extension) and of the direction of the movements relative to the body (e.g., adduction and abduction). Some of these basic movements are illustrated in Figure 7-4, along with the following values for each (as based on a sample of 39 men selected to represent the major physical types in the military services): mean angle (in degrees) and 5th and 95th percentile angles (computed from the standard deviations for the sample). In this, as in other aspects of biomechanics, there are the ever-present individual differences, including the effects of physical condition and the ravages of age.

However, in performing specific activities, as in work, the movements of the body members can be described in more operational terms. There actually are different ways in which movements can be classified. One such scheme is given below:

- *Positioning* movements are those in which the hand or foot moves from one specific position to another, as in reaching for a control knob.

[2]For a more extensive treatment of biomechanics, especially as related to equipment design, the reader is referred to Damon, Stoudt, and McFarland [13, especially pp. 187–252].

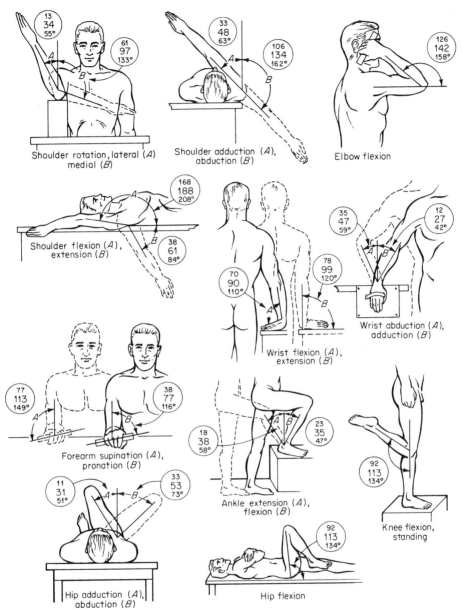

Figure 7-4 Range of certain movements of the upper and lower extremities, based on a sample of 39 men selected to represent the major physical types in the military services. The three values (in degrees) given for each angle are the 5th percentile, the mean, and the 95th percentile, respectively, of voluntary (not forced) movements. [Based largely on data from Dempster, 17, as reanalyzed by Barter et al., 3.]

• *Continuous* movements are those which require muscular control adjustments of some type during the movement, as in operating the steering wheel of a car or guiding a piece of wood through a band saw.

• *Manipulative* movements involve the handling of parts, tools, control mechanisms, etc., typically with the fingers or hands.

- *Repetitive* movements are those in which the same movement is repeated, as in hammering, operating a screwdriver, and turning a handwheel.
- *Sequential* movements are several relatively separate, independent movements in a sequence.
- A *static* adjustment is the absence of a movement, consisting of maintaining a specific position of a body member for a period of time.

As indicated above, various types of movements may be combined in sequence so that they blend one into another. For example, placing the foot on a brake pedal is a positioning movement, but this may be followed by a continuous movement of adjusting the amount of brake pressure to the conditions of the situation. Similarly, a continuous movement may include holding a position (a static adjustment) for a short time.

These operational categories are, of course, rather gross (or, to be more semantically fashionable, we might refer to them as *macromotions*). For certain purposes, categories of a more *micro*motion nature are used. This is especially the case in industrial engineering practices in methods analysis, in which certain elemental motions are identified in work activities. Most elemental motion systems stem from the original concept of therbligs developed by Gilbreth [24] many years ago. A more recent version of these systems is the Methods Time Measurement system [Maynard, 44].

Operational Criteria of Physical Activity

What we will label *operational* criteria include techniques for measuring or otherwise depicting the performance of the body or body members. The most obvious, and most common, operational criteria relate to the performance of body members in making specific types of movements, such performance generally falling into the following groups: *range* of movement, *force* applied during the activity (i.e., strength), *endurance*, *speed*, and *accuracy*. For measuring these, various kinds of gadgetry are used, such as timing devices, motion pictures, strain gauges, and dynamometers. Aside from such criteria of body-member movements, there are certain techniques for depicting, or recording information about, the physical movements involved in activity. Such techniques include motion pictures, the use of interrupted-light photography (chronophotography), and certain electronic and mechanical techniques. One of these, the force platform, will be used as an illustration.

A force platform is a small platform on which a subject stands when carrying out some physical activity. By the use of some sensing elements below the platform (such as piezoelectric crystals) it is possible to sense and then automatically record the forces generated by the subject in each of three planes, namely, vertical, frontal, and transverse. The original force platform was developed by Lauru [39]; other platforms have been used experimentally by Barany [1] and Greene, Morris, and Wiebers [28]. Such devices are sensitive to slight differences in physical movements and can thus lend themselves to use in comparing the three-dimensional forces in different activities. Recordings of the

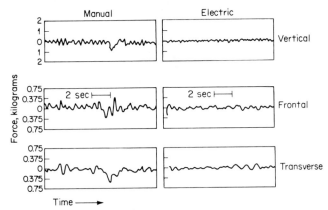

Figure 7-5   Forces in three dimensions (vertical, frontal, and transverse) in the operation of a manual and an electric typewriter, recorded with a force platform. [From Brouha, 5, p. 106.]

forces in the operation of manual and of electric typewriters are shown in Figure 7-5 for comparison.

It has been proposed that such force-time recordings, as possible indices of energy expenditure, are nearly as accurate as metabolic measurements and can thus be used as a measure of physiological cost of a given motion [Brouha, 5, p. 103]. In fact, Brouha presents data for oxygen cost of certain work activities that show high correlations (ranging from .83 to .96) with data from the force platform (force-time *areas*, that is, the area under the curve produced by the platform). Although further data probably would be required to determine the generality of the relationship between force-platform measures of work and strictly physiological measures, the force platform does seem to offer considerable promise as a technique for studying work activities.

ENERGY EXPENDITURE IN PHYSICAL ACTIVITIES

Although human beings are not now used as sources of energy nearly as much as in days gone by, some occupations still require substantial physical effort, at least at certain times or as accumulated over the work day, and in some countries the use of human beings as major sources of energy is almost dictated by economic considerations. When human physical activity in work is potentially dangerous to health and safety, some modification of the work is in order, whether by appropriate redesign of the equipment and work space, by modification of methods, or by reduction of work periods or work pace.

Energy Expenditures of Gross Body Activities

In order to give some "feel" for the numerical values of energy expenditures for different kinds of physical activities, it may be useful to present the physiological costs of certain everyday activities; the following examples are given in kilocalories per minute [Edholm, 19]: sleeping, 1.3; sitting, 1.6; and standing,

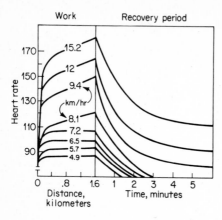

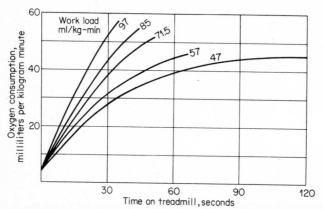

Figure 7-6 Heart rate during and after a march of 1.6 km (1 mile) at various speeds, in kilometers per hour. [From LeBlanc, 40, as presented by Monod, 47.]

2.25. In connection with rate of body movement (as in walking and running), the price per unit of work (in physiological costs) goes up with increasing rate. This is shown quite clearly in Figure 7-6, which shows the heart rate during and after marching a 1600-meter (about a 1-mile) course, when the 18 subjects marched at speeds ranging from 4.9 to 15.2 km/h (about 3.0 to 9.4 mi/h). The increasing energy cost scoots up rather sharply at speeds of 8.1 km/h and above. In addition, recovery time also increases markedly. The energy costs of body exercises of different intensities are shown further in Figure 7-7. The

Figure 7-7 Oxygen consumption from the onset to the end of running on a treadmill with certain work loads that were the consequence of specified speeds and inclines; the energy requirements of the work loads are themselves expressed by an independently derived oxygen-consumption index. [From Margaria et al., 43.]

work loads were the consequence of running on a treadmill at certain combinations of speed and inclines. The oxygen-consumption curves are obviously steeper the higher the work load; the lightest work led to exhaustion in about 3 min, the heaviest in about 30 s. These and other examples clearly indicate the trade-offs in human work, in particular that the physiological price of work—per unit of work—is greater at higher rates of work than at more moderate rates.

Energy Expenditures of Specific Activities

The discussion above dealt simply with the energy expenditures of lugging the body around, as, for example, at different paces, inclines, etc. The energy expenditures of various types of activity vary somewhat for individuals, but estimates of the approximate energy costs for certain specific types of work are given in Figure 7-8. The energy costs for these range from 1.6 to 16.2 kcal/min.

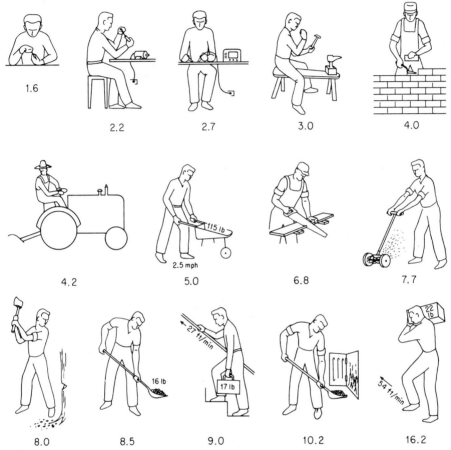

Figure 7-8 Examples of energy costs of various types of human activity. Energy costs are given in kilocalories per minute. [Data from Passmore and Durnin, 50, as adapted and presented by Gordon, 26.]

However, the energy cost for certain types of work can vary with the manner in which the work is carried out. The differential costs of methods of performing an activity are illustrated by the several methods of carrying a load as used in various cultures. Seven such methods were compared by Datta and Ramanathan [14] on the basis of oxygen requirements; these methods and the results are shown in Figure 7-9. The requirement of the most efficient method (the double pack) is used as an arbitrary base of 100 percent. There are advan-

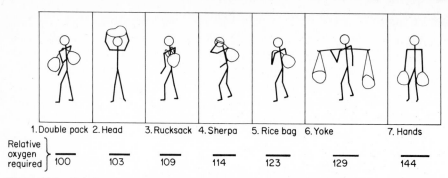

1. Double pack 2. Head   3. Rucksack 4. Sherpa   5. Rice bag 6. Yoke   7. Hands

Relative
oxygen
required
100   103   109   114   123   129   144

Figure 7-9 Relative oxygen consumption of seven methods of carrying a load, with the double-pack method used as a base of 100 percent. This illustrates that the manner in which an activity is carried out can influence the energy requirements. [Adapted from Datta and Ramanathan, 14.]

tages and disadvantages to the various methods over and above their oxygen requirements, but the common denominator of the most efficient methods reported in this and other studies is that of maintenance of good postural balance, one that affects the center of gravity the least.

Energy Expenditures of Different Postures

The posture of workers when performing some tasks is another factor that can influence energy expenditure. In this regard certain agricultural tasks in particular have to be carried out at or near ground level, as in picking strawberries. When performing such work, however, any of several postures can be assumed. The energy costs of certain such postures were measured in a study by Vos [67] in which he used a task of picking up metal tags placed in a standard pattern on the floor. A comparison of the energy expenditures of five different postures is given in Figure 7-10. This figure shows that a kneeling posture with

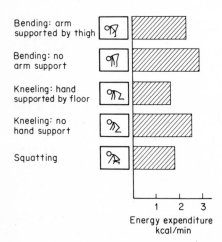

Bending: arm
supported by thigh

Bending: no
arm support

Kneeling: hand
supported by floor

Kneeling: no
hand support

Squatting

1   2   3
Energy expenditure
kcal/min

Figure 7-10 Human energy expenditures (kcal/min) for five postures used in task of picking up light objects from ground level. [Adapted from Vos, 67, fig. 5.]

hand support and a squatting posture required less energy than the other postures. (The kneeling posture, however, precludes the use of one hand and might cause knee discomfort over a period of time.) On the basis of another phase of the study it was shown that a sitting posture (with a low stool) is a bit better than squatting, but a sitting posture is not feasible if the task requires moving from place to place. Although this particular analysis dealt with postures used on ground-level tasks, differences in postures used in certain other tasks also can have differential energy costs.

Energy Expenditure and Rate of Activity

Energy costs are related to the rate or pace of activity, as well as to the type of activity. One indication of this is shown in Figure 7-11, which shows the rela-

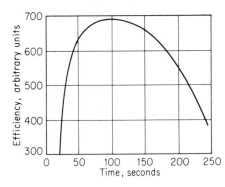

Figure 7-11 Efficiency of stair climbing at different speeds (speed is given by time in seconds to climb the stairs). [From Lupton, as presented by Tuttle and Schottelius, 65.]

tive efficiency of stair climbing at different speeds. However, there is evidence that the optimum varies for different age groups [Salvendy and Pilitsis, 55] and for individuals, and that individuals seem to be able to determine the pace that is most "natural" for themselves, which tends to be the one that involves the minimum energy expenditure for each cycle [Corlett and Mahadeva, 10].

Related to the notion of an optimum pace for any given activity is the question as to whether, with certain types of work (such as assembly operations), the work pace should be set for the worker or controlled by the worker (i.e., paced or self-paced).

Keeping Energy Expenditures within Bounds

If those who are concerned with the nature of human work activities (design engineers, industrial engineers, supervisors, administrators, etc.) want to keep energy costs within reasonable bounds, it is necessary for them to know both what those bounds should be and what the costs are (or would be) for specific activities (such as those shown in Figure 7-8).

Energy Costs of Grades of Work   As a starter toward this, the following definitions of different grades of work may be helpful [Christensen, 9]:

Grade of work	Energy expenditure		Approximate oxygen consumption, liters/min
	kcal/min	kcal/8 h	
Unduly heavy	over 12.5	over 6000	over 2.5
Very heavy	10.0–12.5	4800–6000	2.0–2.5
Heavy	7.5–10.0	3600–4800	1.5–2.0
Moderate	5.0–7.5	2400–3600	1.0–1.5
Light	2.5–5.0	1200–2400	0.5–1.0
Very light	under 2.5	under 1200	under 0.5

In discussing energy expenditures over the period of the conventional working day, Lehmann [41] estimates that the maximum energetic output a normal man can afford in the long run is about 4800 kcal/day; subtracting his estimate of basal and leisure requirements of 2300 kcal/day leaves a maximum of about 2500 kcal/day available for the working day. This figures out to be about 5 kcal/min. But although he proposes this as a maximum, he suggests about 2000 kcal/day as a more normal load, this averaging out to be about 4.2 kcal/min. Edholm [19, p. 91] proposes a somewhat more conservative value, suggesting that the 2000 kcal/day expenditure should be considered as a maximum and that work levels preferably should be kept somewhat below this. Granting some modest differences between these and other physiological standards, we nonetheless get an impression of the general level of physiological costs that should not be exceeded.

Work and Rest   If we accept some ceiling (such as 4 or 5 kcal/min) as a desirable upper limit of the average energy cost of work, it is manifest that if a particular activity per se exceeds that limit, there must be rest to compensate for the excess. In this connection Murrell [48, p. 376] presents a formula for estimating the total amount of rest (scheduled or not scheduled) required for any given work activity, depending on its average energy cost. This formula (with different notations) is

$$R = \frac{T(K - S)}{K - 1.5}$$

in which R is rest required in minutes; T is total working time; K is average kilocalories per minute of work; and S is kilocalories per minute adopted as standard. The value of 1.5 in the denominator is an approximation of the resting level in kilocalories per minute. If we adopt as S a value of 4 kcal/min and want to figure R for a 1-h period ($T = 60$ min), our formula becomes

$$R = \frac{60(K - 4)}{K - 1.5}$$

Applying this to a series of values of K, we can obtain R values shown in the next to the top curve of Figure 7-12 ($S = 4$). The other curves (for values of

$S = 3$, 5, and 6) are given for comparison when lower ($S = 3$ kcal/min) or higher ($S = 5$ or 6 kcal/min) standards of energy expenditure might seem to be appropriate. The lowest curve (for a value of $S = 6$), however, undoubtedly represents a level of activity that probably could not be maintained very long, except possibly by the hardiest among us. This general formulation needs to be accepted with a fair sprinkling of salt, in part because of individual differences in physical condition. Further, although the curves in Figure 7-12 swing

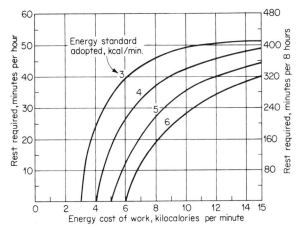

Figure 7-12 Total rest requirements for work activities of varying energy costs, for energy-expenditure standards (ceilings) of 3, 4, 5, and 6 kcal/min; a generally accepted standard is 4 kcal/min. The rest requirements for maintenance of the adopted standard are given per hour (left) and per 8-h day (right). [Based on formulation of Murrell, 48, p. 376.]

down to the zero-rest-required line, we should keep in mind that this formulation deals only with the physiological costs of work. Because of *other* considerations, such as the more psychological factors of boredom and shift in attention, some rest must be provided for virtually any kind of continuous work, even though its physiological costs are nominal.

Work limits of local muscle groups Although the overall energy cost of an activity might be within reasonable limits, it is of course possible for individual muscles or muscle groups to be worn to a frazzle with excessive use. If the rate of contraction of a muscle or muscle group is low enough, it can function almost indefinitely, but at higher rates it can become completely fatigued and cease to function at all. This is shown in the example below from Tuttle and Schottelius [65, table 6].

No. of contractions	Fatigued by	Work done
1 per 1 s	14 contractions	0.912 kg-m
1 per 2 s	18 contractions	1.080 kg-m
1 per 4 s	31 contractions	1.842 kg-m
1 per 10 s	No fatigue (no stress)	Almost indefinite

We can see that the rate of 1 contraction of the muscle per 10 s did not produce complete fatigue and permitted almost indefinite continuation of work; faster rates (1 contraction per 4, 2, or 1 s) produced such stress that the muscle ceased to function after 31, 18, or 14 contractions, respectively.

Discussion If a given type of physical activity is within the reasonable physical ability limits of an individual (and individuals obviously vary in such limits), the outer bounds of the ability of the individual to *continue* the activity are prescribed by the total energy costs and/or the recovery rates of individual muscles or muscle groups. These possible constraints in effect should dictate the work rate that should not be exceeded for continued effectiveness of the individual in performing the activity.

STRENGTH AND ENDURANCE

Strength is the maximal force muscles can exert isometrically in a single voluntary effort, that is, the muscular capacity to exert force under static conditions [Kroemer, 38], and is usually measured by the use of an external device such as a hand dynamometer or a device for measuring the force exerted against some object. The measurement of such force, however, depends not only on the intrinsic muscle strength, but also on the subject's motivation, the experimenter's instructions, and even the measurement index used (e.g., whether use of a peak value or an average of two or three efforts). Since most human activities consist of dynamic efforts rather than static efforts, Kroemer raises serious questions about the use of measures of strength per se, pointing out that it is mechanically difficult, if not impossible, to predict an individual's performance on a dynamic task (for example, turning a crank) from a measurement of his static force capacity (for instance, holding a weight).

Arm Strength

With the above cautions in mind, let us illustrate studies of maximum strength with some data from Hunsicker [32], who tested the arm strength of 55 subjects who made movements in each of several directions, with the upper part of the arm in each of five positions, as illustrated in Figure 7-13. Some of the results are shown in Figure 7-14. This figure shows, for the six movements, the maximum strength of the 5th percentile and the mean maximum strength. It is frequently the practice, in dealing with data relating to strength, to use the 5th percentile value as the maximum force to be overcome by users of equipment being designed, since this would in general ensure that 95 percent of the individuals in question would have that strength level or more.

We can see that pull and push movements are clearly strongest, but that these are noticeably influenced by the position of the hand, with the strongest positions being at angles of 150 and 180°. The differences among the other movements are not great, but what patterns do emerge are the consequence of the mechanical advantages of such movements, considering the levers involved

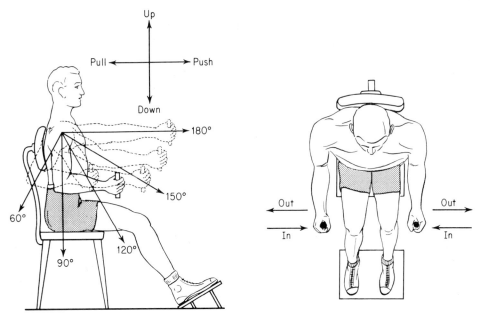

Figure 7-13 Side and top views of subjects being tested for their strength in executing six different movements, namely, push, pull, up, down, in (adduction), and out (abduction). All six movements were made at each of the five arm positions (180, 150, 120, 90, and 60°). See Figure 7-14 for results. [Adapted from Hunsicker, 32.]

and the effectiveness of the muscle contractions in applying leverage to the body members. It might be added that although left-hand data are not shown, the strength of left-hand movements is roughly 10 percent below that of movements of the right hand.

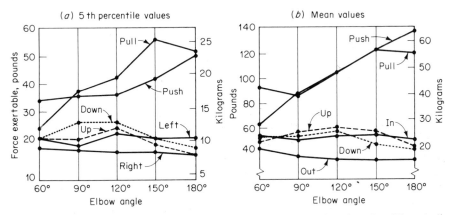

Figure 7-14 Maximum arm strength of movements in various directions for different elbow angles of upper right arm. a shows 5th percentile values, and b the mean values (with compressed vertical scale of pounds) for 55 male subjects. See Figure 7-13 for illustrations of arm positions and of movements. [Based on data from Hunsicker, 32.]

Endurance

If one considers the endurance of people to maintain a given muscular force, we can all attest from our own experience that such ability is related to the magnitude of the force. This is shown dramatically in Figure 7-15, which depicts

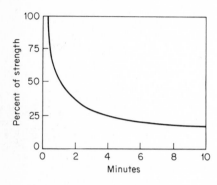

Figure 7-15 Endurance time as a function of force requirements. [From Kroemer, 38, fig. 4, as adapted from various sources.]

the general pattern of endurance time as a function of force requirements of the task. It is obvious that people can maintain their maximum effort very briefly, whereas they can maintain a force of around 25 percent or less of their own maximum for a somewhat extended period (10 minutes or more). (Note that this relationship is based on each individual's *own* measures of strength.) The implication of this relationship is fairly obvious—that if it is necessary to require individuals to maintain force over a period of time, the force required should be well below each individual's own static force capacity.

It should be added that one common-sense concept of endurance refers to the ability to keep up some general body activity over a period of time (rather than the exertion of a given muscle or muscle group). In this frame of reference the "endurance" of individuals would be a function of the total energy cost of the activity and the energy expenditure the individuals can reasonably maintain over time. If the energy costs exceed reasonable limits, rest should be provided to keep total energy requirements within bounds.

Lifting and Moving Loads

Recognizing that exertion of maximum strength should be avoided if at all possible, various individuals and organizations have concerned themselves with determining the maximum loads that individuals should be permitted to lift.

Methods of Lifting   The burden of evidence about methods of lifting loads from the floor (or near floor level) argues for the following technique [adapted from Davies, 15]: (1) feet far enough apart for balanced distribution of weight; (2) knees and hips bent, back reasonably straight; (3) arms as near to body as possible, with load as close to body as possible; (4) wherever possible, use whole hand, not just fingers; (5) lift smoothly with no jerks. The actual lifting is performed largely by an extension of the legs. A posture somewhat akin to this

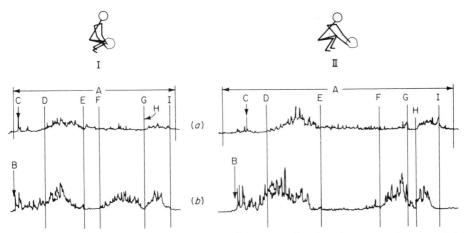

Figure 7-16 Electromyograph recordings of two muscles (a, gluteus maximus, and b, sacrospinalis) made for one subject using two lifting techniques. Technique I incurred less muscle activity than technique II. (The letters and lines represent recordings of corresponding segments of the movements made by the two methods.) [Adapted from Tichauer, 64, fig. 9.]

is shown in Figure 7-16 (example I), along with another lifting technique (example II) as used in a comparative study of these two methods by Tichauer [64] in which he obtained electromyograph recordings for two muscles (the gluteus maximus and the sacrospinalis). Examples of these recordings show less muscle activity for technique I than for technique II, thus tending to support the method recommended by Davies.

Load to Be Lifted   In discussing the various sets of acceptable or permissible loads that have been proposed, Davies [15] refers to "marked variation" between and among them. Although some of the variations can be attributed to the specific experimental conditions and methods used in developing the data, there still are apparently unexplained disparities between and among various sets of data. Although certain sets of recommended standards and of research results are given below for illustrative purposes, these disparities should be kept in mind. One set of recommendations has been proposed by the International Occupational Safety and Health Information Center [72]. These recommendations, which cover reasonable weight limits for occasional lifting by any method, taking into account the relationship between efficiency and age, are given below:

	Weight, lb, for Specified Age Groups (Approximate kilogram equivalents in parentheses)					
	14–16	16–18	18–20	20–35	35–50	Over 50
Male	33 (15)	42 (19)	51 (23)	55 (25)	45 (20)	36 (16)
Female	22 (10)	26 (12)	31 (14)	33 (15)	29 (13)	22 (10)

Recognizing the range of individual differences, including those associated with age, Snook, Irvine, and Bass [62] propose for males the following limits of weights while lifting (these being toward the lower range of the limits shown on page 181):

Lifting range	lb	kg
Floor to knuckle level	37.3	16.7
Knuckle to shoulder level	34.4	15.1
Shoulder to arm-reach level	29.4	13.0

Work Pace and Energy Cost in Lifting When the work activity consists of frequent or virtually continuous lifting, the efficiency of the work—that is, the energy cost per unit of work—is influenced by both the range of the lifting and the work pace. With regard to the range, for example, the energy cost of lifting objects from the floor to about 20 in is about half again as much as lifting the same weight from about 20 to 40 in [Davies, 15]. This is because of the additional effort in lowering and raising the body. Some indication of the relative efficiency of lifting various weights over certain vertical ranges is shown in Figure 7-17 [Frederick, 23, as presented by Davies, 15]. This shows, for each of

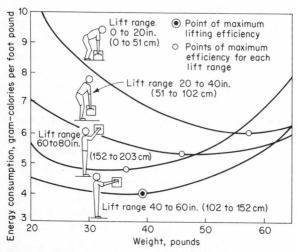

Figure 7-17 Energy consumption in lifting per unit of work for various weights and specified lift ranges. The most efficient lifting is with a weight of about 40 lb for a lift range from about 40 to 60 in. [From Frederick, 23, as presented by Davies, 15, fig. 3.]

four lift ranges, the most efficient weight to be lifted in terms of energy cost per unit of work. The fact that the lift range of 40 to 60 in was the most efficient suggests that workplaces preferably should be designed so that the primary lifting is within that range.

As a further elaboration of the initial study, Frederick developed a formula for estimating the energy cost for any given number of lifts per hour for any given weight and lift range.[3] Alternatively, one can estimate the number of lifts per hour that could be executed and still keep the total energy expenditure within some specified desirable limit, as 200 kcal/h.

Other Aspects of Strength and Endurance

This is not the place to reproduce the many sets of biomechanical data that are available. However, for illustrative purposes here, a few bits and pieces of data and a few unadorned generalizations from certain studies will be recapped below, without the details of the methods or results.

• *Lifting action:* In lifting heavy objects to various levels, markedly heavier weights can be lifted by males from 17 to 32 years of age to a low level of 18 in (42 cm) than to an intermediate level of 42 in (104 cm) or to a higher level, the values for short, medium, and tall males being respectively: low level, 124, 138, and 146 lb (56, 62.5, and 66 kg); intermediate level, 73, 92, and 96 lb (33, 42, and 43.5 kg); high level, 53, 65, and 67 lb (24, 29.5, and 30.5 kg) [Switzer, 63]. However, as indicated above, lifting from the floor is less efficient in energy cost than lifting from an intermediate level.

• *Pushing and pulling:* Forces that can be executed by 90 percent of industrial male workers in pushing and pulling tasks over short distances are approximately as follows [Snook, Irvine, and Bass, 62]:

	Push	Pull
Initial force	58 lb (25.3 kg)	53 lb (24.0 kg)
Sustained force	29 lb (13.1 kg)	32 lb (14.5 kg)

• *Grip strength:* The grip strength of a sample of 552 male industrial workers ranged from 75 to 170 lb (34 to 77 kg), with a mean of 125 lb (57 kg) [Fisher and Birren, 21]. But university students do not do as well, as indicated by the following results [Tuttle et al., 65]: right hand, mean of 108 (49 kg) and standard deviation of 21 lb (9.5 kg); left hand, mean of 95 (43 kg) and standard deviation of 18 (8 kg).

• *Strength of hand turn:* In turning movements with the hand (such as with handles on heavy refrigerator doors or stirrup-type devices on some garden tanks-spraying equipment) the following generalizations can be made [Salter and Darcus, 54]: The forces that people can exert increase in both a

---

[3]The formula for developing such estimates is:

$$\text{kcal/h} = \frac{f \times a \times w \times c}{1000}$$

in which f = no. of lifts/h; a = lifting height in ft; w = weight in lb; and c = energy in gram calories per ft-lb taken from Figure 7-17.

As an example, assuming a 200 kcal/h limit, with a = 2 ft, w = 40 lb, and c = 4 gram calories per ft-lb, f would be 625 lifts per hour [Davies, 15].

pronation movement (turning the hand inward from a palm-up position) and a supination movement (turning the hand outward); in both of these movements the relation between hand position (as it is turned in or out) and force is greater when the elbow is flexed at about 90 or 150°, and least at about 30°.

• *Elbow flexion versus extension:* An elbow flexion action (bending) is about half again as strong as an extension action [Provins and Salter, 53].

• *Elbow versus shoulder forces:* Rotation action of the shoulder is about half again as strong as that of the elbow and has nearly three times as much staying power [Provins, 52].

As indicated earlier, there are certain human variables that are related to muscle strength and endurance. Observations about a few of these, based largely on the discussion by Damon, Stoudt, and McFarland [13], are given below:

• *Age:* Strength reaches a maximum by the middle to late 20s and declines slowly but continuously from then on, until at about age 65 strength is about 75 percent of that exerted in youth. Despite such reduction in strength, however, there are indications that continuous-work capacity of men does not decrease with age, up to about 60, at least in moderate environments [Snook, 60]. Snook suggests, however, that this finding may not apply in hot environments or for intermittent work that requires short durations of high energy expenditure.

• *Sex:* Women's strength is about two-thirds that of men.

• *Body build:* Although body build is related to strength and endurance, the relationships are complicated; for example, athletic-looking individuals generally are stronger than others, but less powerfully built persons may be more efficient; and for rapidly fatiguing, severe exercise, slender subjects are best, with obese subjects worst; and for moderate exercise, those with normal build are best.

• *Exercise:* Exercise can increase strength and endurance within limits, these increases frequently being in the range of 30 to 50 percent above beginning levels.

SPEED AND ACCURACY OF MOVEMENTS

Speed generally is the primary requirement in executing movements that are otherwise not difficult or demanding, such as in applying the brake pedal of an automobile or reaching for parts to be assembled. In turn, accuracy is the primary requirement in executing such movements as those in tracking (in which continuous control is required), in certain positioning actions that require precision and control, and in certain manipulative activities. However, in some circumstances both speed and accuracy may be required.

Response Time

Many movements are triggered by some external stimulus such as a changing traffic light or auditory warning signal. The time to make a movement following such a stimulus actually consists of a combination of delays; the nature of these delays and the range of typical times in milliseconds (ms) required for them have been summarized by Wargo [68] as follows: receptor delays, 1 to 38; neural transmission to the cortex, 2 to 100; central-process delays, 70 to 300; neural transmission to muscle, 10 to 20; and muscle latency and activation time, 30 to 70. These add up to a total ranging from 113 to 528 ms. The total time to make a response following a stimulus frequently is referred to as *reaction time*. However, one can differentiate between the time to initiate a movement (this being a more restrictive definition of reaction time) and the time to make the movement (sometimes called "movement time").

Simple and Choice Reaction Time   Simple reaction time is the time to make a specific response when only one particular stimulus can occur, usually when an individual is anticipating the stimulus (as in conventional laboratory experiments). Reaction time is usually shortest in such circumstances, typically ranging from about 150 to 200 ms (0.15 to 0.20 s), with 200 ms being a fairly representative value; the value may be higher or lower depending on the stimulus modality and the nature of the stimulus (including its intensity and duration), as well as on the subject's age and other individual differences. If there are several possible stimuli, each with its own response, the time goes up, largely because of the additional central-process time required to make a decision on what the correct response should be. Such a choice reaction time (also called *disjunctive* reaction time) is pretty much a function of the number of choices available, as indicated below [summarized from various sources by Damon et al., 13, p. 239]:

Number of choices	1	2	3	4	5	6	7	8	9	10
Approximate reaction time, s	0.20	0.35	0.40	0.45	0.50	0.55	0.60	0.60	0.65	0.65

There is some evidence to suggest that choice reaction time is linearly related to the number of bits of information involved (that is, the logarithm to the base 2 of the number of alternatives available), even up to 10 bits (about 1000 alternatives) [Hilgendorf, 31].

Expectancy   Most data on simple and choice reaction times come from laboratories in which the subject is anticipating a stimulus. (And in some industrial circumstances people actually are waiting for a stimulus.) However, when stimuli occur infrequently or when they are not expected, the ante is raised. This was illustrated, for example, in a study by Warrick et al. [69] in which typists at their regular jobs were asked to press a button whenever a buzzer sounded, the buzzer going off only once or twice a week over a period of six months. Data were also obtained from the same task when the subjects were given a 2- to 5-s warning. The reaction time (the total response time) to the "unexpected" signals averaged about 100 ms above that when the subjects had received an advance warning.

As another example to illustrate this point, Johansson and Rumar [33] collected data in Sweden on the response time of automobile drivers in applying the brake pedal following an auditory signal. They used two methods of investigation. In one method, the response time of 321 drivers was measured when they were anticipating a signal within the next 10 km (about 6 mi). In the other method the cars of five drivers were rigged up with a mechanism that triggered a signal infrequently, such as at intervals of an hour or even several days; this was called a "surprise" condition, although the drivers of course knew of the presence of the mechanism. These drivers were also tested under the first condition, their mean values for these two conditions being 0.73 (surprise condition) and 0.54 (anticipation condition). The ratio of these two (0.73/0.54, or 1.35) was considered to be a correction factor reflecting the approximate ratio of the additional time taken under surprise conditions. The distribution of response times of the 321 drivers under the anticipation condition is shown in Figure 7-18, along with the estimated time under the surprise condi-

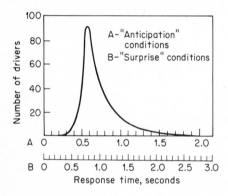

Figure 7-18 Smoothed distributions of actual brake response times of 321 drivers in an "anticipation" condition (A), and of estimated response times under a "surprise" condition (B). The estimation consisted of applying a correction factor of 1.35 as derived from data for a small number of drivers who responded under "surprise" conditions. [Adapted from Johansson and Rumar, 33.]

tion as adjusted by the 1.35 correction factor. Recognizing the roughness of the estimation process, it was estimated that the response time to unexpected events of about half the drivers would be 0.9 s (900 ms) or longer.

Whatever the implications of the increased response time to unexpected events (such as the increased stopping distance in the case of drivers), such increases must be considered as a part of life to be dealt with.

Movement Time The time to effect a movement following a signal would of course vary with the type and distance of movement, but it has been estimated [Wargo, 68] that a minimum of about 300 ms (0.30 s) can be expected for most control activities. Adding this value to an estimated reaction time of 200 ms would result in a total response time of about 500 ms. However, the nature and distance and location of the response mechanism can influence the total time. This is illustrated, for example, by the result of a study by Pattie [51], who investigated the time to activate four types of possible emergency power cutoff devices as they might be used on agricultural tractors. The mean times to activate the devices in response to a buzzer are given below:

	Mean time (ms)
Clutch	613
Toggle switch (underneath steering wheel)	498
Horn rim (on steering wheel)	412
Rim blow (on under edge of steering wheel)	337

Discussion   We can see that the time required to make certain responses can be influenced by a number of variables, such as the nature of the stimulus, the number of choices, the degree of expectancy, and the device used. In some instances the total time required can be of substantial consequence. For example, the response time of the pilot of a supersonic aircraft on a collision course can be as long as 1.7 s, this being the simple addition of 0.3 s for visual acquisition of the other aircraft, 0.6 s for recognition of the impending danger, 0.5 s for selection of a course of action, and 0.3 s for initiation of the desired control response. Add the response time of the aircraft itself, and it would be futile to take any action if the planes were closer than about 4 miles.

When time is, as they say, of the essence, one should not throw up his hands in complete despair of the time lag in human response. There are, indeed, ways of aiding and abetting people in responding rapidly to stimuli. Speed requirements, for example, can be reduced by taking actions such as using sensory modalities with shortest reaction time, using two or more senses to reduce reaction time, presenting stimuli in a clear and unambiguous manner, minimizing the number of alternatives from which to choose, giving advance warning of stimuli if possible, using body members that are close to the cortex to reduce neural transmission time, using control mechanisms that minimize response time, and training the individuals. In more exotic circumstances, one can even bypass the human physical response by the direct use of electrical muscle-action potentials for effecting control responses [Wargo, 68].

Predetermined Time Systems

With particular reference to time involved in making different types of movements, mention should be made of some predetermined time systems that are used in industrial engineering for developing standard times for work operations. Such systems (as the Methods Time Measurement system, MTM, discussed by Maynard [44]) set forth specific time values for specific elemental motions made during an operation; the standard time for the complete operation is based on the sum of the times for the individual elements, usually with certain "allowances" included. Most such systems are predicated on the accumulation of empirical data obtained from extensive time-study observations. It should be noted here that there is evidence that some element times are not strictly additive because of possible interactions among element times. Some predetermined time systems provide for at least a few such interactions by the use of tables of data for a given type of motion for different conditions under which the motion is executed.

Positioning Movements

Positioning movements are made when a person reaches for something or moves something to another location, usually by hand; they are then "travel" movements of the body member. The time and/or accuracy of such movements can be influenced by such factors as the nature of the stimulus that triggers the movement, the distance and direction of the movement, single possible terminal versus alternative terminals, fixed terminal (with automatic stop) versus precise terminal position under control of an individual, and visual versus nonvisual (i.e., "blind") control. Certain positioning movements can be dissected into two or three relatively distinct components, namely, reaction time (the time to initiate a response following the stimulus that triggers it), primary or gross travel time (to bring the body member near the terminal), and a secondary or corrective type of motion to bring the body member to the precise position desired. Where there is an automatic fixed terminal (such as on a typewriter carriage), the secondary, or corrective, component virtually drops out of the picture.

Time and Distance of Movements   In the execution of positioning movements, reaction time is almost a constant value, unrelated to the distance of movement. This is shown, for example, in Figure 7-19, which is based on a pair

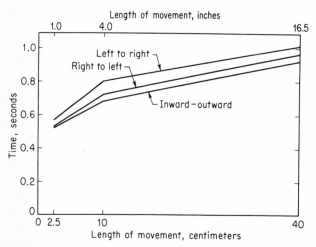

Figure 7-19 Times required for horizontal positioning movements of different lengths. Times given are from the sounding of a buzzer to completion of movement. Reaction time, which was essentially the same for all movements (about 0.25 s), is included in time values. [Adapted from Brown and Slater-Hammel, 6, and Brown, Wieben, and Norris, 7.]

of related studies in which the subjects moved a sliding device to a marked position when a buzzer was sounded [Brown and Slater-Hammel, 6, and Brown, Wieben, and Norris, 7]. Three different distances were used, namely,

2.5, 10, and 40 cm, the movements being left to right, and inward and outward. This figure also shows that movement time is related to distance but is not proportionate to distance. This lack of linear relationship between distance and time probably can be attributed to the time required for accaleration to the maximum speed, and (except where there is a mechanical terminal) the secondary, or corrective, movement in bringing the body member to the precise terminal.

Termination of Movements   The effect on movement time of a mechanical stop at the terminal of the movement was mentioned briefly above; Barnes [2], for example, reported that the use of visual control required about 17 percent more time than did the use of a fixed mechanical terminal. Other variations in the termination of the movement can also have some effect on movement time. For example, in the MTM system mentioned earlier [Karger and Bayha, 36; Maynard, 44], different time units are allowed for *reach* movements of three types, namely, those in which the object to be reached for (1) is in a fixed location, (2) is in a location that may vary from cycle to cycle, and (3) may be jumbled with other objects. These times, based on empirical data, are shown in graphic form in Figure 7-20. We thus see that the movement (reach) time is

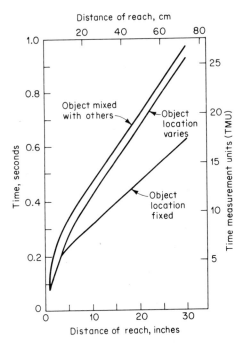

Figure 7-20 "Reach" times allowed under three conditions of terminating movements, for various distances of movement, as based on the MTM (Methods Time Measurement) system of predetermined times. [Reproduced with permission from *Basic Motions of MTM* by William Antis, John M. Honeycutt, Jr., and Edward N. Koch, published by the Maynard Foundation, copyright 1963, 1968, p. 2-2. Reprint of this material is prohibited without express permission of the MTM Association for Standards and Research, Fair Lawn, N. J.]

influenced by the nature of the *termination* of the movement, the variation presumably being essentially a function of the time required to search for or select specific objects in their specific locations.

Direction of Positioning Movements   Differences in time and accuracy of positioning movements made in various directions are a function of the biomechanical functioning of the particular combination of body members and muscles that are brought into the act. The results of certain investigations will illustrate this effect, but we shall skip most of the experimental procedures and, rather, present certain of the results that illustrate the point. As a first example, Briggs [4] carried out a series of experiments with the movement of a stylus back and forth in a horizontal plane between a large (3-in or 8-cm) buzzer and a target. The target could be *in* (in front of the body) or *out* (at different angles and distances from the body), with the buzzer being in the opposite location (out or in). The target circle was drawn on a piece of stretched paper through which the stylus was punched. Accuracy was determined by counting the stylus punches within the target, the targets being $1/4$, $1/2$, $3/4$, or 1 in in diameter. Scores were based on the number of hits at the target within 20-s trials and hence were a combination of accuracy and speed. Figure 7-21 shows the accu-

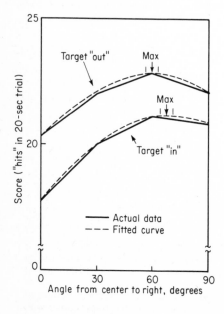

Figure 7-21 Accuracy scores of repeated positioning movements at various angles when the target is in (in front of body) and out (14 in or 37 cm away from central point). [Adapted from Briggs, 4.]

racy scores of repeated 14-in positioning movements at various angles to the right from the front of the body, when the target was in and out. This shows that accuracy is optimum with angles of movement of about 60° right from center, and also that accuracy is generally greater when the target is away from the body (out rather than in).

Keeping this study in mind, let us shift to another investigation of the time required to make 40-cm (15-in) positioning movements in eight different directions from a center starting point [Schmidtke and Stier, 57]. The results, shown in Figure 7-22, indicate that time was generally shorter for movements from lower left to upper right, the same general range of Briggs' 60° optimum. Thus, although these two studies vary in their criteria (time versus accuracy) and in the location of the terminal of the movement (in toward the body

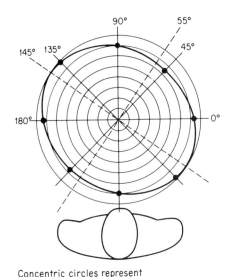

90° 55°

145° 135° 45°

180° 0°

Concentric circles represent
equal time intervals

Figure 7-22 Average times of hand movements made in various directions. Data were available for the points indicated by black dots; the oval was drawn from these points and represents assumed, rather than actual, values between the recorded points. The concentric circles represent equal increments of time to provide a reference for the average movement times depicted by the oval. [Adapted from Schmidtke and Stier, 57.]

versus out away from the body), nonetheless they reveal the same general pattern. This pattern suggests that, in biomechanical terms, controlled arm movements that are primarily a pivoting of the elbow, with fairly nominal upper-arm and shoulder action, tend to be more accurate and to take less time than those with a greater degree of upper-arm and shoulder action.

As we look at two-handed simultaneous positioning movements (such as in reaching for parts in bins), however, a different pattern emerges, as reflected by a study of Barnes and Mundell [2]. The speed of such movements (with mechanical terminals) when made at angles of 0, 30, 60, and 90° from a dead-ahead direction did not vary much with the angle, but errors in moving the hand to the terminal positions that needed visual control increased very markedly from the 0 to 90° angles. The probable explanation for this lies in the requirement for visual control in terminating the movements; the closer the two terminals to the dead-ahead position, the more accurate the visual control of the termination of the movements.

Blind Positioning Movements When visual control of movements is not feasible, the individual needs to depend on his kinesthetic sense for feedback. Probably the most usual type of blind positioning movement is one in which the individual moves his hand (or foot) in free space from one location to another, as in reaching for a control device when the eyes are otherwise occupied. The very well-known study by Fitts [22] probably provides the best available data relating to the accuracy of the *direction* of such movements in free space. He used an arrangement with targets positioned around the subject at 0, 45, 90, and 135° angles left and right, in three tiers, namely, a center (reference) tier, and tiers 45° above and below the center tier. The blindfolded subjects were given a marker with a sharp point, which they pressed against each target when they tried to reach to it. A bull's eye was scored zero, and marks in subsequent circles were scored from 1 to 5, marks outside the circles being scored 6.

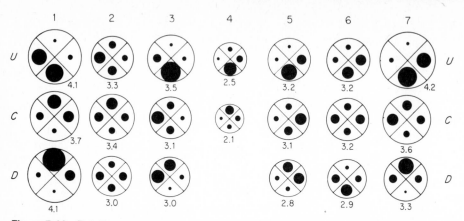

Figure 7-23 Relative accuracy scores for different areas in blind-positioning study by Fitts [22]. The position of the circles represents the location of targets, which ranged from 135° left (number 1) to 135° right (number 7), number 4 being straight ahead. The three tiers represent those up, center, and down. The size of each circle represents the relative number of errors, so small circles indicate greater accuracy. The relative size of the four dark circles within each light one is proportional to the errors in each quadrant of the target. [Courtesy of USAF, AFHRL, Human Engineering Division.]

Figure 7-23 shows the results. Each circle in this figure represents the subjects' accuracy in hitting the target in the corresponding position. The size of the circle is proportional to the average accuracy score for that target; the smaller the size, the better the accuracy. The circles within circles (those in the four quadrants) indicate, relatively, the proportionate number of marks that were made in each quadrant. From this figure it can be seen that blind positioning movements can be made with greatest accuracy in the dead-ahead positions and with least accuracy in the side positions. With regard to the level of the targets, the accuracy is greatest for the lowest tier, average for the middle tier, and poorest for the upper tier. Further, right-hand targets can be reached with a bit more accuracy than left-hand ones.

Thus, in general, in the positioning of control devices or other gadgets that are to be reached for blindly, positions closer to the center and below shoulder height usually can be expected to be reached more accurately than those farther off to the sides or higher up.

In making horizontal hand movements in various directions (such as forward and back, or to the right from a frontal position and back again) without visual control, there is a tendency to overshoot short distances of movement and to undershoot longer distances. This is called the *range* effect.

Continuous Movement

As indicated earlier, continuous movements are those which require some type of muscular control during the movement. There are actually two types of continuous movements: (1) continuous terminal movements, and (2) continuous control movements (as in a tracking task).

Continuous Terminal Movements These are movements that have a terminal, but that require continuous muscular control throughout their span. The variations of such movements in practical circumstances, however, are many, such as in freehand drawing or decorating, controlling the movement of fabric in sewing, threading a needle, or hand engraving. A laboratory example of this type of movement was investigated by Corrigan and Brogden [11] by having subjects move a stylus along a horizontal 35-cm (14-in) narrow track formed between two brass plates on a piece of glass; they were to control the stylus continuously, keeping up with a target of constant velocity that moved in view under the glass along the path. The direction of the path was positioned at 24 different angles around a center at 15° intervals. Errors were recorded automatically when the stylus touched either brass plate. The results are shown in Figure 7-24. This shows the error index at different angles and indicates that errors

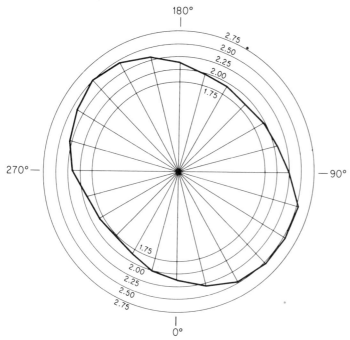

Figure 7-24 Errors in continuous movements (35 cm or 14 in) made at various angles from point of origin. Length of line from the center indicates relative errors at the angle in question. [Adapted from Corrigan and Brogden, 11, table 2.]

are least (and accuracy, therefore, greatest) at around 135 and 315° (1:30 and 7:30 o'clock positions), maximum errors being at around 45 and 225°.

The deviations from the desired path of a movement are produced by tremor of the body member. An interesting approach to the study of tremor during such movements was that carried out by Mead and Sampson [45], in which they had subjects move a stylus (15 in long with a 4-in, 90° bend at the tip) along a narrow groove in any one of the four positions shown in Figure 7-25. As the subject moved the stylus in the directions indicated in that figure, any time the stylus touched the side of the groove an "error" was electronically recorded. These errors, which can be viewed as measures of tremor, are shown in Figure 7-25, and indicate that tremor was greatest during an in-out arm

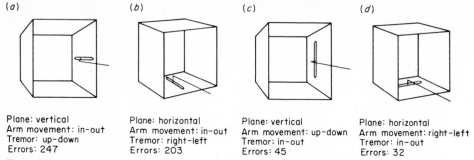

(a)
Plane: vertical
Arm movement: in-out
Tremor: up-down
Errors: 247

(b)
Plane: horizontal
Arm movement: in-out
Tremor: right-left
Errors: 203

(c)
Plane: vertical
Arm movement: up-down
Tremor: in-out
Errors: 45

(d)
Plane: horizontal
Arm movement: right-left
Tremor: in-out
Errors: 32

Figure 7-25 Directions and planes of arm movements with stylus as used in study of hand steadiness, with direction of hand tremor and number of "errors" (number of times the stylus touched the side of the groove) for each condition. [Adapted from Mead and Sampson, 45, fig. 1 and table 1.]

movement in the vertical plane (in which the tremor was up-down), and that it was least during a right-left arm movement in the horizontal plane (in which the tremor was in-out).

It is interesting to note that these findings are reasonably in line with those of Corrigan and Brogden discussed above, in that movements that have an essentially lateral component (lower left to upper right, or left to right, or vice versa) tend to result in less tremor than those of an inward-outward direction (as from lower right to upper left, or in-out, or vice versa). Further, the similarity of Figure 7-24 (from Corrigan and Brogden), which is based on accuracy of continuous movements, and Figure 7-22 (from Schmidtke and Stier), which is based on positioning movements, lends credence to the notion of the generalized biomechanical advantage of an elbow-pivoting movement in contrast to one that requires greater upper-arm and shoulder involvement.

Continuous Control Movements   Continuous control tasks (as in tracking) will be discussed more extensively in the next chapter. However, the results of one study will be mentioned here because they are germane to the previous discussion of the range effect. In this study the subjects tracked a moving stimulus dot with a stylus, the individual (lateral) movements of the dot being 0.25, 0.50, and 1.00 in [Ellson and Wheeler, 20]. The accuracy of the tracking operation was automatically recorded. A recap of some of the results is given below:

	Response	
Movement of stimulus, in	Mean, in	Percent of stimulus
0.25	0.32	127
0.50	0.52	104
1.00	0.96	95

Looking at the response means as compared with the movements of the stimulus dot, we can see here in continuous tracking movements the range effect that we noted previously in positioning movements, that is, an overshooting of short movements and an undershooting of long movements.

Manipulative Movements

Most manipulative movements involve the use of the hand and/or fingers, as in handling items, in assembling parts, or in using hand tools or control devices. (However, the writer has seen in Pakistan and India amazingly facile meat cutters who ply their trade with the knives held between their toes!) Because of the varied nature of manipulative movements, it is not feasible to present any generalizable discussion of them. However, the next chapter will include some discussion of the use of control devices.

Repetitive Movements

The repetitive movements of some class that an individual executes are of course very much alike. However, there is some indication that the specific patterns of such movements are not necessarily the same for all individuals. Such differences were at least evident in the detailed analysis of the operation of a crank by several individuals. In this study, Glencross [25] found, for example, two patterns of wrist movement; one group of subjects reached peak wrist flexion early in the cycle and peak wrist extension later in the cycle, whereas the other group of subjects had the reverse pattern. There were also other individualistic differences, such as in the amount of abduction of the shoulder. Whatever the origin of these differences, they reflect different "subroutines" of muscular response that the individuals have acquired that are automatically executed in accordance with the control of the "executive program" mentioned before. In this task such differences in movement patterns had no effect upon speed of performance, although such effects might occur with other types of tasks.

The information available about different classes of repetitive movements is fairly skimpy, but a few generalizations and bits of data about tapping activities are given below:

- *Rate of tapping:* The maximum rate varies from about 5 to 14 taps per second, with a mean of about 8.4; preferred rates, however, are about 1.5 to 5 taps per second [Miles, 46].
- *Use of various fingers:* Rates of tapping vary considerably for separate fingers, as follows [Dvorak et al., 18]:

	Left hand				Right hand			
	4	3	2	1	1	2	3	4
Taps per 15 s	48	57	63	66	70	69	62	56
Taps per s	3.2	3.8	4.2	4.4	4.7	4.6	4.1	3.7

The rates for the right-hand fingers are systematically higher than for the left-hand fingers, and the rates decrease in order from the index finger, 1, to the little finger, 4.

- *Timing of stimuli:* If individual tapping responses are made to individual signals, the signals should be more than 0.5 s apart; otherwise interference occurs [Vince, 66]. If responses are to be more frequent than about every 0.5 s, it is preferable to have one signal trigger a series of taps, rather than to have a separate signal for each response.
- *Maintenance of tapping movements:* Repeated separate movements can be maintained at a constant rate much more accurately and consistently than continuous movements [Gottsdanker, 27].

Sequential Movements

In most instances sequential movements are of the same general kind, varying in some differentiating feature, as in operating a keyboard. In some instances, however, a potpourri of types of movements may occur in sequence, such as those in starting a car on a rainy night, which might include turning on the ignition, pressing a starter button, turning on the lights, and turning on the windshield wiper. Most of the research on sequential movements relates to the first type (movements of the same type), especially in the use of keyboards. A discussion of keyboards will follow in Chapter 10.

In connection with sequential movements in which the hand moves from one place to another, one particular point should be made, namely, that there are indications that the time required for shifting from one position to another (travel time) can be affected by the nature of the manipulation performed at each position. This interaction was demonstrated, for example, by a comparison of travel time between locations on a panel (the Universal Motion Analyzer) when the manipulations consisted of a switch-turning operation versus a pin-pulling operation [Wehrkamp and Smith, 70], and when the manipulations consisted of three variations in the angle of turn of switches at the various locations (40, 80, and 120°) [Harris and Smith, 29]. A review of research relating to this interaction has led Schnappe [56] to conclude that travel time of a body member (such as a hand) is in fact influenced by the manipulation activities of the terminals of the travel movement, and that both of these are influenced by perceptual factors. Such interactions raise questions about the *additivity* of the times "allowed" for various elemental motions in the use of predetermined time systems for estimating time allowances for complete operations. However, the extent to which the estimate of total times might be adversely affected by such interactions probably is not known. As indicated earlier, however, some predetermined time systems do, to some degree, take such interaction into account in deriving total time allowances for sequences of movements.

Static Reactions

In static reactions, certain sets of muscles typically operate in opposition to each other to maintain equilibrium of the body or of certain portions of it. Thus, if a body member, such as the hand, is being held in a fixed position, the various muscles controlling hand movement are in a balance that permits no net movement one way or the other. The tensions set up in the muscles to bring about this balance, however, require continued effort, as most of us who have attempted to maintain an immobile state for any length of time can testify. In fact, it has been stated that maintaining a static position produces more wear and tear on people than some kind of adjustive posture [Harston, 30]. It was reported, for example, that holding a weight was three to six times more fatiguing than lifting it up and down.

Deviations from static postures are of two types: those called *tremor* (small vibrations of the body member) and those characterized by a gross drifting of the body or body member from its original position.

Tremor in Maintaining Static Position Tremor is of particular importance in work activities in which a body member must be maintained in a precise and immovable position (as in holding an electrode in place when welding). An interesting aspect of tremor, incidentally, is that the more a person tries to control it, the worse it usually is [Young, 71]. There are certain ways in which tremor can be reduced, as reported by Craik [12] and other sources. The following are four conditions that help to reduce tremor:

1 Use of visual reference.
2 Support of body in general (as when seated) and of body member involved in static reaction (as hand or arm).
3 Hand position. (There is less hand tremor if the hand is within 8 in above or below the heart level.)
4 Friction. (Contrary to most situations, mechanical friction in the devices used can reduce tremor by adding enough resistance to movement to counteract in part the energy of the vibrations of the body member.)

DISCUSSION

Many of the sketches and blueprints of systems on the drawing board predetermine the nature of the physical activities that will be required later in the use of the systems, including their energy costs, the range of motions, and their strength, endurance, speed, and accuracy requirements. Some timely consideration given to these affairs during the system design processes frequently will pay handsome dividends in later system performance, and may even relieve someone's aching back or help to keep his bowling arm in good shape.

REFERENCES

1 Barany, J. W.: The nature of individual differences in bodily forces exerted during a simple motor task, *Journal of Industrial Engineering*, 1963, vol. 14, no. 6, pp. 332–341.
2 Barnes, R. M., and M. E. Mundell: *A study of simultaneous symmetrical hand motions*, University of Iowa, Iowa City, Studies in Engineering, Bulletin 17, 1939.

3   Barter, J. T., I. Emanuel, and B. Truett: *A statistical evaluation of joint range data*, USAF, WADC, Technical Note 57–311, 1957.

4   Briggs, S. J.: *A study in the design of work areas*, unpublished doctoral dissertation, Purdue University, Lafayette, Ind., August, 1955.

5   Brouha, L.: *Physiology in industry*, Pergamon Press, New York, 1960.

6   Brown, J. S., and A. T. Slater-Hammel: Discrete movements in the horizontal plane as a function of their length and direction, *Journal of Experimental Psychology*, 1949, vol. 39, pp. 84–95.

7   Brown, J. S., E. W. Wieben, and E. B. Norris: *Discrete movements toward and away from the body in a horizontal plane*, ONR, USN, SDC, Contract N5ori–57, Report 6, September, 1948.

8   Burger, G. C. E.: Heart rate and the concept of circulatory load, *Ergonomics*, 1969, vol. 12, no. 6, pp. 857–864.

9   Christensen, E. H.: "Physiological valuation of work in the Nykroppa iron works," in W. F. Floyd and A. T. Welford (eds.), *Ergonomics Society Symposium on Fatigue*, Lewis, London, 1953, pp. 93–108.

10 Corlett, E. N., and K. Mahadeva: A relationship between a freely chosen working pace and energy consumption curves, *Ergonomics*, 1970, vol. 13, no. 4, pp. 517–524.

11 Corrigan, R. E., and W. J. Brogden: The trigonometric relationship of precision and angle of linear pursuit-movements, *American Journal of Psychology*, 1949, vol. 62, pp. 90–98.

12 Craik, K. J. W.: *Psychological and physiological aspects of control mechanisms with special reference to tank gunnery. Part I*, Medical Research Council (Great Britain), Military Personnel Research Committee, B.P.C. 43/254, August, 1943.

13 Damon, A., H. W. Stoudt, and R. A. McFarland: *The human body in equipment design*, Harvard University Press, Cambridge, Mass., 1966.

14 Datta, S. R., and N. L. Ramanathan: Ergonomics comparison of seven modes of carrying loads on the horizontal plane, *Ergonomics*, 1971, vol. 14, no. 2, pp. 269–278.

15 Davies, B. T.: Moving loads manually, *Applied Ergonomics*, 1972, vol. 3, no. 4, pp. 190–194.

16 Davies, C. T. M.: Cardiac frequency in relation to aerobic capacity for work, *Ergonomics*, 1968, vol. 11, pp. 511–526.

17 Dempster, W. T.: The anthropometry of body action, *Annals of the New York Academy of Sciences*, 1955, vol. 63, pp. 559–585.

18 Dvorak, A., N. I. Merrick, W. L. Dealey, and G. C. Ford: *Typewriting behavior*, American Book Company, New York, 1936.

19 Edholm, O. G.: *The biology of work*, World University Library, McGraw-Hill Book Company, New York, 1967.

20 Ellson, D. G., and L. Wheeler: *The range effect*, USAF, Air Materiel Command, Wright-Patterson Air Force Base, TR 4, Apr. 22, 1947.

21 Fisher, M. B., and J. E. Birren: Standardization of a test of hand strength, *Journal of Applied Psychology*, 1946, vol. 30, pp. 380–387.

22 Fitts, P. M.: "A study of location discrimination ability," in P. M. Fitts (ed.), *Psychological research on equipment design*, Army Air Force, Aviation Psychology Program, Research Report 19, 1947.

23 Frederick, W. S.: Human energy in manual lifting, *Modern Materials Handling*, 1959, vol. 14, no. 3, pp. 74–76.

24 Gilbreth, F.: *Motion study*, D. Van Nostrand Company, Inc., New York, 1911.

25 Glencross, D. J.: Temporal organization in a repetitive speed skill, *Ergonomics*, 1973, vol. 16, no. 6, pp. 765–776.

26 Gordon, E. E.: The use of energy costs in regulating physical activity in chronic disease, *A.M.A. Archives of Industrial Health*, November, 1957, vol. 16, pp. 437–441.

27 Gottsdanker, R. M.: The continuation of tapping sequences, *Journal of Psychology*, 1954, vol. 37, pp. 123–132.

28 Greene, J. H., W. H. M. Morris, and J. E. Wiebers: A method for measuring physiological cost of work, *Journal of Industrial Engineering*, vol. 10, no. 3, May–June, 1959.

29 Harris, S. J., and K. U. Smith: Dimensional analysis of motion: VII. Extent and direction of manipulative movements as factors in defining motions, *Journal of Applied Psychology*, 1954, vol. 38, pp. 126–130.

30 Harston, L. D.: Contrasting approaches to the analysis of skilled movements, *Journal of General Psychology*, 1939, vol. 20, pp. 263–293.

31 Hilgendorf, L.: Information input and response time, *Ergonomics*, 1966, vol. 9, no. 1, pp. 31–37.

32 Hunsicker, P. A.: *Arm strength at selected degrees of elbow flexion*, USAF, WADC, TR 54–548, August, 1955.

33 Johansson, G., and K. Rumar: Drivers' brake reaction times, *Human Factors*, 1971, vol. 13, no. 1, pp. 23–27.

34 Kalsbeek, J. W. H.: "Sinus arrhythmia and the dual task of measuring mental load," in W. T. Singleton, J. G. Fox, and D. Whitfield (eds.), *Measurement of man at work*, Taylor and Francis, London, 1971, pp. 101–113.

35 Kalsbeek, J. W. H.: Do you believe in sinus arrhythmia? *Ergonomics*, 1973, vol. 16, no. 1, pp. 99–104.

36 Karger, D. W., and F. H. Bayha: *Engineered work measurement*, 2d ed., The Industrial Press, New York, 1965.

37 Khalil, T. M.: An electromyographic methodology for the evaluation of industrial design, *Human Factors*, 1973, vol. 15, no. 3, pp. 257–264.

38 Kroemer, K. H. E.: Human strength: terminology, measurement, and interpretation of data, *Human Factors*, 1970, vol. 12, no. 3, pp. 297–313.

39 Lauru, L.: The measurement of fatigue, *The Manager*, 1954, vol. 22, pp. 299–303 and 369–375.

40 LeBlanc, J. A.: Use of heart rate as an index of work output, *Journal of Applied Physiology*, 1957, vol. 10, pp. 275–280.

41 Lehmann, G: Physiological measurements as a basis of work organization in industry, *Ergonomics*, 1958, vol. 1, pp. 328–344.

42 Luczak, H., and W. Laurig: An analysis of heart rate variability, *Ergonomics*, 1973, vol. 16, no. 1, pp. 85–97.

43 Margaria, R., F. Mangili, F. Cuttica, and P. Cerretelli: The kinetics of the oxygen consumption at the onset of muscular exercise in man, *Ergonomics*, 1965, vol. 8, no. 1, pp. 49–54.

44 Maynard, H. B.: *Industrial engineering handbook*, 2d ed., McGraw-Hill Book Company, New York, 1963.

45 Mead, P. G., and P. B. Sampson: Hand steadiness during unrestricted linear arm movements, *Human Factors*, 1972, vol. 14, no. 1, pp. 45–50.

46   Miles, D. W.: Preferred rates in rhythmic response, *Journal of General Psychology*, 1937, vol. 16, pp. 427–469.

47   Monod, Par H.: La validité des mesures de fréquence cardiaque en ergonomie, *Ergonomics*, 1967, vol. 10, no. 5, pp. 485–537.

48   Murrell, K. F. H.: *Human performance in industry*, Reinhold Publishing Corporation, New York, 1965.

49   Passmore, R.: Daily energy expenditure by man, *Proceedings of the Nutrition Society*, 1956, vol. 15, pp. 83–89.

50   Passmore, R., and J. V. G. A. Durnin: Human energy expenditure, *Physiological Reviews*, 1955, vol. 35, pp. 801–875.

51   Pattie, C.: Simulated tractor overturnings: *A study of human responses in an emergency situation*, Ph.D. thesis, Purdue University, May, 1973.

52   Provins, K. A.: Effect of limb position on the forces exerted about the elbow and shoulder joints on the two sides simultaneously, *Journal of Applied Physiology*, 1955, vol. 7, pp. 387–389.

53   Provins, K. A., and N. Salter: Maximum torque exerted about the elbow joint, *Journal of Applied Physiology*, 1955, vol. 7, pp. 393–398.

54   Salter, N., and H. D. Darcus: The effect of the degree of elbow flexion on maximum torques developed in pronation and supination of the right hand, *Journal of Anatomy*, 1952, vol. 86, pp. 197–202.

55   Salvendy, G., and J. Pilitsis: Psychophysiological aspects of paced and unpaced performance as influenced by age, *Ergonomics*, 1971, vol. 14, no. 6, pp. 703–711.

56   Schappe, R. H.: Motion element synthesis: an assessment, *Perceptual and Motor Skills*, 1965, vol. 20, pp. 103–106.

57   Schmidtke, H., and F. Stier: Der aufbau komplexer bewegungsabläufe aus elementarbewegungen, *Forschungsberichte des landes Nordrhein-Westfalen*, 1960, no. 822, pp. 13–32.

58   Shephard, R. J.: Comments on "Cardiac frequency in relation to aerobic capacity for work," *Ergonomics*, 1970, vol. 13, no. 4, pp. 509–513.

59   Singleton, W. T.: "The measurement of man at work with particular reference to arousal," in W. J. Singleton, J. G. Fox, and D. Whitfield (eds.), *Measurement of man at work*, Taylor and Francis, London, 1971, pp. 17–25.

60   Snook, S. H.: The effects of age and physique on continuous-work capacity, *Human Factors*, 1971, vol. 13, no. 5, pp. 467–479.

61   Snook, S. H., and C. H. Irvine, Psychophysical studies of physiological fatigue criteria, *Human Factors*, 1969, vol. 11, no. 3, pp. 291–300.

62   Snook. S. H., C. H. Irvine, and S. F. Bass: *Maximum weights and workloads acceptable to male industrial workers while performing lifting, lowering, pulling, carrying, and walking tasks*, paper presented to American Industrial Hygiene Conference, Denver, May, 1969.

63   Switzer, S. A.: *Weight lifting capabilities of a selected sample of human males*, AMRL, MRL, TDR 62–57, 1962.

64   Tichauer, E. R.: A pilot study of the biomechanics of lifting in simulated industrial work situations, *Journal of Safety Research*, September, 1971, vol. 3, no. 3, pp. 98–115.

65   Tuttle, W. W., and B. A. Schottelius: *Textbook of physiology*, 16th ed., The C. V. Mosby Company, St. Louis, 1969.

66   Vince, M. A.: The intermittency of control movements and the psychological refractory period, *British Journal of Psychology*, 1948, vol. 38, pp. 149–157.

67 Vos, H. W.: Physical workload in different body postures, while working near to, or below ground level, *Ergonomics*, 1973, vol. 16, no. 6, pp. 817–828.

68 Wargo, M. J.: Human operator response speed, frequency, and flexibility: a review and analysis, *Human Factors*, 1967, vol. 9, no. 3, pp. 221–238.

69 Warrick, M. J., A. W. Kibler, and D. A. Topmiller: Response time to unexpected stimuli, *Human Factors*, 1965, vol. 7, no. 1, pp. 81–86.

70 Wehrkamp, R., and K. U. Smith: Dimensional analysis of motion: II. Travel-distance effects, *Journal of Applied Psychology*, 1952, vol. 36, pp. 201–206.

71 Young, I. C.: A study of tremor in normal subjects, *Journal of Experimental Psychology*, 1933, vol. 16, pp. 644–656.

72 *Manual lifting and carrying*, International Occupational Safety and Health Information Centre, Geneva, CIS Information Sheet 3, 1962.

Chapter 8

Human Control of Systems

In the case of most operational systems the human being serves a control function, such control typically being effected by the use of control devices. Such control can range from the very simple chore of starting and stopping a dishwashing machine to the very complicated process of controlling the aircraft around a metropolitan airport or controlling the operation of a petroleum refinery. Whatever the nature of the system, the basic human functions involved are those referred to earlier, namely, information input, information processes leading to a decision, and action or response. The action taken serves as the control input to the system. In the case of some systems that require continuous control, there typically is some form of feedback to the operator.

INPUT AND OUTPUT CHANNELS

A representation of the human sensorimotor system was presented in Chapter 3 in Figure 3-2. That figure can also be viewed as representing the nature of the human involvement in system control, starting with the reception of stimuli by the sense organs and ending with the effectors that are used in the action or response phase.

The primary sensory modalities used are of course vision and audition. However, certain of the other senses also serve as channels of useful input in some circumstances, such as the skin senses (pressure and temperature), the kinesthetic sense, the senses of body movement and equilibrium, and even the sense of smell. The inputs to the sensory receptors should of course consist of stimuli that are relevant to the control of the system in question. As illustrated in Figure 3-1 in Chapter 3 such stimuli may be received *directly* from the original (distal) source, or *indirectly* via some man-made display. (Chapters 4 and 5 dealt with displays of various types, so we need not discuss them extensively in this chapter.)

The primary output channels (i.e., the effectors) used in system control are the motor responses (especially with the hands and feet) and speech (as in aircraft control tower operations). Most typically the motor responses consist of the use of control devices such as levers, cranks, push buttons, and pedals. However, other responses are used in special circumstances, or offer some promise of greater use. These include certain of the physiological measures discussed in Chapter 7 (i.e., EEG, EKG, etc.). In addition, there have been some interesting developments in the use of eye movements in control processes. The choice of the output channel for any given situation would of course depend upon its appropriateness for the purpose at hand.

BETWEEN INPUT AND OUTPUT

The nature and degree of human involvement in control processes naturally hinges on the circumstance—the nature of the system, the level of the relationship of the individual to the system, the extent to which the various phases of control process have been predetermined at higher levels, the designated role of the individual, the type of control mechanism(s) used, etc. For example, in some situations there is an intended one-to-one correspondence between some input information and the output response. In a simple case this could consist of the activation of a machine when an appropriate stimulus is present, or of dialing a telephone number. In other instances, as in a tracking task, the operator is supposed to match his response to a continuously changing input signal. In such instances the individual serves primarily as a transmission link and his primary control function is an internal one of endeavoring to match his output response (physical or communication) to the input information or signal. In other circumstances, certain input information is supposed to undergo some standard, or fairly standard, process before an output response is made (including invariant processes such as classification); such circumstances typically call for some form of information transformation or some form of filtering, such as in interpreting radar signals. And yet in other situations the input information may be varied and unpredictable and require decision processes of varying complexity leading to actions that implement these decisions, such as in driving a car, piloting a plane, or controlling the operation of an electric power station.

Sandwiched between the input and the output is some form of information processing and decision, ranging from the simple (virtually a conditioned response) to the complex. When the input virtually specifies what the output response should be, it would be theoretically possible to so program the system that the process is automated, thus eliminating the need for an individual. This has been done, of course, in many systems. There are, however, limitations on the extent to which this can be now done, including technological constraints, cost, practicality, etc. Perhaps the most severe restriction with many systems, however, is the difficulty or impossibility of programming a system for all possible contingencies that might arise. One of man's rather unique abilities is his adaptability to unexpected or unusual events, and the inclusion of man in many systems is fully justified on the grounds that he has the capability of making decisions on the basis of unexpected or unpredictable events. When man is included within a system, the decision-making processes can be viewed within the framework of decision theory.[1]

To a very considerable degree the specific nature of the information-processing and decision processes required on the part of an operator is prescribed by the nature of the input information (more specifically, of the stimuli that convey information) and of the actions required. Thus, in this chapter we will give attention to the interrelationships between displays and controls as they might influence the intervening mediation processes and the system output, especially in the case of continuous control processes such as tracking tasks. Before discussing this, however, we should elaborate on the concept of compatibility, since it has a critical tie-in with human control processes.

COMPATIBILITY

As discussed in Chapter 3, *compatibility* refers to the spatial, movement, or conceptual relationships of stimuli and of responses, individually or in combination, which are consistent with human expectations. This concept has very definite implications for the design of displays and controls of systems on the grounds that human performance generally is enhanced when the displays and/or controls are in fact compatible with human expectations.

Spatial Compatibility

There are many variations of the theme of spatial compatibility, most of them spanning the gamut of physical similarities in displays and corresponding controls and their arrangement, and the arrangement of any given set of either displays or controls.

Physical Similarity of Displays and Controls   Sometimes there exists the opportunity to design related displays and controls so there is reasonable correspondence of their physical features, and perhaps also of their modes of operation. Such a case is well illustrated by Fitts and Seeger [15]. In this study three different displays and three different controls were used in all possible combi-

---

[1]The reader is referred to Chapter 8 in DeGreene [12] for a discussion of decision theory as it relates to systems.

nations. The displays consisted of lights in various arrangements. As a light would go on, the subject was to move a stylus along a channel (or channels) of the control that was being used to a location at the end of the appropriate channel; when that location was reached, an electric contact would turn the light off. The three displays and controls are illustrated in Figure 8-1. The experimental

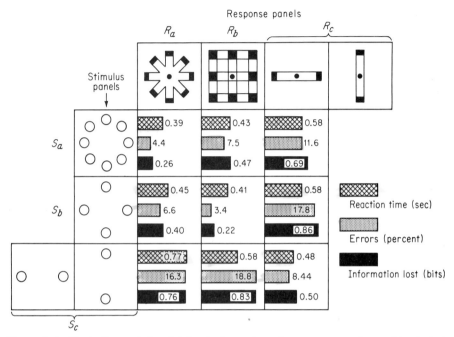

Figure 8-1 Illustrations of signal (stimulus) panels and response panels used in study by Fitts and Seeger [15]. The values in any one of the nine squares are the average performance measures for the combination of stimulus panel and response panel in question. The compatible combinations are S_a-R_a, S_b-R_b, and S_c-R_c, for which results are shown in the diagonal cells.

procedures will not be described, but in general, different groups of subjects used each combination of the three stimulus panels with the three response panels, their performance being measured in reaction-time errors and information lost. The results, also given in Figure 8-1, show that performance with any given stimulus or response panel was better when it was used in combination with its *corresponding* response or stimulus panel (S_a-R_a, S_b-R_b, and S_c-R_c) than when used in combination with a different configuration.

Physical Arrangement of Displays and Controls   Both experiments and rational considerations lead one to conclude that for optimum use, corresponding displays and controls should be arranged in corresponding patterns. This aspect of compatibility was well demonstrated by the down-to-earth example dealing with a gadget used morning, noon, and night (and sometimes in between)—a study dealing with the arrangement of burner controls on a four-burner stove [Chapanis and Lindenbaum, 10]. The burners, in a sense, can be thought of as displays. The four arrangements that were tried out are shown in Figure 8-2. Fifteen subjects each had 80 trials on each design. They were told

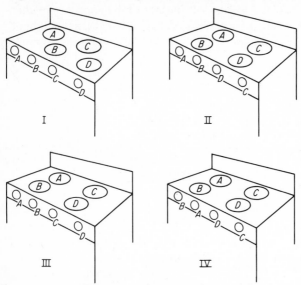

Figure 8-2 Control-burner arrangements of simulated stove used in experiment by Chapanis and Lindenbaum [10].

what burner to turn on, and their reaction time was recorded along with **any** errors they made. The number of errors made on the four designs is **given** below:

Design	No. of errors, out of 1200 trials
I	0
II	76
III	116
IV	129

Design I was also best in reaction time (over the last 40 trials), with design II next best. Clearly, design I was the most compatible.

   An elaboration of essentially the same theme was explored in a study by Chapanis and Mankin [11] in which 10 possible control-display linkages were used experimentally, each linkage consisting of four displays (each with a light) and four push buttons, as illustrated in Figure 8-3. As a display light appeared, the subject was to push a control button to turn it off. The response times are shown in the figure below their respective linkages. The response times for the first four linkages indicate that the arrangements of the displays in those instances were somewhat more compatible with the expectations of the subjects than were the other linkages.

Arrangement of Sets of Similar Devices   Sometimes an assortment of similar devices is used in combination, such as displays or controls that have

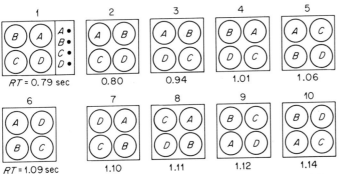

RT = 0.79 sec   0.80   0.94   1.01   1.06

RT = 1.09 sec   1.10   1.11   1.12   1.14

Figure 8-3 Ten control-display linkages used in a study of spatial compatibility. No. 1 shows the basic arrangement of the controls and displays; the controls in the other nine were in the same positions as shown in No. 1. The response time (RT) for each is the average time the subjects took to press the control button when the display light went on. [Adapted from Chapanis and Mankin, 11, figs. 2 and 3.]

some systematic relationship to each other. Here, again, we can capitalize on the concept of compatibility by arranging them in the most natural, or expected, pattern. One such example is the arrangement of push-button telephone key sets [Deininger, 13]. In this example, each push button serves both as a display and as a control. Several different arrangements were experimented with initially, which resulted in the selection of five designs for further analysis. These are shown in Figure 8-4. In the experimental use of these,

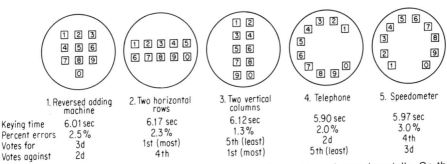

	1. Reversed adding machine	2. Two horizontal rows	3. Two vertical columns	4. Telephone	5. Speedometer
Keying time	6.01 sec	6.17 sec	6.12 sec	5.90 sec	5.97 sec
Percent errors	2.5%	2.3%	1.3%	2.0%	3.0%
Votes for	3d	1st (most)	5th (least)	2d	4th
Votes against	2d	4th	1st (most)	5th (least)	3d

Figure 8-4 Push-button arrangements of telephone key sets used experimentally. On the basis of combined considerations of keying time, errors, and expressed preferences of subjects, the first design was selected for use. [From Deininger, 13. Reprinted by permission of the copyright owner, American Telephone and Telegraph Company.]

various criteria were obtained, including average keying time, errors, and "votes" for and against each arrangement. Although no single design was superior on all criteria, the collective criterion data pointed toward number 1, and this is the design that you find now on push-button telephones.

Compatibility of Movement Relationships

There are several different types of circumstances in which movement relationships in systems can be viewed in a compatibility frame of reference. Following are some examples:

- Movement of a control device to *follow* the movement of a display (as moving a lever to the right to follow a right movement of a blip on a radarscope)
- Movement of a control device to *control* the movement of a display (as tuning a radio to a particular wavelength)
- Movement of a control device that produces a specific system response (as turning a steering wheel to the right to turn right)
- Movement of a display indication without any related response (as the clockwise turn of the hands of a clock)

For most movement relationships there usually is a particular relationship for each that is the most compatible, either because it is intrinsic to the situation (such as turning a steering wheel to the right to turn the car to the right) or because it is a culturally acquired relationship.

However, the compatibility of movement relationships of displays or controls (or both) in some circumstances becomes intertwined with their physical orientation to the user. In this connection, there are two basic orientations of the *display plane* relative to the operator (the display plane may include some controls as well as displays). One of these is a *front-wall* orientation (with the displays on a vertical plane facing the operator), and the other is a *table-top* orientation (with the displays on a horizontal surface). The other two planes are referred to as the *lateral cutting plane* (perpendicular to the display plane and in a lateral relation to it) and the *orthogonal cutting plane* (on a vertical, forward-aft axis). Such surfaces, in a work situation, could be either to the right or to the left of the operator.

Movement Relationships of Displays and Rotary Controls in Same Plane In display-control systems in which a control movement is associated with a movement of some feature of the display (such as a pointer), a generally accepted principle is that with a rotary control in the same plane, a clockwise turn of a control device is associated with an increase in values, as in the left-hand examples of Figure 8-5, although there are certain exceptions. One exception to this is with a fixed horizontal or vertical scale with a moving pointer where a rotary control is beside the scale; in these scales the most compatible relationship has the indicator move in the same direction as the part (or side) of the control device nearest the indicator [Warrick, 32]; such examples are shown in the right-hand examples of Figure 8-5.

For moving scales with fixed pointers, it has been postulated that the following features would be desirable [Bradley, 4]:

1   That the scale rotate in the same direction as its control knob (i.e., that a direct drive exist between control and display)
2   That the scale numbers increase from left to right
3   That the control turn clockwise to increase settings

With the usual orientation of displays, however, it is not possible to incorporate all these three features in a conventional assembly. With the usual

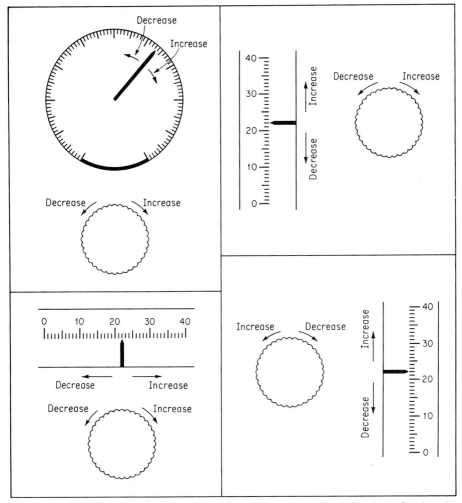

Figure 8-5 Illustration of desirable control-display movement relationships: fixed scales with moving pointer, with control in same plane as display.

display orientation, only the two combinations of these features that are shown in Figure 8-6 as *A*, *B*, *C*, and *D* are possible. In a study relating to moving-display instruments and associated controls, these types and some variations of them were used experimentally, with provision for obtaining the following types of criteria: (1) starting errors (an initial movement in the wrong direction); (2) setting errors (incorrect settings); and (3) rank-order preferences of the subject. Some of the results are given at the bottom of Figure 8-6 (the two or three sets of data for each criterion are from different groups of subjects used in different phases of the study in which only certain assemblies were used).

While assembly *A* incorporated the incompatible feature of a *counter-clockwise* control turn bringing about an *increase* in scale value (rather than a decrease), nonetheless, this assembly was found throughout a series of subex-

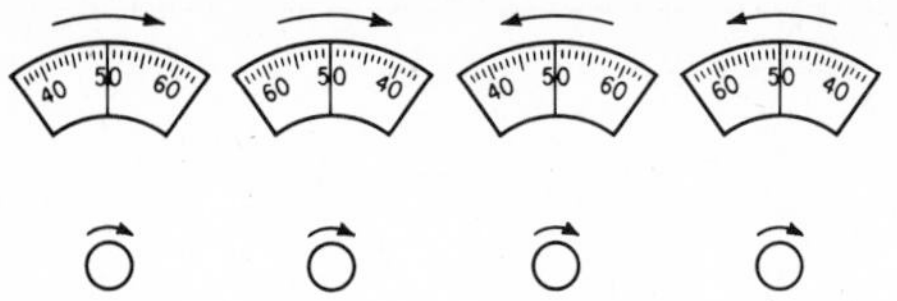

Assembly	A	B	C	D
Drive	Direct	Direct	Reversed	Reversed
Scale numbers increase	Left to right	Right to left	Left to right	Right to left
With clockwise knob movement setting will:	Decrease	Increase	Increase	Decrease

	A	B	C	D
Starting errors	13	11	87	106
	11		116	
Setting errors	0	9	1	8
	7	20		
Preference (number of times ranked "first")	31	22	17.5	1.5
	42		10	
	11	9		

Figure 8-6 Some of the moving-display and control-assembly types used in a study by Bradley [4]. The various features of these related to three desirable characteristics are given below the diagrams; crosshatching indicates an undesirable feature. With the usual display orientation all three desirable features are not possible. Some data on three criteria are given at the bottom of the figure, indicating the general preferability of A.

periments generally to be the best assembly, considering starting errors, setting errors, and preferences. In evaluating the apparent superiority of assembly A over certain others (for example, C), it seemed clear that a *reverse drive* (such as in C and D) tended to give rise to more starting errors than a direct drive.

While, in a sense, all this relates to a fairly incidental feature of display-control design, it probably does have broader implications. If within a situation, there might be different *kinds* of compatibility (as, indeed, there are in this experiment), it may be pertinent to know *what* type of compatibility is the more (or most) critical in case of some possible conflict (for example, *direct-drive* compatibility seems to be more critical than the *clockwise-increase* principle with moving scales).

Movement Relationships of Displays and Rotary Controls in Different Planes In the relationship between movement of display indicators and of control devices, some control devices may be in a different plane from the displays with which they are associated, such as the lateral cutting plane or orthogonal cutting plane. In one such study, the control knobs that were used caused a pointer to move along a straight-line scale [Holding, 17]. The knob and pointer were in different planes, as illustrated in Figure 8-7. More than 700 subjects were used in the study and were asked to "twist the knob" in order to move the pointer. The strongest relationships displayed by the subjects between direction of control movement and direction of pointer movement are shown in Figure 8-7. The results led to the conclusion regarding

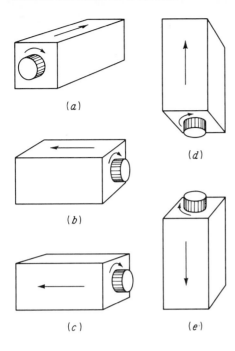

Figure 8-7 Illustration of some of the display-control relationships studied by Holding [17], showing the strongest relationships demonstrated by subjects.

human performance in such situations that people's responses tend to be one of two types: (1) a generalized clockwise tendency; and (2) a helical, or screwlike, tendency in which clockwise rotation is associated with movement away from (as with screws, bolts, etc.), and counterclockwise is associated with movement toward, the individual.

Movement Relationships of Rotary Vehicular Controls   In the operation of most vehicles there is no "display" to reflect the "output" of the system; rather, there is a "response" of the vehicle. In such instances, if the wheel control is in a horizontal plane, the operator tends to orient himself to the forward point of the control, as shown in Figure 8-8(*a*) [Chapanis and Kinkade, 9]. If the wheel is in a vertical plane, the operator tends to orient himself to the top of the control, as shown in Figure 8-8(*b*).

Movement Relationship of Stick-Type Controls   The compatibility of stick-type controls and movements of associated display indicators was investigated in connection with a tracking task [Spragg, Finck, and Smith, 30]. Four combinations of control location and display movement were used, as illustrated in Figure 8-9. This figure also shows average tracking scores of subjects under these four conditions.

For a horizontally mounted stick (on the vertical display plane), the superiority of the *up-up* relationship (control movement up associated with display movement up) over the *up-down* relationship is evident. For a vertically mounted stick (on the horizontal lateral cutting plane) there was less difference

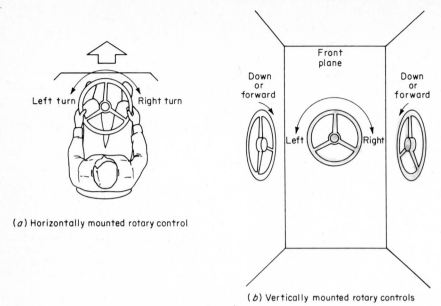

(a) Horizontally mounted rotary control

(b) Vertically mounted rotary controls

Figure 8-8 The most conpatible relationships between the direction of movement of horizontally and vertically mounted rotary controls and the response of vehicles. [Adapted from Chapanis and Kinkade, 9, figs. 8-6 and 8-7.]

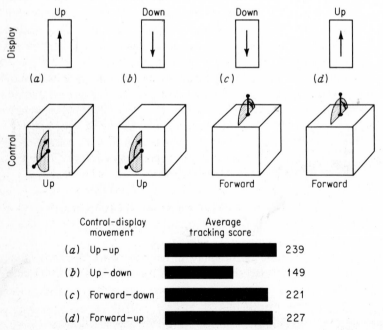

Figure 8-9 Tracking performance with horizontally mounted and vertically mounted stick controls and varying control-display relationships. [Adapted from Spragg, Finck, and Smith, 30; data based on trials 9 to 16.]

between the *forward-up* and *forward-down* relationships, although the forward-up relationship was slightly superior (but not significantly so). The investigators concluded that for the kind of tracking used it was about equally effective to mount the stick in a horizontal or vertical position, provided one *avoids* the up-down relationship with a horizontally mounted stick.

Conceptual Compatibility

Probably the most common variety of conceptual compatibility relates to the associations in the use of coding systems, symbols, or other stimuli; these associations may be intrinsic (i.e., the use of visual symbols that represent things such as airplanes) or they may be culturally acquired. Certain illustrations were given in Chapter 3 and will not be discussed further here.

THE NATURE OF CONTINUOUS CONTROL SYSTEMS

There are many types of systems with continuous control of some process or operation, including the operation of all types of vehicles, tracking operations (as of military targets and aircraft), and certain production processes (as for chemicals and petroleum products).

Inputs and Outputs in Continuous Systems

For continuous systems, the input to the system specifies the system goal, or desired output; this may be constant (e.g., steering a ship at a specified heading or flying a plane at an assigned altitude), or variable (e.g., following a winding road or tracking a maneuvering aircraft). Such input typically is received directly from the environment and sensed by mechanical sensors or by people. If it is sensed mechanically, it may be presented to operators in the form of signals on some display. The input signal is sometimes referred to as a *target* (and in certain situations it actually is a target), and its movement is called a *course*. Whether or not the input signal represents a real target with a course, or some other changing variable such as desired changes in temperature in a production process, it usually can be described mathematically and shown graphically. While in most instances the mathematical or graphic representations do not depict the real geometry of the input (and the input-output relations) in spatial terms, such representations do have utility in characterizing the input. Inputs may change at a constant velocity (ramp), by steps, sinusoidally, or in more complex ways including randomly. Figure 8-10 illustrates some of these possibilities. Whatever the form of the input, it typically specifies the desired output of the system, such as curves in a road specifying the desired path to be followed by an automobile. The output is usually brought about by a physical response with a control mechanism (if by an individual) or by the transmission of some form of energy (if by a mechanical element). In some systems the output is reflected by some indication on a display, sometimes called a *follower* or a *cursor*; in other systems it can be observed by the outward behavior of the system, such as the movement of an

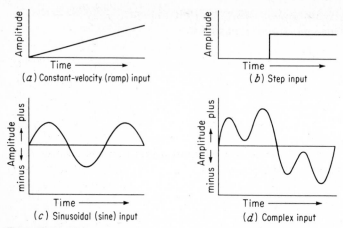

Figure 8-10 Types of inputs of tracking systems. Depending upon the circumstance, the amplitude can represent such variables as direction, distance, angle, voltage, velocity, and acceleration. Complex inputs frequently are the combination of two or more sinusoidal inputs.

automobile; in either case it is frequently called the *controlled element*. In a *compensatory* display, one of the two indications (the target or the controlled element) is fixed and the other moves. When the two are superimposed, the controlled element is *on target*; any difference represents an error, and the function of the operator is that of manipulating the controls to eliminate or minimize that error. The source of any difference (error), however, cannot be diagnosed; whether the target has moved or changed course or whether the tracking has been inaccurate is not shown. An illustration is shown in Figure 8-11. In a pursuit display, both elements move, each showing its own *location* relative to the space represented by the display. With a pursuit display, the

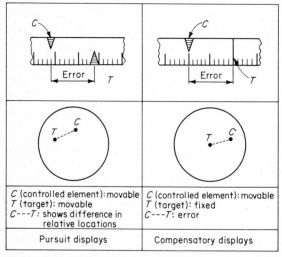

Figure 8-11 Illustrations of compensatory and pursuit tracking displays. A compensatory tracking display shows only the *difference* (error) between the target *T* and the controlled element *C*. A pursuit display shows the location (or other value represented) of both the target and the controlled element.

operator is presented with information about the actual location of both elements, whereas with a compensatory display he knows only the absolute error or difference. However, compensatory displays sometimes have a practical advantage in conserving space on an instrument panel since they do not need to represent the whole range of possible values or locations of the two elements.

Control Loops

Complex systems can be viewed as consisting of a hierarchy of control relationships, each level being called a *loop*. Each loop is related in some fashion to the higher (outer) loops and to the lower (inner) loops, as illustrated in Figure 8-12 for the control of a ship [Kelley, 23]. On the one hand what Kelley calls

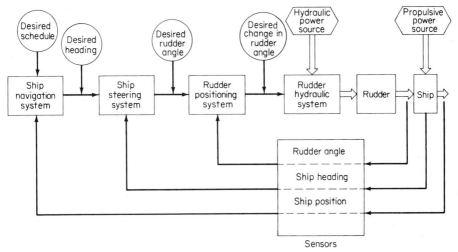

Figure 8-12 Hierarchy of tasks in steering a ship. Note that the desired goal at any given loop (e.g., the desired schedule) specifies the desired goal for the next-lower loop (e.g., desired heading), etc. Also, note the influence of any given loop (e.g., rudder hydraulic system) on the next-higher loop (e.g., rudder). [From Kelley, 23, p. 29.]

the goal-selection and planning aspects of control typically permeate from the outer to the inner loops, specifying in chain-reaction manner the operational goals to be fulfilled. Thus, the desired goal at any given loop (e.g., the desired schedule of the ship) specifies the desired goal for the next-lower loop (i.e., the desired heading), etc. On the other hand, in the actual control process, the output (i.e., the controlled variable) of any given inner loop serves as the input to the next-higher loop, etc. As shown in Figure 8-12, the hydraulic control of the rudder angle controls the ship heading, which in turn controls the ship position. If all goes well in this sequence, the ship position at a given point in time would be that which would put the ship on its desired schedule.

Control Order of Systems

A discussion of the control order of systems involves some familiarity with sinusoidal functions. Thus, let us first discuss sine waves briefly.

Sine Wave A *sine wave* can be defined geometrically from the properties of a rotating vector (or a point on the circumference of a circle rotating at a constant angular velocity). This was illustrated in Chapter 5, Figure 5-1, in connection with the discussion of sound. As the point moves counterclockwise around the circle (starting from horizontal), its projection through time on a vertical axis forms a sine wave, and time is represented along the horizontal base. The sine of the angle, theta, Θ (sometimes called "phase angle"), is characterized as follows (see Figure 5-1 for meaning of *NP* and *OP*):

$$\sin \Theta = \frac{NP}{OP}$$

When $\Theta = 0$, the sine also is 0; as Θ increases up to 90°, the sine increases to unity. Subsequent changes in Θ bring about the characteristic temporal changes of the sine wave depicted, and plus and minus values are above and below the horizontal line.

The changes in amplitude depicted graphically occur in varying physical contexts, such as in vibrations, noise, and the motion of resonating bodies (at their resonating frequencies). These changes in amplitude also are characteristic of changes in velocity and of acceleration of objects in motion. Since changes in amplitude are associated with angular position, it is then possible to use Θ as a measure of amplitude of whatever variable quantity is involved in a system, such as an input or output quantity. It is also possible to give the amplitude of an error (the difference between actual output and desired output) in terms of Θ.

Relationships of Angle, Velocity, and Acceleration For sine functions, there is a specified relation among the variables of angle Θ, velocity, and acceleration, as depicted in Figure 8-13. (That figure also characterizes certain related tracking control "orders," but we shall mention these later.) While these relationships exist in any context in which a sine function is descriptive of continuous change of any type of variable, the relations probably can be most easily comprehended in the context of physical motion. As an example, let us envision a radarscope on which a target is moving up and down on a vertical axis, the motion being that of a sine function. The position on this axis at any point in time is given by the vertical amplitude (up or down) of the angle Θ that designates that point in time. The *angle* curve in Figure 8-13, then, describes *position* through *time*. The *velocity v* curve, in turn, represents the velocity of the target at corresponding points in time; and, in turn, the *acceleration a* curve represents the corresponding rate of change in velocity (i.e., acceleration). Note that the velocity and acceleration curves are similar to the angle (i.e., position) curve, except that they have different phases; the velocity and the acceleration curves are out of phase with the angle by 90 and 180°, respectively.

In terms of derivatives, the phase-angle curve is a zero-order function, the velocity curve is a first-order function, and the acceleration curve is a second-

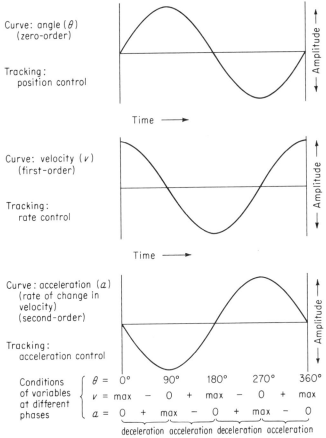

Curve: angle (θ)
(zero-order)

Tracking:
position control

Time ⟶

Curve: velocity (v)
(first-order)

Tracking:
rate control

Time ⟶

Curve: acceleration (a)
(rate of change in
velocity)
(second-order)

Tracking:
acceleration control

Conditions of variables at different phases

	$\theta =$	0°	90°	180°	270°	360°
	$v =$	max	— 0 +	max	— 0 +	max
	$a =$	0 +	max	— 0 +	max	— 0

deceleration   acceleration   deceleration   acceleration

Figure 8-13 Relationship among angle Θ, velocity, and acceleration curves and types of tracking tasks associated with these. The three curves are identical, but at different phases. The velocity curve is out of phase with the angle curve by 90°, and the acceleration curve is at opposite phase, being 180° different.

order function. In other words, velocity v is the first derivative of angle Θ and thus requires one differentiation of Θ. In turn, acceleration a is the first derivative of velocity v, and thus requires one differentiation of velocity, or the second derivative of angle, and thus requires two differentiations of angle. Higher derivatives can be described as higher functions, i.e., third-order, fourth-order, etc.

Control Order as Related to System Control After this diversion into sine waves, let us get back to the matter of the input and output of continuous, closed-loop systems and to the ship we left adrift in Figure 8-12. The control of the system by the helmsman influences the rudder angle and, in turn, the ship heading and its position. But the nature of these influences is complex. Specifically, the *position* of the rudder control (which operates the rudder hydraulic system) produces a *rate of movement* of the rudder, and in a chain-reaction manner, the position of the rudder results in the angular acceleration of the ship and, in turn, the *rate of change of lateral position* with respect to the desired course. This hierarchy of control sequence represents various *control orders.* In continuous control processes that have a series of control loops (such as a ship), the sequence of chain-reaction effects can be described in terms of mathematical functions, such as a change in the *position* of one variable changing the *velocity* (rate) of the next, the *acceleration* of the next, etc. Almost any continuous control operation, then, can be characterized by its *control order* as predetermined by the mathematical derivative of the controlled variable, including zero order (position control), first order (rate or velocity control), second order (acceleration control), etc.

Control Responses in Relation to Control Order   The control order, in effect, specifies the nature of the response that is to be made to various types of inputs. Some such responses are illustrated in Figure 8-14, for sine, step, and

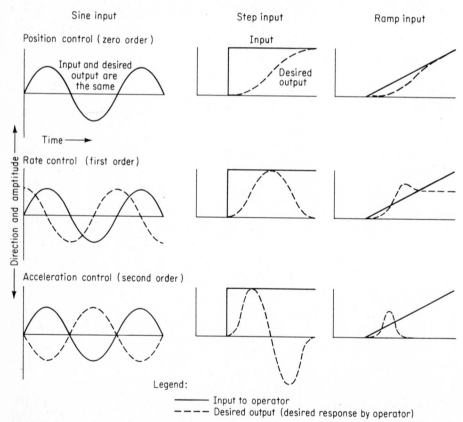

Figure 8-14 Tracking responses to sine, step, and ramp inputs which would be conducive to satisfactory tracking with position, rate, and acceleration control. The desired response, however, is not often achieved to perfection, and actual responses typically show variation from the ideal. In a positioning response to a step input, for example, a person usually overshoots and then hunts for the exact adjustment by overshooting in both directions, the magnitude diminishing until he arrives at the correct adjustment.

ramp inputs when position, rate, and acceleration control systems are used. In each case the dotted line represents the response over time (along the horizontal) that would be required for satisfactory tracking of the input in question. In general, the higher the order of control, the greater is the number of controlled movements that need to be made by an operator in response to any single change in the input, as illustrated in Figure 8-14. This is illustrated further in Figure 8-15 (page 220). If the input, changing over time, follows the pattern (line) shown, and if the control system (whatever it may be) is zero order (i.e., position control), then the movement of the control device by an operator should correspond exactly with that line. But if, instead, the control system is a first-order system (i.e., rate control), then the operator needs to anticipate and make the response movements shown. In turn, the other lines represent the changes that the operator would need to make with a control device if, in fact, the control system were a second-, third-, or fourth-order system. Some of the higher-order systems are, in fact, especially characteristic of certain vehicles, as of our ship. The rudder and flaps of aircraft serve to control the rate of change of heading and elevation and are second-order controls; and submarines typically have at least third-order control.

Procedures for Facilitating Human Control

As we can see from the previous discussion, the particular nature of the desired human response (output) in continuous control systems is predetermined by the type of input to the system (sine, step, ramp, etc.) and the control order of the system. In the case of a zero-order system (position control) as illustrated in the top line of Figure 8-15, the problem of deciding what response to make is relatively simple. In such a case the operator simply has to effect the amount of movement of the control that is required to bring about a specified degree of change in the controlled element (i.e., the system response, such as movement of a vehicle). The mental function in doing this is essentially an amplification process (in particular, the multiplication by a constant that represents the ratio of the input signal to the desired output response). In the case of higher-order systems, however, the mental functions of figuring out what movements to make involve more complex mental gymnastics, including those akin to mathematical differentiation, integration, and algebraic addition. A mental operation analogous to differentiation would be required, for example, in estimating velocity (rate), and to double differentiation in estimating acceleration. An operation analogous to integration would be required in estimating future position (at some specific time) from an estimate of a given velocity, and to double integration for estimating future position from an estimate of acceleration. Algebraic addition is of course the operation of adding two or more values (actually estimates of values), taking into account their sign (+ or −).

In general, people do not do well in these kinds of mental operations, and when one combines two or more operations, such as differentiation, integration, multiplication, and addition, the task becomes virtually impossible to perform.

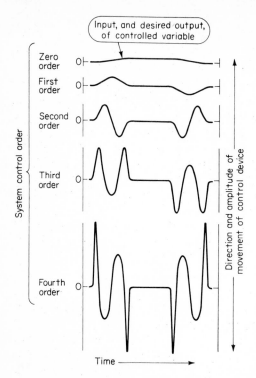

Figure 8-15 Control order illustrated by changes in a controlled variable and in its first four derivatives. Each line represents the changes over time that would have to be made with control systems of various control orders to make the controlled variable correspond to the input. [Adapted from Kelley, 23, p. 31.]

Thus, in the case of complex systems that require high-order controls (such as second-, third-, or fourth-order) some ways and means have to be devised for relieving the operator of the need to perform the mental functions that otherwise would be required, or of "compensating" for the disparity between response demands of the situation and the response capabilities of people. Certain such procedures will be discussed below.

Aiding   One such procedure is the use of aiding. Aiding was initially developed for use in gunnery tracking systems and is most applicable to tracking situations of this general type, in which the operator is following a moving target with some device. Its effect is to modify the output of the control in order to help the tracker. In aided tracking, a single control adjustment usually affects two variables, specifically the position and the rate of the controlled element (and, in some instances, also the acceleration). Let us suppose we are trying to keep a high-powered telescope directed exactly on a high-flying aircraft by using a rate-aided system. When we fall behind the target, our control movement to catch up again would automatically speed up the *rate* of motion of our telescope (and thus, of course, its position). Similarly, if our telescope gets ahead of the target, a corrective motion would automatically slow down its rate (and influence its position accordingly). Such rate-aiding would simplify the problem of quickly matching the rate of motion of the following device to that of the target and would thus improve tracking performance.

The effects of aiding on tracking performance in one experiment are illus-

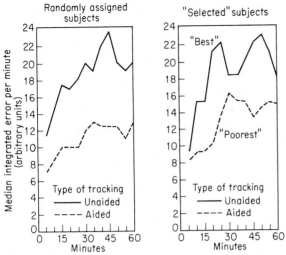

Figure 8-16 Comparison of performance of continuous tracking for 1 h of unaided and of aided acceleration tracking. In the case of the selected subjects, the eight who were poorest on a pretest were assigned to the aided tracking task and the eight who were best were assigned to the unaided task; even the poorest subjects did better with aided tracking than the best ones with unaided. [Adapted from Garvey, 16.]

trated in Figure 8-16. In this, the comparison is between unaided acceleration tracking and aided acceleration tracking. Comparable comparisons have been made of unaided and aided rate tracking.

The operational effect of aiding is to shift from the operator to the control system such functions as differentiation, integration, and algebraic addition which are required in some tracking tasks. In such circumstances the primary function left for the operator is that of amplification. The effects of aiding depend upon a number of factors, including the nature of the input signal. Thus, aiding should be used selectively, in those control situations in which it is definitely appropriate.

Quickening *Quickening* is one form of what is sometimes called *display augmentation* [Kelley, 23], and is particularly relevant for use in vehicular control systems that involve second-, third-, or higher-orders of control. In essence, quickening is a modification of a closed-loop system which reduces the need for the operator to perform analog differentiations or to sense and utilize derivative information separately [Birmingham and Taylor, 3]. It is particularly useful where the dynamics of the system are such that the apparent response of the system to control actions is delayed. The quickening is usually accomplished by appropriate modification of the information going into the display, so the displayed information is more easily and rapidly translatable in terms of the consequences of the operator's control actions. It indicates, in effect, what control action to take to achieve a desired system output.

By its nature, a quickened system is most appropriate where the consequences of the operator's actions are not immediately reflected in the behavior of the system, but rather have a delayed effect, the delay frequently caused by the dynamics of the system, as in aircraft and submarines. Refer back again to Figure 8-15 and consider the responses that would be required for a second-order or third-order system (i.e., the lower curves of that figure); in quickening, the operator would still have to make those responses, but he would be shown what responses to make and would not have to through the mental gymnastics of figuring what those complex movements should be (which, incidentally, he could not do).

A comparison of tracking performance with a quickened versus an unquickened system is reported by Birmingham, Kahn, and Taylor [2]. The six Navy enlisted men who served as subjects used, at different times, three nonquickened systems with one or two joy sticks to be moved in one or two dimensions, and a quickened system with two joy sticks in two dimensions. A summary of the results, shown in Figure 8-17, indicates a clear superiority of

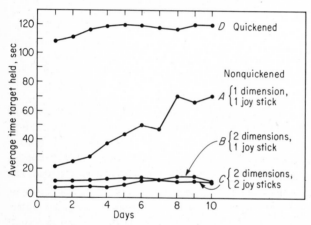

Figure 8-17 Comparison of tracking performance with a simulated quickened versus unquickened system. [Adapted from Birmingham, Kahn, and Taylor, 2.]

the quickened system over the others in time on target (TOT). In fact, four of the six subjects made perfect scores with the quickened display.

Although quickening generally simplifies and improves some tracking tasks, it has certain possible disadvantages and limitations. For example, in a typical quickened system the operator is not provided with information regarding the current condition of the system, since his display shows primarily what control action he should take. It should also be kept in mind that quickening does not have any appreciable advantage in very simple systems, or in systems where there is no delay in the system effect from the control action and where there is already immediate feedback of such system response.

Use of Predictor Displays Still another procedure for use in simplifying the control of high-order systems (especially large vehicles such as submarines) is the use of predictor displays, as proposed primarily by Kelley [21, 22, 23]. However, such displays are essentially in the experimental and development stage. In effect, predictor displays use a fast-time model of the system to predict the future excursion of the system (or controlled variable) and display this excursion to the operator on a scope or other device. The model repetitively computes predictions of the real system's future, based on one or more assumptions about what the operator will do with his control, e.g., return it to a neutral position, hold it where it is, or move it to one or another extreme. The predictions so generated are displayed to the operator to enable him to reduce the difference between the predicted and desired output of the system.

An illustration of what a predictor display might look like is given in Figure 8-18. This shows the actual and future predicted position of a ship that is com-

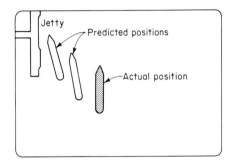

Figure 8-18 Illustration of the type of information that might be displayed on a predictor display to show the actual and predicted position of a ship approaching a jetty for berthing. [Brigham, 6, fig. 4.]

ing in to a jetty for berthing, the predicted positions shown being those for two chosen time intervals ahead [Brigham, 6]. Such a display would permit corrective action to be taken in the eventuality of impending contact with the jetty.

In some situations in real life we do have such "advance" information, as in driving a car on a winding road when we can see the road ahead and can then "predict" what control actions would be required. In the case of a blinding snowstorm or heavy fog, we do not have that advance information. Predictor displays would provide the relevant predictive information when it is not otherwise available.

Predictor displays offer particular advantages for complex control systems in which the operator needs to anticipate several seconds in advance, such as with submarines, aircraft, and spacecraft. The advantages in such situations have been demonstrated by the results of experiments such as the one by Dey [14] simulating the control of a VTOL (vertical takeoff and landing) aircraft. One of the measures of the performance of the subjects was the mean square value of one index of the deviations from a desired "course." The values with, and without, the use of a predictor display were 2.48 and 7.92, respectively. The evidence from this and other experiments shows a rather consistent enhancement of control performance with a predictor display.

DISPLAY FACTORS THAT INFLUENCE SYSTEM CONTROL

The effectiveness of the human control of systems is influenced by a wide variety of factors, some of which are associated with particular features of the displays that are used. In discussing a few such display factors, however, it should be kept in mind that the nature of the relationship between displays and controls frequently has an impact on control performance.

Specificity of Displayed Error in Tracking

In certain compensatory tracking systems the error (the difference between input and output) can be presented in varying degrees of *specificity*. Some such variations are shown in Figure 8-19 as they were used in a tracking experiment

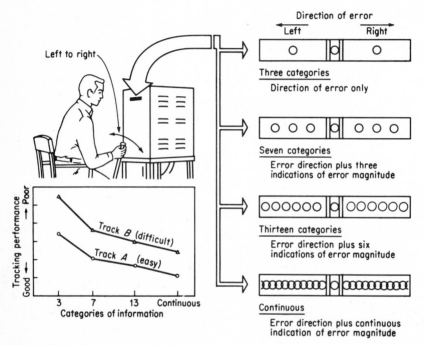

Figure 8-19 Compensatory displays used in study of the effects of specificity of feedback of error information, and tracking performance using such displays. The feedback of error was presented by the use of lights (3, 7, 13, and continuous). [Adapted from Hunt, 19.]

by Hunt [19], these including: 3 categories of specificity (left, on target, and right); 7 categories; 13 categories; and continuous. The accuracy of tracking performance under these conditions (for two levels of task difficulty) indicates quite clearly that performance improved with the number of categories of information (greater specificity), this improvement taking a negatively accelerated form for tasks of both levels of difficulty. Although the results of other studies are not entirely consistent with these, the evidence suggests that in a tracking task, the mediating control functions are facilitated by the presentation of more specific, rather than less specific, display information.

Visual Noise in Tracking Displays

Noise in a visual display (such as a CRT) is a visual disturbance in the display, as in the background or in the target or response signals. Either variety of noise typically causes degradation in tracking performance. The effect on tracking performance of background noise on a CRT display was investigated by Wolf and Green [34] by varying the density of white blips (somewhat like targets) on the black background. On the basis of their investigation it is apparent that, although increasing noise levels (i.e., density) generally tend to cause performance degradation, degradation was particularly noticeable for short exposure times: when the exposure time is long, the time for scanning is apparently sufficient to overcome some of the adverse effects of noise.

The effects of visual noise in the form of disturbances of the display signals are illustrated by the results of a study by Howell and Briggs [18] and are shown in Figure 8-20. This investigation was concerned with perturbations in

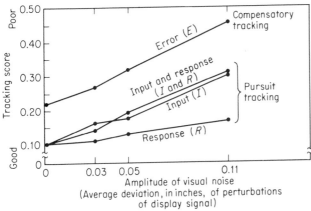

Figure 8-20 Effects on tracking performance of visual noise in the form of disturbances of error signals *E* in compensatory tracking and of input *I* and response *R* signals in pursuit tracking. Lines show tracking performance when the signal indicated (*E, I,* or *R*) was caused to deviate around its "true" position. [Adapted from Howell and Briggs, 18. Copyright 1959 by the American Psychological Association and reproduced by permission.]

pursuit tracking of the input signal *I*, the response signal *R*, and both *I* and *R*, and in compensatory tracking of the error *E*. The amplitude of the noise perturbations generated was such that, for three noise levels used, the average deviations of the signals produced by the noise were 0.03, 0.05, and 0.11 in, respectively. The most obvious thing about Figure 8-20 is that tracking performance was not noticeably affected when the visual noise was of the response *R* display signal only; apparently visually coded feedback information (the response signal) is less critical to operator performance than input information is. It has been suggested that experienced subjects in position control tasks can rely in part upon proprioceptive cues (pressure and movement) as a source of feedback and thereby maintain a reasonable degree of performance.

Type of Display: Compensatory and Pursuit

The comparisons that have been made of tracking performance by using compensatory and pursuit displays have not demonstrated consistent superiority of one over the other. Rather, it seems that the human control performance with these two types of displays tends to vary with such variables as the control order and the nature of the course being tracked. In an investigation with certain of these relationships, the following variables were varied [Obermayer, Swartz, and Muckler, 25]: (1) compensatory and pursuit tracking and (2) control order, namely, position, rate, and acceleration. Three of the criteria of performance used were particularly pertinent, namely, average absolute error (AAE, average difference in inches between the input and output positions on the display), root mean square (rms, square root of the integrated square display error, in inches), and time on target (TOT, the time in seconds that the cursor was within $1/10$ in of the target).

Some indication of the interactions of tracking mode and tracking control orders is shown in Figure 8-21. The results for the three criteria are relatively

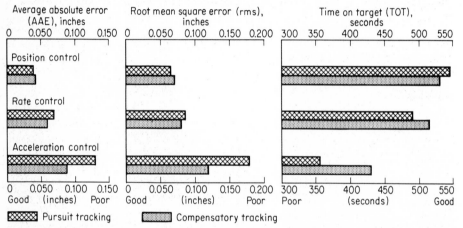

Figure 8-21 Tracking performance, by three criteria, on pursuit and compensatory tracking tasks with position, rate, and acceleration control orders. [Adapted from Obermayer, Swartz and Muckler, 25, fig. 1. Copyright 1961 by the American Psychological Association and reproduced by permission.]

comparable—the *rank order* of performance of the six conditions was essentially the same for all three criteria, with performance generally being best with position control, next best with rate control, and poorest with acceleration control. But although there was no significant difference in performance for either position control or rate control between pursuit tracking and compensatory tracking, for acceleration control compensatory tracking was significantly better than pursuit tracking.

The complexities of tracking are increased by throwing a couple of other variables into the mix, in particular, input *frequency* (i.e., the frequency of changes in the course of the target, especially in the case of sine-wave inputs) and input *amplitude*. However, we will not go into the specific nature of the interactions of such variables as they influence performance.

In pulling together the results of studies with pursuit and compensatory tracking, it seems that under most experimental conditions, pursuit displays are preferable, but the actual choice of a particular display must be made in the light of the kind of information required by the operator. For example, if the inputs are of very low frequencies, compensatory displays may have an advantage.

CONTROL FACTORS THAT INFLUENCE SYSTEM CONTROL

While it is recognized that the combinations of displays and controls interact to influence the human control of systems, there are certain specific features of controls that have a fairly direct bearing upon performance, either because of the physical and psychomotor requirements of their operation or because of the nature of the feedback that they produce.

Feedback

The operation of control mechanisms can provide either of two types of feedback to the individual. What we will call *intrinsic* feedback is that which the individual senses directly from the operation of the control device, such as the amplitude of movement of the control or pressure required to effect the movement, including the "feel" of the control. The second type of feedback—what we will call *extrinsic*—is that which is sensed from some "external" source that indicates the consequences of the control action, such as observing the system output on a display or observing it directly (such as the movement of an automobile), or picking up auditory cues (such as from a machine or vehicle). Some forms of intrinsic feedback received from the operation of controls can provide useful cues, whereas other forms can provide misinformation and result in degradation of system performance. The problem for the designer is to select or design those displays that provide the types of feedback that would enhance the operation of the system—or at least that would not have serious adverse effect. Although it is not feasible here to include a comprehensive treatment of such feedback, certain aspects will be discussed as they are related to control characteristics.

Types of Controls and Their Feedback

The degree of the "effect" of control devices on the output they control can be a function of either of two types of input to the device. One of these is the amount of *displacement* of the control (such as the amount of depression of a brake pedal or the degree of turn of a steering wheel). The other is the amount of *force* applied to the control. Some displacement controls (also called "position controls") have virtually no resistance; they are free-positioning or isotonic controls, and the only type of feedback they provide is the amount of movement of the body member as sensed primarily by the kinesthetic senses. Other displacement controls have some form of resistance and require some degree of force to be operated. Such controls provide varying combinations of amount of movement and pressure feedback. An example of a free-positioning (isotonic) joy stick control is shown in Figure 8-22(a). An example of a force or

(a)

(b)

Figure 8-22 Illustration of a two-axis isotonic (i.e., free-positioning or displacement) joy stick control (*a*), and an isometric (i.e., force or pressure "stiff-stick") joy stick control (*b*). [Photographs courtesy of Measurement Systems, Inc., Norwalk, Conn.]

pressure (isometric) control is shown in Figure 8-22(*b*). In such controls the output is related to the amount of force or pressure that is applied to the control, and the release of the control reduces the effect to zero. The feedback to the operator from force-operated (isometric) controls is from the "feel" of the resistance of the control as varying amounts of force are applied to it. In practice, most control devices partake of both types, since they typically do involve some movement and some force to overcome resistance in the controls.

Resistance in Controls

All control devices have some resistance, although in the case of free-positioning (isotonic) controls this is virtually negligible. Resistance can of course serve as a source of feedback, but the feedback from some types of resistance can be useful to the operator while the feedback from other types of resistance can have a negative effect. However, in some circumstances the designer can design or select those controls that have resistance characteristics that will minimize any possible negative effects and that possibly can enhance performance. This can be done in a number of ways, such as with servomechanisms, with hydraulic or other power, and with mechanical linkages.

The primary types of resistance are illustrated in Figure 8-23 and are described further below.

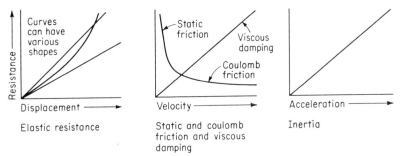

Figure 8-23 Illustration of certain types of resistance in control devices as related to movement variables.

- *Static and coulomb friction:* Static friction, the resistance to initial movement, is maximum at the initiation of a movement but drops off sharply. Coulomb (sliding) friction continues as a resistance to movement, but this friction force is not related to either velocity or displacement. With a couple of possible exceptions, static and coulomb friction tend to cause degradation in human performance. This is essentially a function of the fact that there is no systematic relation between such resistance and any aspect of the control movement such as displacement, speed, or acceleration; thus it cannot produce any meaningful feedback to the user of the control movement. On the other hand, such friction can have the advantages of reducing the possibility of accidental activation and of helping to hold the control in place.
- *Elastic resistance:* Such resistance (as in spring-loaded controls) varies with the *displacement* of a control device (the greater the displacement, the greater the resistance). The relationship may be linear or nonlinear. A major

advantage of elastic resistance (such as in spring-loaded controls) is that it usually (but not always) serves as a source of useful feedback, presumably because the displacement feedback has some systematic relation to resistance. By so designing the control that there is a distinct gradient in resistance at critical positions (such as near the terminal), additional cues can be provided. In addition, such resistance permits sudden changes in direction and automatic return to a neutral position.

* *Viscous damping:* This is caused by a force operating opposite to that of the output but proportional to the output velocity. It generally has the effect of helping to execute smooth control, especially in maintaining a prescribed *rate* of movement. The feedback, however, being related to *velocity* and not displacement, probably is not readily interpretable by operators. Such resistance, however, does minimize accidental activation.

* *Inertia:* This is the resistance to movement (or change in direction of movement) caused by the mass (weight) of the mechanism involved. It varies in relation to *acceleration*. It generally aids in smooth control and minimizes the possibility of accidental activation. Because of the forces required to overcome inertia, however, it complicates the process of changing directions. Its use probably is particularly warranted when heavy friction is involved in a control. In such a situation, inertia presumably compensates in part for friction drag.

The above relationships are illustrated in Figure 8-23. In the case of a spring-centered joy stick (i.e., a control with elastic resistance), the amount of *displacement* of the stick is directly proportional to the amount of force applied (i.e., the resistance that is overcome). In a viscously damped control stick, the force applied will produce an angular *velocity* of the stick that is proportional to the force and, in turn, a displacement of the stick that is proportional to the time integral of the force. With a high inertia control, the *acceleration* of the stick will be proportional to the force applied; thus its displacement will be proportional to the second integral of the force applied. We can see, then, that certain control devices, by reason of their interaction with the forces applied to them, can bring about the effects of integration when using them.

Ability of People to Judge Resistance   Regardless of what variety of resistance is intrinsic to a control mechanism, if it is to be used as a source of useful feedback, the meaningful differences in resistance have to be such that the corresponding pressure cues can be discriminated. Such discriminations were the focus of an investigation by Jenkins [20], with three kinds of pressure controls, namely, a stick, a wheel, and a pedal (like a rudder control in a plane). After some training and practice in reproducing specified forces, a series of trials was made by each subject. Measures of actual pressure exerted were then compared with the pressures that the subjects attempted to reproduce. The difference limens by pounds of pressure for the various types of controls are shown in Figure 8-24. Since the difference limen is the average difference that can just barely be detected, two pressures have to differ by an amount greater than the limen to be detected as being different. The systematic drop-off of all the curves between 5 and 10 lb implies that if differences in pressure are to be

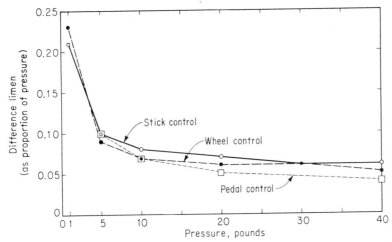

Figure 8-24 Difference limens for three control devices for various pressures that were to be reproduced. (*Limen* is the standard deviation divided by the standard pressure.) [From Jenkins, 20.]

used as feedback in operation of control devices of the types used, the pressures used preferably should be around or above these values (and perhaps more for pedals because of the weight of the foot). The experimenter suggested that if varying levels of pressure discrimination are to be made, the equipment should provide a wide range of pressures up to 30 to 40 lb. Beyond these pressures, the likelihood of fatigue increases, and also the likelihood of slower operation.

Pressure and Displacement as Feedback Cues

When vision is not to be used as a feedback source (or at least not as the dominant source), it would be helpful for the designer to know the relative usefulness of pressure and of displacement cues, singly or in combination. The evidence regarding their combination, however, is not entirely consistent. In reviewing the use of different types of controls in *tracking* tasks, Zeigler and Chernikoff [35] cite certain studies in which tracking errors were lower with pressure types of controls than with displacement (e.g., free-moving) controls. In their own study Zeigler and Chernikoff found a pressure type of lever to result in a somewhat lower error rate than a displacement type of lever, in a third-order tracking task. Similar results also were reported by Burke and Gibbs [7] in a position-tracking (zero-order) task; however, in an analysis of errors in relation to *gain* (i.e., displacement-force ratios which were varied), it was found that errors were *fewer* with *low* ratios, that is, when *greater* forces were associated with a given displacement. This finding, of course, ties in with the results reported by Jenkins [20] as shown above in Figure 8-24, which indicate that differences between *small* forces cannot be differentiated very reliably with only pressure feedback.

On the other hand the results of certain other studies have shown that displacement (amplitude of movement) as a source of feedback tends to be superior to pressure. Such studies, however, generally have dealt with *positioning* tasks (in which the control serves to move the controlled element to a particular position or location) rather than with tracking tasks. For example, in a study by Weiss [33] of blind positioning movements, it was found that the distance the lever moved (amplitude) provided much better cues for positioning the lever accurately than pressure (force) cues did. Positioning errors were greater for short movements than for long ones, however. Another investigation [Bahrick, Bennett, and Fitts, 1] tended to confirm these inklings by demonstrating that changes in torque (increases in torque as the movement continues) aided in accurate positioning for the short movements (the angular excursions of 17.5 and 35°) but not for the longer movements (70°).

In still another study it was found that changes in amplitude of movement significantly affected performance only when force cues were high—not when force cues were low; and changes in force cues affected performance only when amplitude was high—not when it was low [Briggs, Fitts, and Bahrick, 5]. It was proposed that the combination effect of displacement and force cues is expressed by the following ratio:

$$\frac{\Delta F}{F(\Delta D)}$$

where ΔF is the force associated with a given displacement change ΔD and F is the terminal force required for attaining the displacement. In turn, Burrows [8] points to the very complicated effects of the dynamic characteristics of the controlled element on the feedback that the operator would receive (at least in the case of complex systems) as a potentially limiting factor in the use of such feedback in control operations.

It is thus evident that there are numerous missing and confusing pieces of this jigsaw puzzle. In sifting through some of the evidence, however, a few conclusions seem to be warranted. To begin with, it seems evident that there is no clear and consistent superiority of one mode of feedback over the other, which suggests that each may have utility in certain kinds of control circumstances. For example, as indicated above, pressure controls may be more useful in certain types of continuous-tracking tasks than in positioning tasks (but this is not to say that all tracking tasks should be executed with a pressure type of control). And, in any event, pressure controls presumably should be used only when the pressure feedback cues have a tolerably systematic relationship to changes in the controlled variable. If a pressure-stick control is to be used, however, the force to be overcome needs to be of some consequence (that is, the gain, or displacement-force ratio, needs to be low) for the feedback to be of greatest use. Since such forces might induce fatigue (if the controls are used continuously), one might follow the suggestion of Burke and Gibbs [7] to use a nonlinear control with lower force required for large displacements.

On the other side of the coin, if moving controls are to be used, one needs to be forewarned that distance of movement is not a very reliable source of feedback when the amount of displacement is limited. All of this adds up to the notion that for some conventional positioning control purposes (as contrasted with tracking tasks) a combination of force and displacement feedback may have some advantage, using dominantly pressure cues for limited movements of the controlled element and displacement cues for larger movements. Keeping in mind Burrows' observations [8] on the abysmal lack of knowledge about the effects of various types of control feedback, it probably can be said that, barring complicated pressure cues, neither type of feedback (pressure or displacement) in combination with the other seems to affect performance adversely. Although there may still be some question as to whether, with long movements, the combination actually facilitates performance over either one independently, the reference of pilots and other vehicle operators to the *feel* of the control device seems to imply the desirability of at least moderate feel in such mechanisms.

Response Lag in Systems

In the operation of virtually every system there is some lag in the response of the system following the input from the operator. Actually, there are at least a couple of types of lag. *Transmission* lag refers to a situation in which there is a constant delay between input and output; the output is identical with the input, but simply follows it, temporally, after a constant time interval. *Exponential* lag, on the other hand, refers to the situation where the output follows essentially an exponential function following a *step* input.

Although some studies of the effects of lag on tracking tasks have indicated that lag in appreciable amounts brings about degradation in performance [Levine, 24; Wallach, 31], such effects are not universal. In fact, in some circumstances it has been found to have a facilitating effect, these differential effects being particularly related to the control-display ratio being used. The control-display ratio (C/D ratio) is the ratio of the amount of movement of the control device associated with the amount of corresponding movement of the display. Figure 8-25 illustrates such relationships, that figure showing data on the relationship between time delay and average TOT (time on target) for eight subjects. This shows that with high C/D ratios (1:6 and 1:3, limited display movement relative to control movement), a greater C/D time lag apparently causes degradation in compensatory tracking performance, whereas with low C/D ratios (considerable display movement relative to control movement), longer time lags are not serious and, in fact, were even associated with improved performance (especially with a 1:30 ratio).

While there has perhaps been a tendency to consider any delay between control and display as undesirable, such results have suggested that for exponential time delay, a more appropriate principle would be: The optimum delay between control and display depends, among other things, on the magnitude of display change produced by a given control input [Rockway, 27].

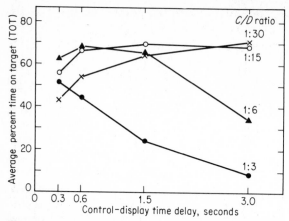

Figure 8-25 Relationship between control-display (C/D) time delay and average percentage of time on target (TOT) scores for subjects using various C/D ratios on a compensatory tracking task. Under long (as opposed to short) delays, high C/D ratios (1:3 and 1:6) result in performance degradation, whereas low C/D ratios (1:15 and 1:30) result in tracking improvement. [From Rockway, 27.]

Anticipation of Input   In some circumstances, the operator can anticipate input changes, as by seeing ahead the road to be followed when driving a vehicle. In other circumstances, where there may be no advance information available as such to the operator, he may be able to deduce the nature of future signals from past experience, as when a person learns the time interval between a warning signal and a subsequent signal to which he is to react, or when (in a tracking task) there is a systematic input such as a sine wave. This type of circumstance has been referred to by Poulton [26] as *perceptual anticipation*.

Backlash

Backlash in a control system is a tendency for the system response to be reversed when a control movement is stopped; and typically it cannot be coped with very well by operators. This effect was illustrated by the results of an investigation by Rockway and Franks [29], using a control task under varying conditions of backlash and of display gain (which is essentially the reciprocal of C/D ratio). The results show that performance deteriorated with increasing backlash for all display gains, but was most accentuated for high gains. The implications of such results are that if a high display gain is strongly indicated (as in, say, high-speed aircraft), the backlash needs to be minimized in order to reduce system errors; or conversely, if it is not practical to minimize backlash, the display gain should be as low as possible—also to minimize errors from the operation of the system.

Deadspace

Deadspace in a control mechanism is the amount of control movement that results in no movement of the device being controlled. It is almost inevitable that some deadspace will exist in a control device. Deadspace of any consequence usually affects control performance, but here, again, the amount of effect is related to the sensitivity of the control system. This is indicated in Figure 8-26 [Rockway, 28]. It can be observed that tracking performance de-

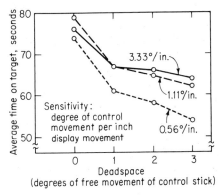

Figure 8-26 Relationship between deadspace in a control mechanism and tracking performance for various levels of control sensitivity. [Adapted from Rockway, 28.]

teriorated with increases in deadspace (in degrees of control movement that produced no movement of the controlled device). But the deterioration was less with the less-sensitive systems (higher C/D ratios) than with more sensitive systems (lower C/D ratios). This, of course, suggests that deadspace can, in part, be compensated for by building-in less-sensitive C/D relationships.

REFERENCES

1 Bahrick, H. P., W. F. Bennett, and P. M. Fitts: Accuracy of positioning responses as a function of spring loading in a control, *Journal of Experimental Psychology*, 1955, vol. 49, pp. 437–444.

2 Birmingham, H. P., A. Kahn, and F. V. Taylor: A *demonstration of the effects of quickening in multiple-coordinate control tasks*, USN, NRL, Report 4380, June 23, 1954.

3 Birmingham, H. P., and F. V. Taylor: Why quickening works, *Automatic Control*, April, 1958, vol. 8, no. 4, pp. 16–18.

4 Bradley, J. V.: *Desirable control-display relationships for moving-scale instrument*, USAF, WADC, TR 54–423, September, 1954.

5 Briggs, G. E., P. M. Fitts, and H. P. Bahrick: Effects of force and amplitude cues on learning and performance in a complex tracking task, *Journal of Experimental Psychology*, 1957, vol. 54, pp. 262–268.

6 Brigham, R. R.: Ergonomic problems in ship control, *Applied Ergonomics*, 1972, vol. 3, no. 1, pp. 14–19.

7 Burke, D., and C. B. Gibbs: A comparison of free-moving and pressure levers in a positional control task, *Ergonomics*, 1965, vol. 8, no. 1, pp. 23–29.

8 Burrows, A. A.: Control feel and the dependent variable, *Human Factors*, 1965, vol. 7, no. 5, pp. 413–421.

9 Chapanis, A., and R. G. Kinkade: "Design of controls," in H. P. Van Cott and R. G. Kinkade (eds.), *Human engineering guide to equipment design*, U.S. Government Printing Office, Washington, D.C., 1972, chap. 8.

10 Chapanis, A., and L. Lindenbaum: A reaction time study of four control-display linkages, *Human Factors*, November, 1959, vol. 1, no. 4, pp. 1–7.

11 Chapanis, A., and D. A. Mankin: Test of ten control-display linkages, *Human Factors*, 1967, vol. 9, no. 2, pp. 119–126.

12 De Greene, K. B.: *Systems psychology*, McGraw-Hill Book Company, New York, 1971.

13 Deininger, R. L.: Human factors engineering studies of the design and use of pushbutton telephone sets, *Bell System Technical Journal*, July, 1960, vol. 39, no. 4, pp. 995–1012.

14 Dey, D.: The influence of a prediction display on a quasi-linear describing function and remnant measured with an adaptive analog-pilot in a closed loop, in *Proceedings of Seventh Annual Conference on Manual Control*, NASA SP-281, National Aeronautics and Space Administration, Washington, D.C., 1972.

15 Fitts, P. M., and C. M. Seeger: *S-R* compatibility: spatial characteristics of stimulus and response codes, *Journal of Experimental Psychology*, 1953, vol. 46, pp. 199–210.

16 Garvey, W. D.: *The effects of "task-induced stress" on man-machine system performance*, USN, NRL, Report 5015, Sept. 9, 1957.

17 Holding, D. H.: Direction of motion relationships between controls and displays in different planes, *Journal of Applied Psychology*, 1957, vol. 41, pp. 93–97.

18 Howell, W. C., and G. E. Briggs: The effects of visual noise and locus of perturbation on tracking performance, *Journal of Experimental Psychology*, 1959, vol. 58, pp. 166–173.

19 Hunt, D. P.: The effect of the precision of informational feedback on human tracking performance, *Human Factors*, 1961, vol. 3, pp. 77–85.

20 Jenkins, W. O.: The discrimination and reproduction of motor adjustments with various types of aircraft controls, *American Journal of Psychology*, 1947, vol. 60, pp. 397–406.

21 Kelley, C. R.: *A predictor instrument for manual control*, paper read before the Eighth Annual Office of Naval Research Human Engineering Conference, September, 1958, Ann Arbor, Mich., revised January, 1962.

22 Kelley, C. R.: Predictor instruments look to the future, *Control Engineering*, March, 1962, pp. 86f.

23 Kelley, C. R.: *Manual and automatic control*, John Wiley & Sons. Inc., New York, 1968.

24 Levine, M.: *Tracking performance as a function of exponential delay between control and display*, USAF, WADC, TR 53–236, October, 1953.

25 Obermayer, R. W., W. F. Swartz, and F. A. Muckler: The interaction of information displays with control system dynamics in continuous tracking, *Journal of Applied Psychology*, 1961, vol. 45, pp. 369–375.

26 Poulton, E. C.: Perceptual anticipation in tracking with two-pointer and one-pointer displays, *British Journal of Psychology*, 1952, vol. 43, pp. 222–229.

27 Rockway, M. R.: *The effect of variations in control-display ratio and exponential time delay on tracking performance*, USAF, WADC, TR 54–618, December, 1954.

28 Rockway, M. R.: *Effects of variations in control deadspace and gain on tracking performance*, USAF, WADC, TR 57–326, September, 1957.

29 Rockway, M. R., and P. E. Franks: *Effects of variations in control backlash and gain on tracking performance*, USAF, WADC, TR 58–553, January, 1959.

30 Spragg, S. D. S., A. Finck, and S. Smith: Performance on a two-dimensional following tracking task with miniature stick control, as a function of control-display movement relationship, *Journal of Psychology*, 1959, vol. 48, pp. 247–254.

31 Wallach, H. C.: *Performance of a pursuit tracking task with different time delays inserted between the control mechanism and the display cursor*, USA Ordnance Human Engineering Laboratories. Technical Memorandum 12–61, OMS Code 50 10.11.841 A, August, 1961.

32 Warrick, M. J.: "Direction of movement in the use of control knobs to position visual indicators," in P. M. Fitts (ed.), *Psychological research on equipment design*, Army Air Force, Aviation Psychology Program, Research Report 19, 1947.

33 Weiss, B.: *Building "feel" into controls: the effect on motor performance of different kinds and amounts of feedback*, USN/ONR, SDC, TR 241–6–11, Oct. 14, 1953.

34 Wolf, Alice K., and B. F. Green, Jr.: *Tracking studies: II. Human performance on one-target displays*, M I T, Lincoln Laboratories, Lexington, Mass., Group Report 38–31, Nov. 20, 1957.

35 Ziegler, P. N., and R. Chernikoff: A comparison of three types of manual controls on a third-order tracking task, *Ergonomics*, 1968, vol. 11, no. 4, pp. 369–374.

Chapter 9

Controls, Tools, and Related Devices

The human being has demonstrated amazing ingenuity in designing devices for accomplishing things with less wear and tear upon himself. For example, various types of hand tools have been traced back to prehistoric man. Hand tools (from then and now) are essentially extensions of the upper extremities. Some of the machines of more recent times perform more efficiently the functions that previously were performed by hand (or by hand with the use of hand tools), whereas others accomplish things that previously could not be accomplished. In either event, most machines require control with some type of control device such as wheels, push buttons, or levers. Such devices, however, are not "extensions" of the extremities (as is the case with hand tools), but rather they require various types of psychomotor action on the part of the operators. In the case of either tools or control devices, the design should be such as to fulfill two related objectives. In the first place the tool or device needs to be capable of performing its function effectively when used by the intended user(s). In the second place it needs to be suitable for human use in terms of the sensory, psychomotor, and other abilities and the anthropometric characteristics of the user(s). Usually these two objectives are compatible with the same design, since the device that can best be used by people usually is the one that best fulfills its function.

In considering particularly the psychomotor abilities and the anthropometric characteristics of people as they might be relevant to the design of tools and controls, however, the designer may, in different circumstances, consider these factors from two different points of view. In some circumstances one may be primarily interested in the range of individual differences in some particular characteristic, as in designing for, say, the extreme individuals (such as the smallest, the largest, the weakest, or the slowest). In other circumstances, however, the designer may be primarily interested in mean differences, such as in the mean performance of groups of people in using one device versus another (such as a short versus a long lever), or a device in one location versus that in another location. Comparison of such mean differences would aid in the selection of the feature in question (i.e., type of device, location, etc.) that generally seems best in terms of human considerations.

The following discussion will touch on various types of controls and their uses, certain principles and concepts involved in their use, and a few illustrations of research investigations relating to the use of controls and tools. It will not be the intent to recite a comprehensive set of recommendations regarding specific design features of such devices, although a partial set of such recommendations is given in Appendix B.

FUNCTIONS OF CONTROLS

Controls are devices which transmit control information to some mechanism or system. The type of information so transmitted can be characterized in terms of classes of information related to displays, as discussed in Chapter 3. The types of information related to various control functions are given below:

Type of control function	Type of related information
Activation (usually on-off)	Status (dichotomous)
Discrete setting (at any separate, discrete) position	Status (discrete indications) Quantitative Warning and signal
Quantitative setting (setting of control at any position along quantitative continuum)	Quantitative
Continuous control	Quantitative Qualitative Representational
Data entry (as in typewriters, computers, pianos)	Alphanumeric Symbolic

The information associated with any given control function may be presented in a display, or it may be manifested in the nature of the system response. It might be noted in passing that system response and control relationships have a parallel to the chicken-and-egg conundrum, since in some circumstances the control action is made *following* indications of the system response (as in adjusting the steering wheel if a car is getting too close to the edge of the road), whereas in other circumstances the control action is made to *bring about* some predetermined desirable system response (as in setting the thermometer control on an oven). It might be added that the what-follows-what conundrum relationship sometimes becomes strained because some controls double in brass by also serving as displays to indicate the system response, as in the case of selector switches and volume controls.

Types of Controls as Related to Functions

Certain control functions can be executed more effectively by some type(s) of controls than by others. Table 9-1 lists some of the more common types of con-

Table 9-1 Common Types of Controls and the Control Functions They Can Serve

Type of control	Activation	Discrete setting	Quanti- tative setting	Con- tinuous control	Data entry
Hand push button	X				
Foot push button	X				
Toggle switch	X	X			
Rotary selector switch		X			
Knob		X	X	X	
Thumbwheel		X	X	X	
Crank			X	X	
Handwheel			X	X	
Lever			X	X	
Pedal			X	X	
Keyboard					X

trols, showing the control functions for which each type tends to be suitable. Certain of these types of controls are illustrated in Figure 9-1, along with the control functions for which they can be used. Although a general type might be considered most appropriate for a particular function, the specific utility of a particular variant of that type for a specific application can be influenced by such features (if relevant) as ease of identification, location, size, control-display ratio, resistance, lag, backlash, rate of operation, and distance of movement. Certain of these features were touched on in the preceding chapter. Some of the others will be discussed below.

IDENTIFICATION OF CONTROLS

Although the correct identification of controls is not really critical in some circumstances (as in operating a pinball machine), there are some operating circumtances in which their correct and rapid identification is of major consequence—even of life and death. For example, McFarland [24, pp. 605–608] cites cases and statistics relating to aircraft accidents that have been attributed

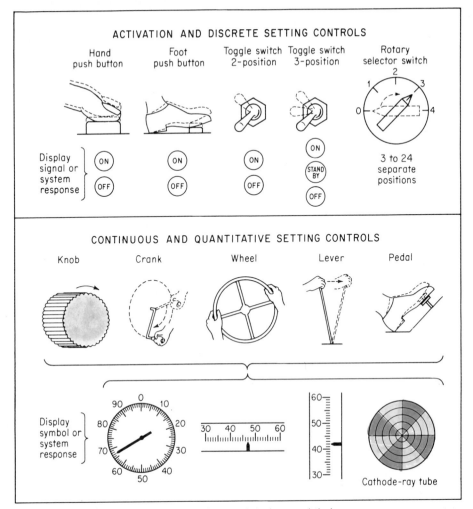

Figure 9-1 Examples of some types of control devices and their uses.

to errors in identifying control devices. For example, confusion between landing gear and flap controls was reported to be the cause of over 400 Air Force accidents in a 22-month period during World War II. It is with these types of circumstances in mind that consideration of control identification becomes important.

The identification of controls is essentially a coding problem, the primary coding methods including shape, texture, size, location, operational method, color, and labels. (Since shape, texture, size, color, and alphanumeric codes were discussed in Chapters 3 and 4, they will not be covered in this chapter.) The utility of these methods typically is evaluated by such criteria as the number of discriminable differences that people can make (such as the number of shapes they can identify), bits of information, accuracy of use, and speed of use.

Location Coding of Controls

Whenever we shift our foot from the accelerator to the brake, feel for the light switch at night, or grasp for a machine control that we cannot see, we are responding to *location coding*. But if there are several similar controls from which to choose, the selection of the correct one may be difficult unless they are far enough apart so that our kinesthetic sense makes it possible for us to discriminate. Some indications about this come from a study by Fitts and Crannell as reported by Hunt [14]. In this study blindfolded subjects were asked to reach for designated toggle switches on vertical and horizontal panels, the switches being separated by 1 in (2.5 cm). The major results are summarized in Figure 9-2, which shows the percentage of reaches that were in error by

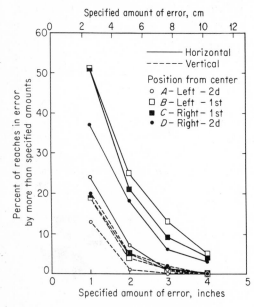

Figure 9-2 Accuracy of blind reaching to toggle switches (nine in a row on switch box) with switch box positioned horizontally and vertically at four locations. [Adapted from Fitts and Crannell, as presented by Hunt, 14.]

specified amounts when the panels were in horizontal and vertical positions, left and right from center. The curves indicate quite clearly that accuracy was greatest when the toggle switches were arranged vertically. For the vertically arranged locations probably a 5-in (13-cm) difference would be desirable for reasonably accurate positioning. Since errors drop quite low at about $2^1/_2$ in (6 cm), and since these errors are in both directions from the control, the central range of errors is double this distance, or 5 in (13 cm). For horizontally arranged controls there should be 8 in or more between them if they are to be recognized by location.

Operational Method of Coding Controls

In the operational method of coding controls, each control has its own unique method for its operation. For example, one control might be of a push-pull variety, and another of a rotary variety. Each can be activated *only* by the movement that is unique to it. It is quite apparent that this scheme would be inappropriate if there were any premium on time in operating a control device and where operating errors are of considerable importance. When such a method is used, it is desirable that compatibility relationships be utilized, if feasible. By and large, this method of coding should be avoided except in those individual circumstances in which it seems to be uniquely appropriate.

Discussion of Coding Methods

In the use of codes for identification of controls, two or more code systems can be used in combination. Actually, combinations can be used in two ways. In the first place, *unique combinations* of two or more codes can be used to identify separate control devices, such as the various combinations of texture, diameter, and thickness mentioned before [Bradley, 4]. And, in the second place, there can be completely *redundant codes*, such as identifying each control by a distinct shape *and* by a distinct color. Such a scheme probably would be particularly useful when accurate identification is especially critical. In discussing codes, we should be remiss if we failed to make a plug for standardization in the case of corresponding controls that are used in various models of the same type of equipment, such as automobiles and tractors. When individuals are likely to transfer from one situation to another of the same general type, the same system of coding should be used if at all possible. Otherwise, it is probable that marked "habit interference" will result [Weitz, 33] and that people will revert to their previously learned modes of response. In connection with the use of individual control coding methods, a few general principles can be set forth, as evolving from both research and experience. Some of these are given below, with the usual words of caution about the usual exceptions to general principles:

I Shape and texture
    A Desirable features. (1) Useful where illumination is low or where device may be identified and operated by feel, without use of vision; (2) can supplement visual identification; (3) useful in standardizing controls for identification purposes.
    B Undesirable features. (1) Limitation in number of controls that can be identified (fewer for texture than for shape); (2) use of gloves reduces human discrimination.
II Location
    A Desirable features. (1) Same advantages as for shape and texture.
    B Undesirable features. (1) Limitation in number of controls that can be identified; (2) may increase space requirements; (3) identification may not be as certain (may be desirable to combine with other coding scheme).
III Color
    A Desirable features. (1) Useful for visual identification; (2) useful for standardizing controls for identification purposes; (3) moderate number of coding categories possible.
    B Undesirable features. (1) Must be viewed directly (but can be combined with some other coding method, such as shape); (2) cannot be

used under poor illumination; (3) requires people who have adequate color vision.

IV Labels

    A Desirable features. (1) Large number can be identified; (2) does not require much learning.

    B Undesirable features. (1) Must be viewed directly; (2) cannot be used under poor illumination; (3) may take additional space.

V Operational method

    A Desirable features. (1) Usually cannot be used incorrectly (control usually is operable in only one way); (2) can capitalize on compatible relationships (but not necessarily).

    B Undesirable features. (1) Must be tried before knowing if correct control has been selected; (2) specific design might have to incorporate incompatible relationships.

CONTROL-DISPLAY RATIO

In continuous control tasks or when a quantitative setting is to be made with a control device, the ratio of the movement of the control device to the movement of the display indicator (i.e., the controlled element) is called the *control-display ratio* (C/D ratio). The movement may be measured in *distance* (in the case of levers, linear displays, etc.) or *angle* or *number of revolutions* (in the case of knobs, wheels, circular displays, etc.). When there *is* no display, the display movement is some measure of system response (such as angle of turn of an automobile). A very "sensitive" control is one which brings about a marked change in the controlled element (display) with a slight control movement; its C/D ratio would be low (a small control movement is associated with a large display movement). Examples of low and high C/D ratios are shown in Figure 9-3.

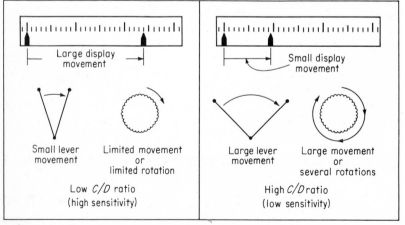

Figure 9-3 Generalized illustrations of low and of high control-display ratios (C/D ratios) for lever and rotary controls. The C/D ratio is a function of the linkage between the control and display.

C/D Ratios and Control Operation

The performance of human beings in the use of continuous or quantitative-setting control devices which have associated display movements is distinctly affected by the C/D ratio. This effect is not simple, but rather is a function of the nature of human motor activities when using such controls. In a sense, there are two types of human motions in such tasks. In the first place, there is essentially a gross adjustment movement (travel time or a slewing movement) in which the operator brings the controlled element (say, the display indicator) to the approximate desired position. This gross movement is followed by a fine adjustment movement, in which the operator makes an adjustment to bring the controlled element right to the desired location. (Actually, these two movements may not be individually identifiable, but there is typically some change in motor behavior as the desired position is approached.)

Optimum C/D Ratios

The determination of an optimum C/D ratio for any continuous control or quantitative-setting control task needs to take into account these two components of human motions. The general nature of the relationships is reflected by the results of studies by Jenkins and Connor [15] and is illustrated in Figure 9-4. That figure illustrates the essential features of the relationships, namely,

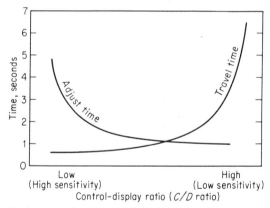

Figure 9-4   Relationship between C/D ratio and movement time (travel time and adjust time). While the data are from a study by Jenkins and Connor [15], the specific C/D ratios are not meaningful out of that context, so are omitted here. These data, however, depict very typically the nature of the relationships.

that travel time drops off sharply with decreasing C/D ratios and then tends to level off, and that adjustment time has the reverse pattern. The optimum, then, is somewhere around the point of intersection. Within this general range the combination of travel and adjust time usually would be minimized. The numerical values of the C/D ratios are of course a function of the physical nature and sizes of the controls and displays; the C/D ratio that would be applicable in controlling, say, a crane would be very different from that for a radar set.

There are no formulas for determining what C/D ratio would be optimum for given circumstances. Rather, this ratio should be determined experimentally for the control and display being contemplated.

DESIGN ASPECTS OF SPECIFIC CONTROLS

The aspects of control devices discussed above predominantly have some general implications in control design (considerations of identification, resistance, lag, etc.), rather than a relation to specific types of controls (although some of the aspects dealt with are more relevant to tracking controls than to other types). It is neither relevant nor feasible to discuss or illustrate the many control mechanisms that ingenious people have concocted. It may be useful, however, to illustrate with examples the principle that the specific design features, sizes, and locations of certain types of such gadgets have a direct bearing on how adequately people can use them for their intended purpose.

Selector Switches

As a case in point, in one investigation a comparison was made of the use of four types of rotary selector switches for making settings of three-digit numbers, and, separately, for reading three-digit numbers already set into the four switches. One type of switch (1) was a fixed-scale model with a moving pointer, as shown in Figure 9-5; the others were moving-scale models with

Figure 9-5 Illustration of fixed-scale, moving-pointer selector switches used in study by Kolesnik [22]. See text for discussion. [Reproduced by permission of Autonetics Division, North American Rockwell Corporation.]

fixed pointers with 10, 3, or 1 of the 10-scale positions visible. Some of the results of the study are given below [Kolesnik, 22].

Type of switch	Average setting time, s	Preference, rank order	Reading errors
Fixed scale	4.5	1	287
Moving scale, 10 digits shown	5.4	3	135
Moving scale, 3 digits shown	5.8	2	101
Moving scale, 1 digit shown	6.3	4	92

In the task of making *settings* of specified three-digit numbers, there was little difference in accuracy among the four styles, but the fixed scale resulted in lowest setting times and was first in preferences of the users. However, in the task of *reading* three-digit values at which the switches were set, it had the most errors. These differences, of course, need to be considered in the light of the *use* to which selector switches are to be put.

Concentrically Mounted Knobs

Space restrictions and other considerations sometimes argue for the use of concentrically mounted (or "ganged") knobs, such as those illustrated in Figure 9-6. Granting situational advantages for such controls, there are also some pos-

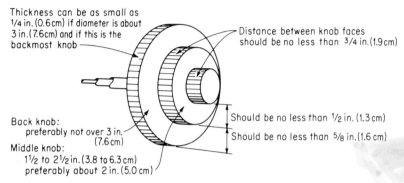

Thickness can be as small as 1/4 in. (0.6 cm) if diameter is about 3 in. (7.6 cm) and if this is the backmost knob

Distance between knob faces should be no less than 3/4 in. (1.9 cm)

Back knob:
  preferably not over 3 in. (7.6 cm)
Middle knob:
  1 1/2 to 2 1/2 in. (3.8 to 6.3 cm)
  preferably about 2 in. (5.0 cm)

Should be no less than 1/2 in. (1.3 cm)
Should be no less than 5/8 in. (1.6 cm)

Figure 9-6 Dimensions of concentrically mounted knobs that are desirable in order to allow human beings to differentiate knobs by touch. [Adapted from Bradley, 5.]

sible disadvantages associated with them, especially the possibility of inadvertent operation of adjacent knobs. If the knobs are too thin, the fingers may scrape against and operate the knob behind, and if the diameter distance is too small, the fingers may overlap and operate the knob in front. The problem, then, is one of identifying the optimum dimensions of such knobs, to minimize such possibilities. In a study dealing with this, Bradley [5] used various combinations of such knobs, and various performance criteria. He found that the dimensions shown in Figure 9-6 were optimum.

Cranks and Handwheels for Moving Objects

Cranks and handwheels frequently are used as a means of applying force to perform various types of functions, such as moving a carriage or a cutting tool or lifting objects. Let us see to what extent performance with such control devices is affected by their size, friction, direction of rotation, location, and use of preferred versus nonpreferred hand.

Size of Cranks and Handwheels In a study by Davis [11] several different sizes of cranks and handwheels were used to control the position of a pointer on a dial. In general, about one revolution of the crank was required in making the settings. The cranks and handwheels were mounted so that the plane of rotation was parallel to the frontal plane of the body. By using torques of 0, 20, 40, 60, and 90 in lb, it was possible to determine the time required to make settings under different friction-torque conditions that typically occur in the operation of such controls.

Figure 9-7 shows average times for certain torque levels by size of crank or handwheel. A couple of points are illustrated by this figure. In the first place, times for making settings under 0 torque were shortest for small sizes of cranks

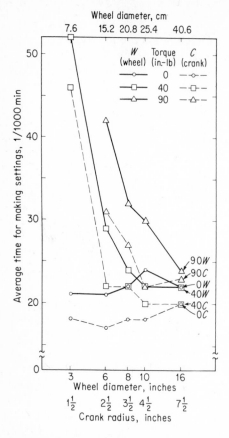

Figure 9-7 Average times for making settings with vertically mounted handwheels and cranks of various sizes under different torque conditions. [Adapted from Davis, 11.]

and wheels, whereas for torques of 40 and 90 in lb, the small cranks were definitely inferior to the larger ones. In the second place, the average times for cranks were, in general, lower than those for handwheels. In some cases, however, the time differences were not great, especially with the larger sizes.

Work Output with Different Cranks   The work output of subjects using three sizes of cranks and five resistant torque loads was investigated by Katchmer [19], with a view to identifying fairly optimum combinations. The cranks were adjusted to waist height of each subject, and the subjects were instructed to turn the crank at a rapid rate until they felt they could no longer continue the chore (or until 10 min had passed). A summary of certain data from the study is given in Figure 9-8. In total foot-pounds of work (shown by the bars), it is apparent that the 30 and 50 in lb of torque load resulted in higher values than the lower or higher loads did. The 5-in (13-cm) and especially the 7-in (18-cm) radius cranks were generally superior on this score to the 4-in (10-cm). In average time that subjects could "take it," the lower loads of course were highest, but it is apparent that they could operate the 7-in (18-cm) cranks longer under moderate loads that the shorter cranks. While average horsepower (hp) per minute was highest for the 90 in lb torque (average for the period actually worked), the durability of the subjects for this heavy load was short (about

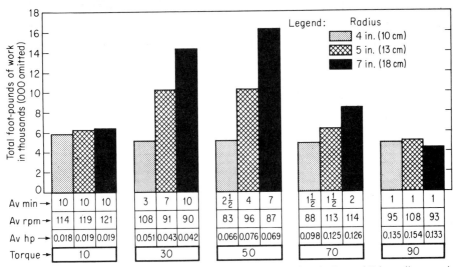

Figure 9-8 Average performance measures for subjects using 4-, 5-, and 7-in-radius cranks under 10, 30, 50, 70, and 90 in-lb torque resistance. [Adapted from Katchmer, 19.]

1 min). These data illustrate emphatically the point that *different* design features might be preferable if different criteria are used (in this case the different criteria being time tolerance, i.e., average minutes; speed, i.e., rpm; and average horsepower per minute when working).

Rotary Controls in Tracking Tasks

There are various human factors aspects of rotary controls (such as cranks) as used in tracking tasks. As examples, Swartz, Norris, and Spragg [31] investigated the relationship between size of cranks and their location in a two-handed tracking task, as related to the performance by their subjects. Some of the results are shown in Figure 9-9, that figure showing the time on target (TOT)

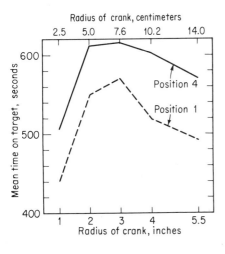

Figure 9-9 Accuracy of pursuit tracking with various sizes of cranks. The two curves show the relationship between radius of crank and time on target (TOT) for two combinations of crank positions (1, both cranks vertical and facing body; 4, left crank same as 1 but right crank vertical and at right angles to body). [Adapted from Swartz, Norris, and Spragg, 31, fig. 1.]

for various sizes of cranks and for two combinations of positions of the two cranks. The results indicate that TOT was greatest for the 2- and 3-in cranks (5.0 and 7.6 cm) with 4-in (10.2-cm) cranks being next best, and that the position of the cranks also influenced tracking performance.

Stick-Type Controls

As indicated earlier, there is evidence that the ratio of lever movement to display-indicator movement apparently is a fairly critical feature of stick control devices. In one study it was found that length of joy stick (12, 18, 24, or 30 in or 30.5, 45.7, 61.0, or 76.2 cm) was relatively unimportant in both speed and accuracy of making settings, as long as the C/D ratio was around 2.5 or 3.0 [Jenkins and Karr, 16].

In a tracking task (as opposed to a task of making a particular setting), however, it was found that joy-stick length had at least a moderate effect upon tracking performance [Hartman, 13]. Within the range of stick lengths from 6 to 27 in (15 to 69 cm), those around 18 in (46 cm) were relatively optimum, although the advantage of that length over others was not great—about 10 percent. It was proposed that, where it would not interfere with other more critical design requirements, the length of such sticks be about 18 in (46 cm). It was further proposed that especially long sticks (say, 27 in or 69 cm) be avoided, since they tend to interfere with comfortable positioning of the operator.

Control Devices for Cranes   An interesting series of applied studies dealing with human factors has been carried out by the Ergonomics Research Section of the British Iron and Steel Institute. One of these studies was concerned, in part, with the design of *master controllers* for cranes used in steel mills [Box and Sell, 3; Sell, 30]. In the pertinent part of the study, lever-type hand controls with the following features were used experimentally over three weeks by three subjects:

• Speed steps: the number of different speeds for which specific settings could be made by the lever, specifically 3, 4, and 6 speeds on either side of the off position
  • Arc of movement of the handle: 20, 35, and 50° in each direction
  • Handle length: 9, 12, and 15 in (23, 30.5, and 38 cm)

Granting that the investigation was highly situational and used only three subjects, the following conclusions may still be of interest:

1.   Performance was significantly better when using four speed steps than when using either three or six.
2.   With either three or four speed steps, handle lengths of 9 to 12 in (23 to 30.5 cm) were better than the larger one (15 in); there was no difference with the 15-in (38-cm) handle for six speed steps.
3.   With six speed steps an arc of 20° movement in each direction was significantly worse than arcs of 35 or 50°.

Foot Controls

When foot control devices are used, they typically are used to control a single function or perhaps two or three. Commenting on the use of the foot, Kroemer [23] makes the point that the general tenor of human factors handbooks implies that the feet are slower and less accurate than the hands, but hastens to add that this belief is neither supported nor discredited by experimental results.

Pedal Design Considerations   Pedals used for various and sundry purposes vary in terms of several features, such as: whether they require a thrust with or without ankle action; the location of the fulcrum (if the pedal is hinged); the angle of the foot to the tibia bone of the leg; the load (the force required); and the operational requirements (such as reaction time, travel time, speed of operation, and precision). For illustrative purposes we will present certain of the results of an experiment by Ayoub and Trombley [1] in which one of the factors which they varied was the location of the fulcrum of the pedal. As one variation in the experimental procedure they had the subjects activate the pedal through a prescribed arc of 12° (constant angle), and separately for a distance of 0.75 in (1.9 cm) (constant distance) measured at the ball of the foot. (Although the *distance* of movement of the ball of the foot would be constant with different fulcrum locations, the angle of movement would increase as the fulcrum is moved forward toward the ball of the foot.)

The mean travel times for various fulcrum locations are shown in Figure 9-10. It can be seen that for a constant angle of movement (of 12°), the op-

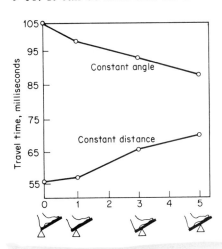

Figure 9-10 Mean travel time in pedal movement as related to location of the fulcrum for conditions of constant angle of movement (12°) and constant distance of movement (0.75 in or 1.9 cm). [Adapted from Ayoub and Trombley, 1, fig. 13. Copyright 1967 by American Institute of Industrial Engineers, Inc., 25 Technology Park, Atlanta, Norcross, Georgia 30071 and reproduced with permission.]

timum location of the fulcrum is forward of the ankle (about a third of the distance between the ankle and the ball of the foot), whereas for a constant distance of movement the optimum location of the fulcrum is at the heel. On the basis of other results not reported here, they found that reaction time (the time to initiate a movement) was independent of the location, but that both reaction time and travel time increased with increasing resistance of the pedal. However, as Kroemer points out [23], the somewhat inconsistent results of various experiments with pedals (in part based on differences in experimental conditions) preclude any general statements as to what pedal allows the fastest activation or highest frequency in repetitive operation under a variety of specified conditions.

Foot Controls for Discrete Control Action   Foot control mechanisms usually are used to control one or a couple of functions. There are some indications, however, that the feet can be used for more varied control functions. In one investigation, for example, Kroemer [23] developed an arrangement of 12 foot positions (targets) for subjects as illustrated in Figure 9-11. Using

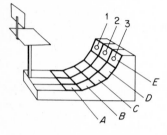

Figure 9-11   Arrangement of foot targets used in experiment by Kroemer in studying the speed and accuracy of discrete foot motions. [From Kroemer, 23, fig. 7.]

procedures that need not be described here, he was able to measure the speed and accuracy with which subjects could hit the various targets with their feet. By an large, he found that after a short learning period the subject could perform the task with considerable accuracy and with very short travel time (averaging about 0.1 s). Although forward movements were slightly faster than backward or lateral movements, these differences were not of any practical consequence. His results strongly suggest that it might be possible to assign control tasks to the feet which heretofore have been considered in the domain of the hands.

Automobile Brake and Accelerator Pedals   The most commonly used foot controls are the brake and accelerator controls of automobiles. Aside from their individual characteristics, an important factor in their use is that of their relative positions. In most automobiles the accelerator is lower than the brake pedal, thus requiring the lifting of the foot from the accelerator, its lateral movement, and then the depression of the brake. It has been clearly demonstrated by Davies and Watts [10] that such an arrangement (with the brake pedal 6 in or 15 cm higher than the accelerator) increases movement times as shown below:

	Movement times, s	
Location of brake	Male	Female
6 in (15 cm) above accelerator	313	309
Even with accelerator	155	194
Difference	158	115

Such differences in activation of the brake could reduce stopping distances by approximately 10 to 22 ft (3.0 to 6.7 m) when traveling at 60 to 100 mi/h (97 to 161 km/h).

Data Entry Processes

The insatiable appetite for information that is part and parcel of this world has brought about a proliferation of mechanisms whose inputs and outputs deal with information, such as typewriters, computers, calculating machines, bookkeeping machines, and telephones.

The data that serve as the input to such mechanisms usually are alphabetical, numerical, or of other symbolic form, and in most instances the data are entered by the use of data entry devices such as keyboards. Some exploratory work is being carried out, however, with the use of voice and eye movements as the basis for data entry.

Nature of Data Input As would be expected on the basis of common sense, data entry speed and accuracy are generally greater when the data presented to the operator are clear and legible and are presented in an appropriate format. There are, however, certain other aspects of the form of the input data that can influence speed and accuracy of the operator, as discussed by Seibel [29], as follows:

- *Readability of source documents.* In the case of written material, reading rates are higher with conventional upper- and lower-case characters than with all-capital characters.
- *Redundancy.* "Redundancy" refers to various forms of repetition or the "expectancy" of items. In the present context, it can be said that words in a meaningful sequence are more redundant than a sequence of random words. In general, input rates are higher with redundant material than with nonredundant material.
- *Chunking of material.* There is some indication that, for short-term memory, people tend to memorize material in "chunks," especially in the case of long messages. Thus, in some circumstances it is useful to present the material in subunits or "chunks." In presenting strings of digits, such as telephone numbers, for example, presenting groupings of three or four digits is optimum in terms of short-term memory.
- *Exposure of characters.* In the case of redundant material (such as sentences), accuracy tends to be greater if the operator can "see ahead" what is to be entered (rather than being able to see the characters one at a time). In the case of nonredundant (i.e., random) characters, there is apparently no advantage in being able to see more than three characters ahead.

Data Entry Devices There are of course many types of data entry devices, and variations of each type. Most such devices are *sequential* in the sense that individual characters are to be entered in a specific sequence (such as a girl friend's telephone number), there being a specific "key" or other device for every character. On the other hand, *chord* keyboards are those in which a single stimulus symbol (as a letter or number) requires the simultaneous activation of two or more keys.

A comparison of the effectiveness with which four different types of sequential data entry devices are used is given below [Miner and Revesman, 27], the task in question being the entry of numerical data:

Type of device	Average time to enter 10 digits, s	Percent of error
10-key keyboard, as on hand calculators	12	0.6
10 by 10 matrix keyboard	13	1.2
10-lever device, one for each digit	17	2.3
10 rotary knobs, one for each digit	18	2.3

It is clear from this comparison that certain types of devices lend themselves to more effective use (in terms of speed and accuracy) than others, in this particular instance the 10-key keyboard (such as that used on hand calculators) being the best.

Chord keyboards are not nearly as common as sequential keyboards, but are used with Stenotype machines, pianos, and mail-sorting machines. Figure 9-12 shows the Burroughs chord key-board that is used in a few post offices for

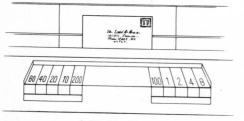 

(a) Design of keys

(b) Keyboard in use

Figure 9-12 Burroughs chord keyboard used in letter sorting in some post offices. Individual numbers can be reproduced by simultaneous activation of certain combinations of keys, following the binary system. For example, 3 is activated by pressing 1 and 2, and 5 by pressing 1 and 4. [Courtesy of the Burroughs Corporation.]

semiautomatic mail sorting. The theoretical ceilings of both types of keyboards can be expressed in terms of information bits per stroke. As pointed out by Seibel [28], for any given number of keys, there is an approximate ratio of about 2:1, in terms of such information, in favor of the chord keyboard. The actual limits, however, are influenced by the coding system and are constrained by human performance capabilities. For example, the information rates of stenotyping versus conventional typing are in the ratio of 2:1, and the stroke rates per second for good operators are in the ratio of 3:8.3 per second, which gives an advantage to stenotyping of 5.6:1 (in terms of information input per stroke).

An actual comparison of the two types of keyboards was made by Bowen and Guinness [2] in the context of a simulated semiautomatic mail-sorting task. In this study the *encoding* of mail by memory into various classes (each with a three-digit numerical code) was done by some subjects with a regular typewriter and by other subjects with a small (12-key) chord keyboard (requiring the use of one to four keys simultaneously) and a large (24-key) chord keyboard (requiring use of one to three keys). The digitation patterns used represented the fingering difficulty and frequency that would be used for a 500-separation mail-sorting scheme. Some of the results are summarized below.

	No. sorted per minute	
Type of keyboard	Correct	Incorrect
Sequential, typewriter	40.4	4.2
Chord, small	55.3	7.8
Chord, large	49.0	3.3

Both chord-keyboard tests resulted in higher numbers of items sorted correctly than resulted from the typewriter tests. The superiority of the large over the small chord keyboard in reduced errors probably can be attributed to the fact that it did not require such difficult finger patterns. Some generally confirming evidence comes from a study by Conrad and Longman [7], who report that for subjects working over a period of weeks, those using a chord keyboard had higher keystroke-per-minute rates than those using a typewriter. Bowen and Guinness [2] suggest that the apparent advantage of chord over sequential encoding in the context of their study may be attributable to the fact that when using a learned memory code, one pairs it immediately with a *unit* response—not a response *spread out over time*, as called for in sequential keying.

Although chord keyboards seem to offer advantages over sequential keyboards, however, chord keyboards generally are more difficult to learn—in particular, it is more difficult to learn their coding system (i.e., the combinations of keys that represent individual items).

Arrangement of Keyboards Various alternative arrangements of the conventional typewriter keyboard have been proposed, one such having been proposed years ago by Dvorak [12]. This and other arrangements have been suggested as being more efficient in the combined use of the fingers, considering the frequency of use of various letters individually and in combination. One feature of the Dvorak keyboard provided that all of the vowels be on the middle row at the left, with the most common consonants at the right, as follows:

A O E U I   D H T N S

This arrangement would tend to capitalize on the common alternative sequences of vowels and consonants in the English language, thus providing for common alternative use of the fingers of the two hands.

The question relating to the advantage of the conventional arrangement (the QWERTY keyboard, so called because of the sequence of letters in the top line) versus other arrangements has never been firmly resolved, as reflected by the disparate observations of Klemmer [20] and Martin [25]. In any event, Seibel [29] offers, as a *guess*, an estimate that the *upper limit* of the daily production advantage of a modified keyboard would be about 10 percent.

The two most common numerical data entry keyboards consist of three rows of three digits with the zero below, although there are two different arrangements of the numerals. One arrangement is used on many adding machines and the other is used on push-button telephones. These two arrangements are shown below:

Adding machine	Telephone
9 8 7	7 8 9
6 5 4	4 5 6
3 2 1	1 2 3
0	0

Since the numerical arrangements of these two are reversed, one would be curious about the accuracy and speed with which the two might be used. Conrad and Hill [6] provide some data to satisfy this curiosity. The subjects used in their experiment were housewives, one group using the adding-machine arrangement, another group using the telephone arrangement. The telephone arrangement resulted in a slightly greater rate of data entry than the adding-machine arrangement (7.8 versus 7.4 eight-digit entries per minute). There was a more clear-cut difference, however, in terms of accuracy, as shown below:

Arrangement	Percent errors
Adding machine	8.2
Telephone	6.4

It is probable that the superiority of the telephone arrangement can be attributed to its greater compatibility with our expectations of digits increasing from left to right, and with our practice of reading from top to bottom.

Teleoperators

The technological explosion of recent decades has made it possible to explore outer space and the depths of the ocean, thus exposing man to new and hazardous environments. These and other developments require some type of "extension" of man's control functions across distance or through some physical barrier, as in the control of lunar vehicles or of underwater activities. The control systems for performing such functions have been called "teleoperators" [Johnson and Corliss, 17]. The technology for making such control possible also opens up other possible applications, such as the remote manipulation of radioactive nuclear fuel elements, the control of prosthetic devices such as artificial arms, and the lifting and handling of tremendous weights. The common denominator of teleoperators is that they represent some augmentation of man's innate physical skills via the mechanism that performs the activities in question—whatever they may be. One example is shown in Figure 9-13, this being

Figure 9-13 Experimental remote-handling equipment used in various studies. [Courtesy USAF, AFHRL.]

an experimental remote-handling setup used in a series of studies at the United States Air Force Human Resources Laboratory, Wright-Patterson Air Force Base. Another type of teleoperator is the model of HardiMan shown in Figure 9-14. This is an exoskeletal device which, with an accompanying power source, would make it possible for a person to "amplify" his own physical movements so he could lift a weight of up to 1500 lb (680 kg).

Feedback to Teleoperators The engineering design of various types of teleoperators must of course take completely into account the nature of human physical responses and of man's anthropometric characteristics in order that the device can simulate the human motions as they are executed by an individual. In this regard, however, particular consideration needs to be given to the nature of the feedback to the individual, since appropriate feedback would be necessary for effective operation. In the use of such devices the individual is bereft of his normal direct physical contact with that which is being controlled; thus he loses his usual kinesthetic feedback, and in some cases he may also lose direct visual feedback. In the control of a remote-handling unit (specifically, a moveable cart), however, Kama [18] reports that one does not really need direct viewing of the object being manipulated. Somewhat in line with this, Knowles [21] expresses the opinion that the normality or naturalness of the situations, movements, and feedback information is in all probability irrelevant. (After all, we mortals frequently perform activities in ways that are not natural,

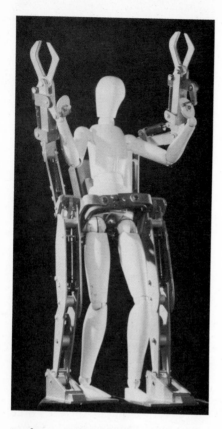

Figure 9-14 A model of HardiMan, a walking machine that mimics the control responses of a human operator and that is capable of amplifying those responses by a ratio of 25:1 up to a force of 1500 lb (680 kg). [Courtesy of the General Electric Company Research and Development Center, Schenectady, N.Y.]

such as unplugging the kitchen drain with a plumber's snake.) Knowles points out that what *is* needed for remote-handling devices is a situation in which *usable inputs* are combined with *usable control devices*—or what has been referred to as the "verified body image"; in simpler terms this means that the operator needs to have some meaningful *identification* or *integration* with his equipment.

Perhaps it is because of this need for integration with the equipment that operators of remote-handling controls tend to do better with those that operate on a pantograph principle (in which the operator's response is translated directly into corresponding three-dimensional movements) as contrasted with some *rectilinear* models (in which electric-powered manipulators must be controlled independently in each of the three dimensions). A manifestation of this principle comes from the study by Crawford [8] in which he found that in the remote handling of disks to be placed in holes in a form board (by using a variation of the device shown in Figure 9-13), subjects did better with a joy stick than with multiple-lever controls. These results are shown in Figure 9-15. Manipulations of the single joy stick actually controlled six different control movements, simulating shoulder pivot, shoulder rotation, elbow pivot, wrist pivot, wrist rotation, and grip. This single joy stick provides a higher degree of integration of the control with the operator than the independent use of six separate multiple

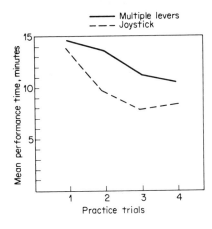

Figure 9-15 Comparison of remote-handling performance as a function of practice with a single joy stick (which governed six control functions) versus six multiple levers. Although practice resulted in improvement with both types of controls, performance with the joy stick was consistently superior. [From Crawford, 8.]

levers. (This probably is, in a sense, another manifestation of the principle of compatibility.)

TOOLS AND OTHER DEVICES

There are probably a couple of interrelated considerations (criteria, if you will) that are (or should be) dominant in the design or selection of the many hand tools and other hand devices that people use (ranging from eyebrow tweezers to carving knives, to billiard sticks, to hacksaws). Certainly the devices need to be capable of performing their function (such as pulling eyebrows), but in addition they need to be usable by people, and this means that they need to be compatible with the anthropometric and biomechanical performance characteristics of people. (And in some cases aesthetic considerations may also become relevant, for better or worse; if the purpose of a knife is to cut the steak, there is nothing like a plain butcher knife, but if the purpose is to impress one's guests, a solid-silver place setting probably is indicated.) To illustrate the relevance of anthropometric and biomechanical considerations (if not the aesthetic aspects), let us cite two or three different types of examples.

Pliers

The Western Electric Company has been a forerunner in the application of biomechanical principles to the design of tools and work places. As an example, in a certain wiring operation the operators used a conventional set of pliers that required the operator to work with the wrist bent in the manner shown in Figure 9-16*a*. As Tichauer [32] points out, improperly designed hand tools can induce biomechanical stress. For example, these pliers exert strong bending forces on the wrist. Further, the configuration of the hand favors ulnar drift of the extensor tendons, and it tends to induce compressive stress between these tendons and ancillary structures. In addition, the fact that the axis of rotation of the tool does not coincide with the normal axis of the forearm can also impose some stress on the shoulder and elbow. An anatomically correct design

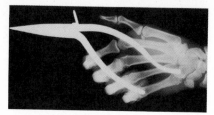

(*a*) Conventional design

(*b*) Redesigned pliers

Figure 9-16 X-rays of hand using conventional pliers in a wiring operation, *a*, and in using a redesigned model, *b*. The redesigned model is more anatomically correct. [See Damon, 8, and Tichauer, 32, for discussion; photographs courtesy of Western Electric Company, Kansas City.]

was developed which essentially eliminates the deficiencies in the original design. That model is shown in Figure 9-16*b*.

A Redesigned Surgical Instrument

Some problems in the use of the conventional design of bayonet forceps as used by neurosurgeons led to the development and evaluation of a new design [Miller, Ransohoff, and Tichauer, 26]. The original and new designs are shown in Figure 9-17. The problems with the original design included a tendency for it to

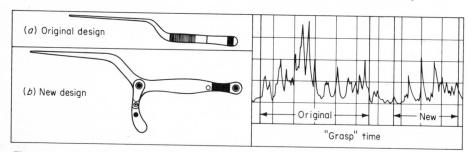

Figure 9-17 Examples of electromyograms of the finger flexor muscle during the "grasp" phase of using an original and an improved design of a bayonet forceps as used by neurosurgeons. [Adapted from Miller, Ransohoff, and Tichauer, 26, figs. 1 and 6.]

roll in the fingers while in use, and for fatigue to develop in the flexor muscles of the fingers and thumb. The new design was intended to overcome these disadvantages. The two designs were evaluated in use by four experienced surgeons, each instrument being tested during approximately 25 procedures. The comparisons included electromyogram recordings of the flexor muscle of the finger, one illustrative comparison being shown in Figure 9-17. Other comparisons are given below:

	Original forceps	New forceps
Stabilizing time, s	1.30	0.98
Myograph values:		
Flexor muscle of finger	0.84	0.52
Thumb opposer muscles	3.96	2.27

These comparisons indicated that the stabilizing time (the time to grasp the instrument securely) was reduced by about 25 percent, and the work loads on the fingers and thumbs were reduced by about 38 and 42 percent, respectively, thus indicating the general superiority of the improved design.

DISCUSSION

In bygone centuries man used various types of tools and hand devices to accomplish certain objectives, such as making things or performing certain tasks (even to club one's attackers). Various types of hand tools and other hand devices (albeit of improved design) still serve—and will continue to serve—many purposes. The industrial revolution brought about the use of machines for producing things we use and for performing many other functions, such as transportation. The advent of the machine typically required that people exercise control, usually by some type of hand or foot control mechanisms. (As we have seen in this chapter, the designs of such mechanisms can influence the effectiveness with which they are used.) Although we can expect that such physical control of machines will continue to be important, there are some interesting developments—so far largely at the experimental level—with other modes of human control. These include, among others, the use of head movements, eye movements, physiological signals, neural signals, and muscular signals as possible inputs to control mechanisms. The use of such methods for control of artificial limbs would be one of the most useful applications, but many other possibilities come to mind, such as in tracking tasks.

REFERENCES

1 Ayoub, M. M., and D. J. Trombley: Experimental determination of an optimal foot pedal design, *Journal of Industrial Engineering*, 1967, vol. 17, pp. 550–559.

2 Bowen, H. M., and G. V. Guinness: Preliminary experiments on keyboard design for semiautomatic mail sorting, *Journal of Applied Psychology*, 1965, vol. 49, no. 3, pp. 194–198.

3 Box, A., and R. G. Sell: Ergonomic investigations into the design of master controllers, *Journal of the Iron and Steel Institute*, October, 1958, vol. 90, pp. 178–187.

4 Bradley, J. V.: Tactual coding of cylindrical knobs, *Human Factors*, 1967, vol. 9, no. 5, pp. 483–496.

5 Bradley, J. V.: Desirable dimensions for concentric controls, *Human Factors*, 1969, vol. 11, no. 3, pp. 213–226.

6 Conrad, R., and A. J. Hull: The preferred layout for numerical data-entry keysets, *Ergonomics*, 1968, vol. 11, no. 2, pp. 165–173.

7 Conrad, R., and D. J. A. Longman: Standard typewriter versus chord keyboard —an experimental comparison, *Ergonomics*, 1965, vol. 8, no. 1, pp. 77–88.

8 Crawford, B. M.: Joy stick vs. multiple levers for remote manipulator control, *Human Factors*, 1964, vol. 6, no. 1, pp. 39–48.

9   Damon, F. A.: The use of biomechanics in manufacturing operations, *The Western Electric Engineer*, October, 1965, vol. 9, no. 4.

10 Davies, B. T., and J. M. Watts, Jr.: Further investigations of movement time between brake and accelerator pedals in automobiles, *Human Factors*, 1970, vol. 12, no. 6, pp. 559–561.

11 Davis, L. E.: Human factors in design of manual machine controls, *Mechanical Engineering*, October, 1949, vol. 71, pp. 811–816.

12 Dvorak, A.: There is a better typewriter keyboard, *National Business Education Quarterly*, 1943, vol. 12, pp. 51–58.

13 Hartman, B. O.: *The effect of joystick length on pursuit tracking*, U.S. Army Medical Research Laboratory, Fort Knox, Ky., Report 279, November, 1956.

14 Hunt, D. P.: *The coding of aircraft controls*, USAF, WADC, TR 53–221, August, 1953.

15 Jenkins, W. L., and M. B. Connor: Some design factors in making settings on a linear scale, *Journal of Applied Psychology*, 1949, vol. 33, pp. 395–409.

16 Jenkins, W. L., and A. C. Karr: The use of a joy-stick in making settings on a simulated scope face, *Journal of Applied Psychology*, 1954, vol. 38, pp. 457–461.

17 Johnson, E. G., and W. R. Corliss: *Teleoperators and human augmentation*, NASA SP-5047, December, 1967.

18 Kama, W. N.: *Effect of augmented television depth cues on the terminal phase of remote driving*, USAF, AMRL, TR 65–6, April, 1965.

19 Katchmer, L. T.: *Physical force problems: 1. Hand crank performance for various crank radii and torque load combinations*, USA, Human Engineering Laboratory, Aberdeen Proving Ground, Technical Memorandum 3–57, March, 1957.

20 Klemmer, E. T.: Keyboard entry, *Applied Ergonomics*, 1971, vol. 2, no. 1, pp. 2–6.

21 Knowles, W. B.: *Human engineering in remote handling*, USAF, MRL, TDR 62–58, August, 1962.

22 Kolesnik, P. E.: *A comparison of operability and readability of four types of rotary selector switches*, Autonetics Division, North American Aviation, Inc., T5–1187/3111, June, 1965.

23 Kroemer, K. H. E.: Foot operation of controls, *Ergonomics*, 1971, vol. 14, no. 3, pp. 333–361.

24 McFarland, R. A.: *Human factors in air transport design*, McGraw-Hill Book Company, New York, 1946.

25 Martin, A.: A new keyboard layout, *Applied Ergonomics*, 1972, vol. 3, no. 1, pp. 48–51.

26 Miller, M., J. Ransohoff, and E. R. Tichauer: Ergonomic evaluation of a redesigned surgical instrument, *Applied Ergonomics*, 1971, vol. 2, no. 4, pp. 194–197.

27 Minor, F. J., and S. L. Revesman: Evaluation of input devices for a data setting task, *Journal of Applied Psychology*, 1962, vol. 46, pp. 332–336.

28 Seibel, R.: Data entry through chord, parallel entry devices, *Human Factors*, 1964, vol. 6, no. 2, pp. 189–192.

29 Seibel, R.: "Data entry devices and procedures," in H. P. VanCott and R. G. Kinkade (eds.), *Human engineering guide to equipment design*, U.S. Government Printing Office, Washington, D.C., 1972, chap. 7.

30 Sell, R. G.: Letter to the editor: Ergonomic investigations into the design of master controllers, *Journal of the Iron and Steel Institute*, February, 1960.

31 Swartz, P., E. B. Norris, and S. D. S. Spragg: Performance on a following tracking

task as a function of radius of control cranks, *Journal of Psychology*, 1954, vol. 37, pp. 163–171.

32 Tichauer, E. R.: Some aspects of stress on forearm and hand in industry, *Journal of Occupational Medicine*, February, 1966, vol. 8, no. 2, pp. 63–71.

33 Weitz, J.: "The coding of airplane control knobs," in P. M. Fitts (ed.), *Psychological research on equipment design*, Army Air Force, Aviation Psychology Program, Research Report 19, 1947.

Part
Four

Work Space and Arrangement

Chapter 10

Applied Anthropometry and Work Space

Every day of our lives we use some physical facilities that have (or should have!) some relationship to our basic physical features and dimensions—facilities such as chairs, seats, tables, desks, workplaces, and clothing. As we know from universal experience, the comfort, physical welfare, and performance of people can be influenced for better or worse by the extent to which such facilities "fit" people.

ANTHROPOMETRY

Anthropometry and the closely related field of biomechanics deal with the measurement of the physical features and functions of the body, including linear dimensions, weight, volume, range of movements, and the like. Although we do not want to reproduce here the voluminous anthropometric data that people have accumulated over the years, we shall at least illustrate some such data:[1] In general terms, the measurements of body dimensions fall into two classes, namely, structural dimensions and functional dimensions.

[1]The interested reader is referred particularly to the excellent compilation of such data by Damon, Stoudt, and McFarland [5], Hansen and Cornog [1], Hertzberg, Daniels, and Churchill [15], and Public Health Service Publication No. 1000, series 11, no. 8 [25].

Structural Body Dimensions

Structural body dimensions are taken with the body of the subjects in fixed (static), standardized positions. Many different body features can be measured, of course. In one survey [Herzberg, et al., 15], for example, 132 different features were measured of 4000 Air Force flying personnel. Measurements of different body features could have some specific application, whether in designing chest protectors for baseball umpires, earphones, or pince-nez glasses. However, measurements of certain body features probably have rather general utility, and summary data on some of these features will be presented for illustrative purposes. These data come from a survey by the United States Public Health Service [25] of a representative sample of 6672 adult males and females. The specific body features measured are shown in Figure 10-1; for each of these (plus weight), data on the 5th, 50th, and 95th percentiles are given in Table 10-1. It should be pointed out that these values cover ages from 18 to 79 and

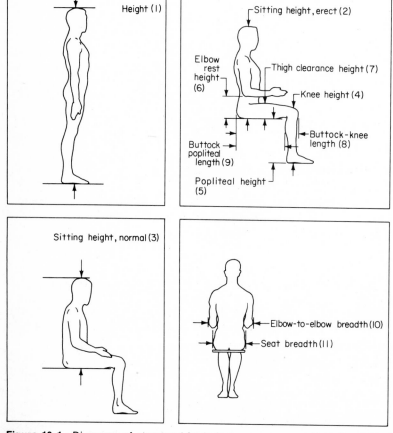

Figure 10-1 Diagrams of structural body features measured in National Health Survey of anthropometric measurements of 6672 adults [25]. See Table 10-1 for selected data based on the survey.

Table 10-1 Selected Structural Body Dimensions and Weights of Adults*

Body feature	Dimensions: in						Dimensions: cm‡					
	Male, percentile			Female, percentile			Male, percentile			Female, percentile		
	5th	50th	95th	5th	50th	95th	5th	50th	95th	5th	50th	95th
1 Height	63.6	68.3	72.8	59.0	62.9	67.1	162	173	185	150	160	170
2 Sitting height, erect	33.2	35.7	38.0	30.9	33.4	35.7	84	91	97	79	85	91
3 Sitting height, normal	31.6	34.1	36.6	29.6	32.3	34.7	80	87	93	75	82	88
4 Knee height	19.3	21.4	23.4	17.9	19.6	21.5	49	54	59	46	50	55
5 Popliteal height	15.5	17.3	19.3	14.0	15.7	17.5	39	44	49	36	40	45
6 Elbow-rest height	7.4	9.5	11.6	7.1	9.2	11.0	19	24	30	18	23	28
7 Thigh-clearance height	4.3	5.7	6.9	4.1	5.4	6.9	11	15	18	10	14	18
8 Buttock-knee length	21.3	23.3	25.2	20.4	22.4	24.6	54	59	64	52	57	63
9 Buttock-popliteal length	17.3	19.5	21.6	17.0	18.9	21.0	44	50	55	43	48	53
10 Elbow-to-elbow breadth	13.7	16.5	19.9	12.3	15.1	19.3	35	42	51	31	38	49
11 Seat breadth	12.2	14.0	15.9	12.3	14.3	17.1	31	36	40	31	36	43
12 Weight*	120	166	217	104	137	199	58	75	98	47	62	90

*Weight given in pounds (first six columns) and kilograms (last six columns).

‡Centimeter values rounded to whole numbers.

Source: From Weight, height, and selected body dimensions of adults: 1960–1962. Data from National Health Survey, USPHS Publication 1000, series 11, no. 8, June, 1965.

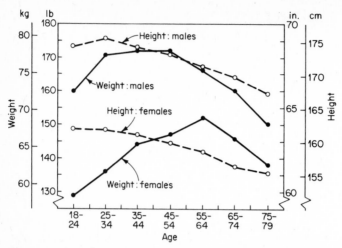

Figure 10-2 Average weight (left scale) and height (right scale) of 6672 adults, showing changes by age. Other physical characteristics also are related somewhat to age. [Based on data from National Health Survey, 25, tables 1 and 2.]

that most of the measurements given did vary somewhat by age, particularly weight and height, as illustrated in Figure 10-2. Further, corresponding data from surveys of other samples can vary from data from this survey. And as a word of caution, measurements of personnel wearing special gear, such as arctic clothing or heavy work clothing, can add inches to the persons' space requirements.

Functional Body Dimensions

Functional body dimensions are taken from body positions that result from motion. Although structural body dimensions are useful for certain design purposes, functional dimensions are probably more widely useful for most design problems. In most circumstances in life people are not inert (not even when sleeping). Rather, in most work and nonwork situations people are functional—whether operating a steering wheel, assembling a mouse trap, or reaching across the table for the salt. Figure 10-3 illustrates the difference in the application of structural versus functional body dimensions to the design of a vehicle cab.

The central postulate of the emphasis on use of functional dimensions relates to the fact that in performing physical functions, the individual body members normally do not operate independently, but rather in concert. The practical limit of arm reach, for example, is not the sole consequence of arm length; it is also affected in part by shoulder movement, partial trunk rotation, possible bending of the back, and the function that is to be performed by the hand. These and other variables make it difficult, or at least very risky, to try to resolve all space and dimension problems on the basis of structural body dimensions.

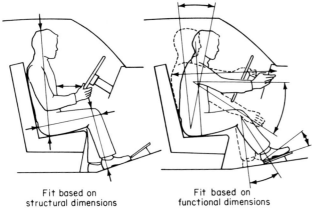

Fit based on   Fit based on
structural dimensions   functional dimensions

Figure 10-3 Illustration of the difference in the application of structural versus functional body dimensions in the context of vehicular cab design. The use of structural dimensions tends to focus on clearances of body dimensions with the surroundings, whereas the use of functional measurements tends to focus on the functions of the operations involved. [Adapted from Damon et al., 5.]

THE USE OF ANTHROPOMETRIC DATA

As indicated above, anthropometric data can have a wide range of applications in the design of physical equipment and facilities. In the use of such data, however, the designer should select data from samples of people who are reasonably similar to the ones who will actually use the facilities in question.

Principles in the Application of Anthropometric Data

In the application of anthropometric data there are certain principles that may be relevant, each one being appropriate to certain types of design problems.

Design for Extreme Individuals   In the design of certain aspects of physical facilities there is some "limiting" factor that argues for a design that specifically would accommodate individuals at one extreme or the other of some anthropometric characteristic, on the grounds that such a design *also* would accommodate virtually the entire population. A *minimum* dimension, or other aspect, of a facility would usually be based on an *upper* percentile value of the relevant anthropometric feature of the sample used, such as the 90th, 95th, or 99th. Perhaps most typically a minimum dimension would be used to establish clearances, such as for doors, escape hatches, and passageways. If the physical facility in question accommodates large individuals (say, the 95th percentile), it also would accommodate all those smaller in size. The minimum weight carried by supporting devices (a trapeze, rope ladder, or other support) is another example. On the other hand, *maximum* dimensions of some facility would be predicted on *lower* percentiles (say, the 1st, 5th, or 10th) of the distribution of people on the relevant anthropometric feature. The distance of control devices from an operator is an example; if those with short functional arm reach can reach a control, persons with longer arm reach generally could also do so. In setting such maximums and minimums it is frequently the practice to use the 95th and 5th percentile values, if the accommodation of 100 percent would incur trade-off costs out of proportion to the additional benefits to be derived. To take an absurd case, we do not build $8^1/_2$-ft doorways for the rare eight-footers, or dining-room chairs for the rare 400-pound guest. There are circumstances, however, in which designs that accommodate all people can be achieved without appreciable trade-off costs.

Design for Adjustable Range   Certain features of equipment or facilities preferably should be adjustable in order to accommodate people of varying sizes. The forward-backward adjustments of automobile seats and the vertical adjustments of typists' chairs are examples. In the design of adjustable items such as these, it is fairly common practice to do so for the range of cases from the 5th to the 95th percentiles. The example given in Figure 10-4 illustrates

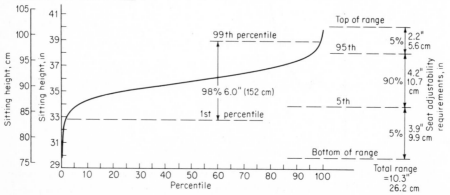

Figure 10-4   Illustration of the relationship between percentiles of cases and anthropometric measurements. This figure shows, specifically, the cumulative percentiles (along the base line) of people whose sitting height is at or below the values on the left vertical scale. The right vertical scale shows the corresponding seat adjustability requirements for various percentile groups. The differences in seat adjustability requirements are disproportionate at the extremes to the additional percentiles that would be accommodated. [From Hertzberg, 13.]

seat adjustability requirements (i.e., the range of adjustment that should be provided for in a seat) to accommodate different segments of the population according to their sitting height. This illustrates the point that the amount of seat adjustment required to accommodate the extreme cases (such as below the 5th percentile and above the 95th) is disproportionate to the additional numbers of individuals who would be accommodated.

In this connection, as in other contexts, trade-off considerations may be in order. The military practice of rejecting persons who are extremely short or tall is dictated, in part, by the fact that the requirement for smaller or larger items of clothing, shoes, etc., would impose an additional administrative load in supplying such items; this is trading off the possible utility of such men in the service for a certain degree of administrative simplicity.

Design for the Average While we frequently hear of the "average" or "typical" man, this is, in one sense, an illusive, will-of-the wisp concept. In the domain of human anthropometry there are few, if any, people who would really qualify as average—average in each and every respect. In connection with this, Hertzberg [12] indicates that in a survey of over 4000 Air Force personnel there were *no* men who fell within the (approximately) 30 percent central (average) range on all 10 of a series of measurements. Since the concept of the average man is then something of a myth, there is some rationale for the common proposition that physical equipment should not be designed for this mythical individual. Recognizing this, however, we would like to make a case here for the use of "average" values in the design of *certain* types of equipment or facilities, specifically those for which, for legitimate reasons, it is not appropriate to pitch the design at an extreme value (minimum or maximum) or feasible to provide for an adjustable range. As an example, the checkout counter of a supermarket built for the average customer probably would discommode customers less in general than one built either for the circus midget or for Goliath. This is not to say that it would be optimum for all people, but that, collectively, it would cause less inconvenience and difficulty than one which might be lower or higher.

WORK-SPACE DIMENSIONS

Human *work space* can consist of many different physical situations, including that of the plumber working under a stopped-up sink, the astronaut in his capsule, the assembler at his position on the assembly line, the flagpole painter, and the minister in his pulpit. Since we cannot here work out the space problems of the plumbers or flagpole sitters, we shall consider certain of the more conventional work locations.

Work Space for Personnel When Seated

There are millions of people whose work activities are carried out while seated in a fixed location. The space within which such an individual works is sometimes referred to as the "work-space envelope." This envelope should, of course, be designed on a situational basis, considering the particular activities to be performed and the types of people who are to use the space. To illustrate the types of data that would be relevant in designing specific work-space envelopes, however, the results of a couple of anthropometric studies will be shown.

Grasping Reach The first of these deals with the measurement of the grasping reach of a sample of 20 Air Force personnel [Kennedy, 16]. "Grasping reach" would of course impose outer-bound constraints on the space within which seated personnel could conveniently perform certain manual functions. The subjects of this study were presented with a vertical rack of measuring staves, each pointing toward the approximate joint center of the right shoulder, each with a knob at the end; the subject grasped each rod between the thumb and forefinger and moved it out until the arm was fully extended without pulling the shoulder away from the seat back. This was done with the staves at positions 15° apart around an imaginary vertical reference line beginning at a seat reference point behind the subject. As each subject positioned each stave, its distance from the reference line was recorded. Figure 10-5a illustrates the general seat arrangement, and b presents the resulting curves for the 5th and 95th percentiles of the subjects for each of certain horizontal "slices" of the three-dimensional space. The 5th percentile values for certain angular positions of these and other horizontal slices are given in Table 10-2. Data such as these

Table 10-2 Fifth Percentile of Grasping Reach, in Inches, to Selected Horizontal Planes above Seat Reference Level*

		Inches above seat reference level (SRL)					
Angle	SRL	10	20	25	30	40	45
L 135							7.75
L 90						12.15	7.25
L 45			19.50	20.00	19.00	14.00	8.50
L 30			21.50	22.50	21.50	15.50	9.50
0			25.50	26.25	25.50	19.00	12.75
R 30	17.50	27.00	30.00	30.25	29.00	22.75	17.50
R 45	19.50	28.25	31.00	31.00	30.25	24.75	19.00
R 90	19.50	29.25	32.25	32.25	31.25	26.25	21.00
R 135	16.50	26.25					20.00
180							12.75

*Sample: 20 male USAF personnel.
Source: Adapted from Kennedy [16].

naturally cannot be extrapolated to different types of people, to other seating arrangements, or to markedly different manual tasks.

Effects of Manual Task on Work-Space Envelope   The confounding effects upon the work-space envelope of the nature of the manual task being performed are illustrated by the results of some anthropometric research by Dempster [6]. Some of his research involved the analysis of photographic traces of contours of the hand as it moved over a series of frontal planes spaced at 6-in intervals. Eight different hand grasps were used, in which the hand, grasping a handlelike device, was in one of the eight fixed orientations (supine, prone, inverted, and at specified angles); but the hand was free to be moved over the plane in question. The mean data for 22 male subjects were summarized in several ways and presented in different forms in order to characterize different functional areas of the three-dimensional space of the subjects. *Kinetospheres* were developed for each type of grasp, showing graphically the

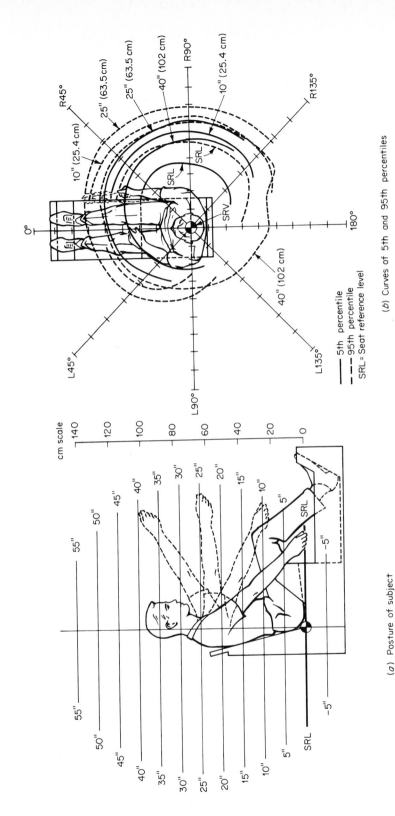

(a) Posture of subject

(b) Curves of 5th and 95th percentiles

— 5th percentile
- - - 95th percentile
SRL = Seat reference level

Figure 10-5 Part a illustrates the physical arrangement used in an anthropometric study of the three-dimensional space envelope of seated subjects (male United States Air Force personnel); the grasping reach was measured at different positions relative to the seat reference level (SRL) and at every 15° around the subject. Part b presents the curves of the 5th and 95th percentiles for each of four horizontal "slices" of the space, namely, at the SRL and at 10, 25, and 40 in (25.4, 63.5, and 102 cm) above that level. [Adapted from Kennedy, 16.]

275

mean contours of the tracings as photographed from each of three angles, top (transverse), front (coronal), and side (sagittal). Although the kinetospheres for the several types of grip will not be illustrated, they were substantially different. These were, however, combined to form *strophospheres* as shown in Figure 10-6. The shaded areas define the region that is common to the hand motions made with the various hand grips.

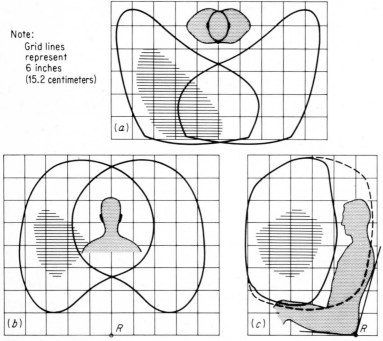

Note:
  Grid lines
  represent
  6 inches
  (15.2 centimeters)

Figure 10-6 Strophosphere resulting from superimposition of kinetospheres of range of hand movements with a number of hand-grasp positions in three-dimensional space. The shaded areas depict the region common to all hand motions (prone, supine, inverted, and several different angles of grasps), probably the optimum region, collectively, of the different types of hand manipulations. [Adapted from Dempster, 6.]

Minimum Requirements for Restricted Spaces

People sometimes find themselves working in, or moving through, some restricted and sometimes awkward spaces (such as an astronaut crawling through an escape hatch). For certain types of restricted spaces dynamic anthropometric data have been derived that provide minimum values. Some such data are given in Figure 10-7 for illustration. Note that the dimensions given include those applicable to individuals with heavy clothing. In most cases such clothing adds 4 to 6 in (10-15 cm), and in the case of a vertical escape hatch it adds 10 in (25 cm) to the requirements.

WORK SURFACES

Within the three-dimensional envelope of a work space, more specific considerations of work-area design relate to horizontal (dimensions, contours, height,

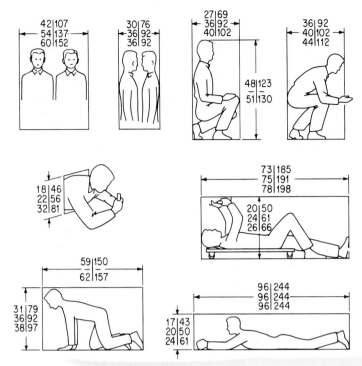

Figure 10-7 Clearances for certain work spaces that individuals may be required to work in or pass through. *Note:* The three dimensions given (in at left, cm at right) are (from top to bottom in each case) minimum, best (with normal clothing), and with heavy clothing (such as arctic). [Adapted from Rigby, Cooper, and Spickard, 19.]

etc.), vertical, and sloping work surfaces (dimensions, positions, angles, etc.). These features of the work situation also preferably should be determined on the basis of anthropometric considerations of the people who are to use the facilities in question.

Horizontal Work Surface

Many types of manual activities are carried out on horizontal surfaces such as workbenches, desks, tables, and kitchen counters. For such work surfaces, the *normal* and *maximum* areas have been proposed by Barnes [2] based on the measurements of 30 men. These two areas are shown in Figure 10-8 and have been described as follows:

    1 *Normal area:* This is the area that can be conveniently reached with a sweep of the forearm, the upper arm hanging in a natural position at the side.
    2 *Maximum area:* This is the area that could be reached by extending the arm from the shoulder.

    Related investigations by Squires [21], however, have served as the basis for proposing a somewhat different work-surface contour that takes into account the dynamic interaction of the movement of the forearm as the elbow

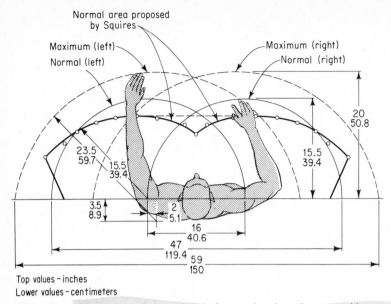

Figure 10-8 Dimensions (in in and cm) of normal and maximum working areas in horizontal plane proposed by Barnes, with normal work area proposed by Squires superimposed to show differences. [From Barnes, 2, and Squires, 21.]

also is moving. The area that is so circumscribed[2] is superimposed over the area proposed by Barnes in Figure 10-8. The fact that the normal work area proposed by Barnes has gained wide acceptance probably indicates that it is quite adequate, although the somewhat shallower area proposed by Squires probably tends to correspond somewhat better with dynamic anthropometric realities.

Work-Surface Height: Seated

The wide range of tasks performed by seated personnel at tables, desks, workbenches, etc., plus the range of individual differences, obviously precludes the establishment of any single, universal height that would be appropriate for such surfaces. However, considering body structure and biomechanics, at least one can set forth a guiding principle that should be applied, namely, that the work surface (or, really, the location of the devices or objects to be used continuously) should be at such a level that the arms can hang in a reasonably natural, relaxed position from the shoulder, with the elbow having, as Floyd and Roberts [8] refer to it, a "satisfactory" relationship to the working surface. Generally this would mean that the forearm would be approximately horizontal or sloping down slightly when performing most simple manual tasks. When the

[2]The resulting curve is known as a *prolate epicycloid*. The following data characterize certain points of this curve; the first value of each pair given is the distance in inches from the center (right or left), and the second value, in parentheses, is the distance in inches from the table edge of the curve at that position: 0 (10.04), 1.8 (11.57), 6.2 (12.49), 9.6 (12.86), 13 (12.65), 16.2 (11.88), 19.2 (10.57), 21.2 (8.67), and 24.5 (6.59).

work surface requires the upper arm to be raised above the relaxed elbow height, the metabolic costs of work tend to be increased [Tichauer, 22].

Such a principle would suggest that work surfaces generally should be lower than is reflected by common practice. With respect to desk heights, Bex [3], on the basis of a European survey, reports that the most common heights have in fact been reduced from about 30 in (76 cm) in 1958 to about $28\frac{1}{2}$ in (72 cm) in 1970. But on the basis of his own and other anthropometric data he argues for a further reduction of fixed desk heights to about 27 in (68.5 cm). However, he urges that, when feasible, adjustable desk heights be provided, the range to be from about 23 to 30 in (58 to 76 cm).

The case for adjustable work-surface heights is supported on the basis of three factors, as follows: individual differences in physical dimensions (expecially seated elbow height); individual differences in preference; and differences in the tasks being carried out. In this regard, for example, Ward and Kirk [24] carried out a survey of the preferences of British housewives with respect to working-surface heights when performing three different types of tasks. The percents preferring work surfaces at certain levels (relative to the elbow) when performing these tasks are given below:

	Level relative to elbow		
Type of task	Lower	Even	Above
A. Working above surface (peeling vegetables, slicing bread, etc.)	54	14	32
B. Working on surface (spreading butter, chopping ingredients, etc.)	16	11	73
C. Exerting pressure (ironing, rolling pastry, etc.)	41	9	50

The mean preferred work-surface heights were: A—23.7 in (60.2 cm); B—25.3 in (64.3 cm); and C—24.2 in (14.7 cm), but it is clear that there were individual differences in preferences as to relative level for each type of task, presumably because of the nature of the muscular involvement of the tasks.

The implications of the nature of the task with respect to work-surface height are emphasized further by Ayoub [1], who offers the following guidelines for three types of tasks, these being based on average anthropometric dimensions:

	Male		Female	
Type of task	in	cm	in	cm
a. Fine work (e.g., fine assembly)	39.0–41.5	99–105	35.0–37.5	89–95
b. Precision work (e.g., mechanical assembly)	35.0–37.0	89–94	32.5–34.5	82–87
c. Writing, or light assembly	29.0–31.0	74–78	27.5–29.5	70–75
d. Coarse or medium work	27.0–28.5	69–72	26.0–27.5	66–70

The work surface that would be most appropriate for individuals, or for people generally, however, is closely tied in with seat height (to be discussed later), the thickness of the surface, and the thickness of the thigh, as shown in Figure 10-9. The combinations of variables involved make it almost impossible

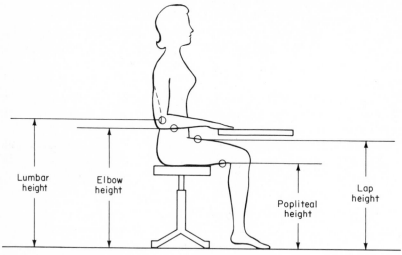

Figure 10-9   Illustration of the relationships of certain body dimensions and of work surface and seat height. If feasible, the work surface height should be adjustable for the individual's dimensions and preferences. [From Bex, 3, fig. 4.]

to design a fixed work-surface and seat arrangement that would be fully suitable for people of all sizes. Thus, where feasible, some adjustable features should be provided, such as seat height, foot position (as by the use of a footrest), or work-surface height. In this connection, an interesting innovation has been developed by the Western Electric Company, as illustrated in Figure 10-10. The work surface is adjustable in height by an electrically activated control. The seat also has adjustments for height and backrest (height, angle, degree of rigidity, etc.). (The frame on the surface holds the chassis of the electrical equipment to be wired by an operator, and can be rotated to the desired position.)

Work-Surface Height: Standing

Some experimental evidence relating to work-surface height for persons working in a standing posture comes from a study by Ellis [7]. Using a manipulation test of turning over wooden disks, he varied the work-surface height in relation to the distance from floor to fingertip of each subject in inches as follows: 25.9; 31.3; 36.6; 42.0; 47.4; and 52.7 (66 through 134 cm). The 42-in (107-cm) difference between the floor and fingertip was optimum for speed of performance, with the 36.6-in (93-cm) difference being nearly as satisfactory. These represented distances below elbow height of 2.8 and 8.2 in (7.0 and 20.8 cm), respectively, for the 42- and 36.6-in distances, and lead to the conclusion (which

Figure 10-10   Illustration of adjustable work place and chair. The height of the work surface can be adjusted by use of an electrically controlled device. The chassis (to be wired by the operator) can be rotated on its frame to any desired position. Various features of the chair also can be adjusted. [Photograph courtesy of Western Electric Company, Kansas City.]

is supported from other investigations and by experience) that for standing, a work surface normally should be a bit below elbow height. Barnes proposes 2 to 4 in or 5 to 10 cm below elbow for at least light assembly or similar manipulatory tasks. In terms of height above the floor, these would represent a range of *average* values for males from about 42 or 41 in (107 or 104 cm) to 36 in (81 cm), and for females from about 38 in (97 cm) to 33 in (84 cm).

This range for females jibes reasonably well with the preferences of working heights expressed by the sample of British women mentioned earlier [Ward and Kirk, 24]. The mean preferred standing work-surface heights for the same three types of task mentioned before are given below:

Type of task	Preferred standing work-surface height
A	34.6 in (87.9 cm)
B	35.8 in (90.9 cm)
C	34.5 in (87.6 cm)

These values are all a few inches below the mean elbow height for these women of 39.3 in (100 cm). In a subsequent study Ward [23] used four methods for assessing work-surface heights of women performing various kitchen tasks, the heights being 30, 33, 36, and 39 in (76, 84, 91, and 99 cm). The methods used were electromyography, anthropometry, "center of weight" determination, and expressed preferences. On the basis of data obtained from women of three size groups (small, medium, and large) she proposed the following ranges of heights for six different tasks:

	At sink		At work top		At stove	
	Washup	Peel potatoes	Iron	Slice potatoes	Frying	Boiling
in →	36–42	36–42	33–39	36–39	33–39	33–39
cm →	90–105	90–105	85–100	90–100	85–100	85–100

These ranges—and corresponding ranges for performing other tasks in other circumstances—are of course very much a function of individual differences, which are always with us.

Further indication of the fact that the nature of the activity influences the desirable work-surface height is reflected by the guidelines proposed by Ayoub [1] for three types of tasks as follows (as based on average dimensions):

	Male		Female	
Type of task	in	cm	in	cm
a. Precision work, elbows supported	43.0–47.0	109–119	40.5–44.5	103–113
b. Light assembly work	39.0–43.0	99–109	34.5–38.5	87–98
c. Heavy work	33.5–39.5	85–101	31.0–37.0	78–94

Although many work-surface heights do not lend themselves to on-the-spot adjustments in height, there sometimes are ways and means of matching facilities to people, as in selecting or building facilities for individuals (as counter tops, workbenches, etc.), placing blocks under legs of benches or tables, having mechanically adjustable legs, or having low platforms (a few inches high) for people to stand on.

THE SCIENCE OF SEATING

Whether at work, at home, at horse races, on buses, or elsewhere, the members of the human race spend a major fraction of their lives sitting down. As we know from experience, the chairs and seats we use cover the gamut of comfort; they can also vary in their influence on the performance of people who use them when carrying out some types of work activities.

Principles of Seat Design

The relative comfort and functional utility of chairs and seats are, of course, the consequence of their physical design in relationship to the physical structure and biomechanics of the human body. The uses of chairs and seats (from TV lounge chairs to stadium bleachers) obviously require different designs, and the range of individual differences complicates the design problem. Granting that compromises sometimes are necessary in the design of seating facilities, there nonetheless are certain general guidelines that may aid in the selection of designs that are sufficiently optimum for the purposes in mind. Some such guidelines have been expressed by Floyd and Roberts [8] and Kroemer and Robinette [17], including most of those discussed below.

Weight Distribution   Various seating studies have led to the conclusion that people are generally most comfortable when the weight of the body is borne primarily by the ischial tuberosities. These are the bony structures of the buttocks and in their anatomical features seem well suited to their weight-bearing responsibilities. Figure 10-11 shows what is considered to be a desirable

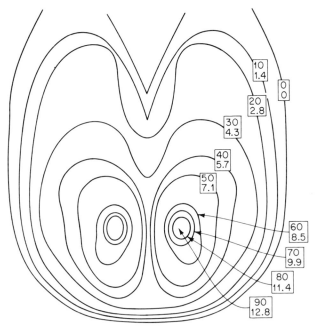

Figure 10-11   Representation of what is considered to be a desirable distribution of weight on the buttocks, showing equal-pressure contours from the ischial tuberosities to the periphery. The upper value for each contour is grams/cm² and the lower value is lb/in². [Adapted from Rebiffé, 18, fig. 7, p. 256.]

distribution of weight for a person when driving a vehicle, each of the lines representing an equal pressure contour, these ranging from the pressure on the ischial tuberosities of 90 grams/cm² (12.8 lb/in²) to the outer contours of 10 grams/cm² (1.4 lb/in²).

Seat Height   To avoid excessive pressure on the thigh (toward the front of the seat), the front of the seat should be no higher than the distance from the floor to the thigh when seated (i.e., popliteal height). This dimension generally should be selected to accommodate, say, all individuals from the 5th percentile up. Referring back to Table 10-1, the 5th percentiles for males and females are

15.5 and 14.0 in (39 and 36 cm), respectively. Fixed seat heights of such values, however, can complicate the mechanics of sitting for the taller members of the human clan by a chain reaction starting with the knee angle that can cause such an individual to sit with his lumbar back area in a convex rather than concave posture. By taking into account the fact that heels typically add an inch or more to the 5th percentile values (and more for females), it has become fairly common practice to use seat heights of around 17 in (43 cm). This jibes quite well with the recommendation by Grandjean et al. [10] of 43 cm (16.8 in) for multipurpose chairs which have a slanting seat. Where feasible, of course, adjustable seat heights (perhaps from 15 to 19 in or 38 to 48 cm) should be provided to accommodate persons of various heights.

Seat Depth and Width   The preferable depth and width of seats would depend in part on the type of seat (whether a multipurpose chair, typing chair, lounge chair, etc.). In general terms, however, the depth should be set to be suitable for small persons (to provide clearance for the calf of the leg and to minimize thigh pressure) and the width should be set to be suitable for large persons. On the basis of comfort ratings for chairs of various designs, Grandjean et al. [10] recommend that for multipurpose chairs depths not exceed 43 cm (16.8 in) and that the width of the seat surface be not less than 40 cm (15.7 in), although such a seat width (or perhaps one a bit wider, say 17 in) would do the trick for individual seats; if people are to be lined up in a row, or seats are to be adjacent to each other, elbow-to-elbow breadth values need to be taken into account, with even 95th-percentile values around 19 and 20 in producing a moderate sardine effect. (And for bundled-up football observers the sardine effect is amplified further.) In any event these are approximate minimum values for chairs with arms on them (and for your well-fed friends, you should have even wider lounge chairs).

Trunk Stabilization   The stabilization of the trunk is facilitated in large part by designs which provide for the primary weight to be borne by the area around the ischial tuberosities. In this regard, the seat angle and back angle play important roles, along with the curvature of the back of the seat. Such features, however, are intertwined with the function of the seat. For example, in the case of an office seat, as illustrated in Figure 10-12, the recommended seat angle is about 3° and the back angle (the angle between the back and the seat) is 100°. For resting and reading, however, Grandjean et al. [9] found that most people prefer greater angles (as reported in a later discussion of chairs for resting and reading).

Trunk stability can also be aided by the use of armrests and even by resting the arms on desks or work-surface areas, *but* these should also be at levels that would make it possible for the arms to hang freely, and for the elbows to rest in a natural position.

Postural Changes   Although some seats have been tested by the postural changes that people make in them (such as the number of fidgets), this does not

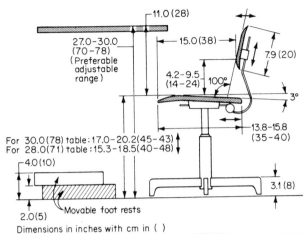

Dimensions in inches with cm in ( )

Figure 10-12 Recommended dimensions (in inches) of adjustable features of office seat. Note ranges of adjustability of seat height in relation to two table heights, 30 and 28 in (78 and 71 cm). To maintain approximately 11 in (28 cm) between seat height and work surface, the seat should have a range of adjustability that would depend on the work-surface height, as indicated. With short persons, footrests may be required if work-surface height is high. (Data in inches converted from centimeters.) [Adapted from Burandt and Grandjean, 3.]

mean that the goal of seat design should be to reduce mobility to zero. Generally a chair or seat should permit moderate mobility and changes in posture.

Seat Designs for Varied Purposes

Since the specific features of seats need to be determined for the particular use, we will illustrate a few selected examples.

Office Seats On the basis of a substantial amount of data relative to comfort reported by people using office seats, Burandt and Grandjean [4] have proposed the design features shown in Figure 10-12. This also shows the relationship between the dimensions of the seat and the height of the work surface.

Multipurpose chairs In the study by Grandjean et al. mentioned earlier [10], 50 men and women were asked to rate the comfort of 11 parts of the body when testing 12 different designs of multipurpose chairs. In addition, each subject compared every chair with every other one and rated the overall comfort by the paired-comparison method. The contours of the two most preferred chairs are shown in Figure 10-13, along with design recommendations made on the basis of the analysis of the results of all the data. The recommendations include foam rubber of 2 to 4 cm (about 0.75 to 1.5 in) on the entire seat.

Chairs for Resting and Reading The desirable features of chairs for relaxation and reading are of course different from those of chairs used for more active use. A study was carried out by Grandjean et al. [9] in which they used a "seating machine" for eliciting judgments of subjects about the comfort of various seat designs. The "seating machine" consisted of features that could be

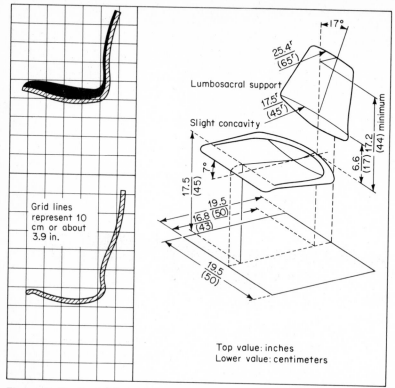

Figure 10-13 Contours of the two multipurpose chairs (of 12) judged to be most comfortable by 50 subjects, and the design features recommended for multipurpose chairs based on the study. [From Grandjean et al., 10, figs. 2, 6, 13.]

adjusted to virtually any profile. Without summarizing all of the results, they found that the following angles and dimensions were preferred by more subjects than others for the two purposes of reading and resting:

	Reading	Resting
Seat inclination, deg	23–24	25–26
Backrest inclination, deg	101–104	105–108
Seat height, cm	39–40	37–38
Seat height, in	15.3–15.7	14.6–15.0

Profiles for two such chairs are shown in Figure 10-14. Note in these the distinct angles of the backs and seats to provide full back support with particular support for the lower (lumbar) sections of the spine.

Automobile Driver Seats The desirability of adequate back support—for whatever activity is involved—is further illustrated in the case of automobile driver seats, as illustrated by the sketches in Figure 10-15. With unsatisfactory

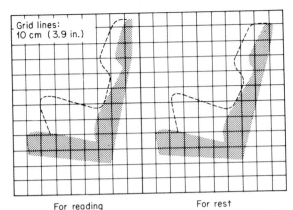

For reading   For rest

Figure 10-14 Profiles of seats proposed for reading and resting. The dotted lines correspond to the armrests and a possible outer contour. The shaded area shows the surface of the seat, including upholstering to be 6 cm (2.5 in) thick. [From Grandjean et al., 9, fig. 2, p. 310.]

(a)

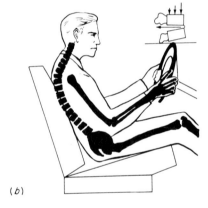

(b)

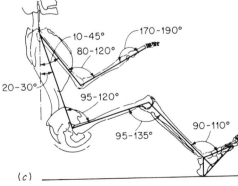

(c)

10-45°
80-120°
170-190°
20-30°
95-120°
95-135°
90-110°

Figure 10-15 Illustrations of anthropometric considerations in designing seats for automobile drivers. Parts (a) and (b) illustrate desirable and undesirable postures as related to the spine. (Note in the inset of b the uneven pressures between the vertebrae caused by the angle between them.) Part (c) illustrates the angles of body members that would provide a reasonably satisfactory posture while driving. [a and b from Rizzi, 20, fig. 5, p. 230; c from Rebiffé, 18, fig. 1, p. 247.]

support (as shown in *b* of that figure), the angles between the vertebrae (shown in the inset) can generate discomfort and conceivably cause spinal complications. The angles of the various body joints shown in *c* of that figure are those which are proposed by Rebiffé [18] as providing the basic driving posture which he considers to be desirable in terms of anthropometric considerations.

DISCUSSION

It must be granted that tables of anthropometric data do not make for very enthralling bedtime reading (although they might help those who have insomnia). But when we consider the application of such data to the design of the physical facilities and objects we use, we can see that such intrinsically uninteresting data play a live, active role in the real, dynamic world in which we live and work.

In the application of anthropometric data to specific design problems there can be no nicely honed set of procedures to follow, because of the variations in the circumstances in question and in the types of individuals for whom the facilities would be designed. As a general approach, however, the following suggestions are offered [based in part on those of Hertzberg, 14]:

1 Determine the body dimensions important in the design (e.g., sitting height as a basic factor in seat-to-roof dimensions in automobiles).
2 Define the population to use the equipment or facilities. This establishes the dimensional range that needs to be considered (e.g., children, housewives, United States civilians, different age groups, world-wide populations, different races, etc.).
3 Determine what "principle" should be applied (e.g., design for extreme individuals, for an adjustable range, or for the "average").
4 When relevant, select the percentage of the population to be accommodated (e.g., 90 percent, 95 percent) or whatever is relevant to the problem.
5 Locate anthropometric tables appropriate for the population, and extract relevant values.
6 If special clothing is to be worn, add appropriate allowances (some of which are available in the anthropometric literature).

REFERENCES

1 Ayoub, M. M.: Work place design and posture, *Human Factors*, 1973, vol. 15, no. 3, pp. 265–268.
2 Barnes, R. M.: *Motion and time study*, 5th ed., John Wiley & Sons, Inc., New York, 1963.
3 Bex, F. H. A.: Desk heights, *Applied Ergonomics*, 1971, vol. 2, no. 3, pp. 138–140.
4 Burandt, V., and E. Grandjean: Sitting habits of office employees, *Ergonomics*, 1963, vol. 6, no. 2, pp. 217–228.
5 Damon, A., H. W. Stoudt, and R. A. McFarland: *The human body in equipment design*, Harvard University Press, Cambridge, Mass., 1966.

6   Dempster, W. T.: *Space requirements of the seated operator*, USAF, WADC, TR 55–159, July, 1955.

7   Ellis, D. S.: Speed of manipulative performance as a function of work-surface height, *Journal of Applied Psychology*, 1951, vol. 35, pp. 289–296.

8   Floyd, W. F., and D. F. Roberts: Anatomical and physiological principles in chair and table design, *Ergonomics*, 1958, vol. 2, no. 1, pp. 1–16.

9   Grandjean, E., A. Boni, and H. Krestzschmer: The development of a rest chair profile for healthy and notalgic people, *Ergonomics*, 1969, vol. 12, no. 2, pp. 307–315.

10 Grandjean, E., W. Hunting, G. Wotzka, and R. Shärer: An ergonomic investigation of multipurpose chairs, *Human Factors*, 1973, vol. 15, no. 3, pp. 247–255.

11 Hansen, R., and D. Y. Cornog: *Annotated bibliography of applied physical anthropology in human engineering*, USAF, WADC, TR 56–30, May, 1958.

12 Hertzberg, H. T. E.: *Some contributions of applied physical anthropology to human engineering*, USAF, WADD, TR 60–19, January, 1960.

13 Hertzberg, H. T. E. : Dynamic anthropometry of working positions, *Human Factors*, August, 1960, vol. 2, no. 3, pp. 147–155.

14 Hertzberg, H. T. E.: "Engineering anthropology," in H. P. Van Cott and R. G. Kinkade (eds.), *Human engineering guide to equipment design*, Washington, D.C., U.S. Government Printing Office, 1972, chap. 11.

15 Hertzberg, H. T. E., G. S. Daniels, and E. Churchill: *Anthropometry of flying personnel—1950*, USAF, WADC, TR 52–321, September, 1954.

16 Kennedy, K. W.: *Reach capability of the USAF population: Phase I. The outer boundaries of grasping-reach envelopes for the shirt-sleeved, seated operator*, USAF, AMRL, TDR 64–59, 1964.

17 Kroemer, K. H. E., and J. C. Robinette: *Ergonomics in the design of office furniture: A review of European literature*, USAF, AMRL, TR 68–80, 1968.

18 Rebiffé, P. R.: Le siége du conducteur: son adaptation aux exigences fonctionnelles et anthropométriques, *Ergonomics*, 1969, vol. 12, no. 2, pp. 246–261.

19 Rigby, L. V., J. I. Cooper, and W. A. Spickard: Guide to integrated system design for maintainability, USAF, ASD, TR 61–424, October, 1961.

20 Rizzi, V. M.: Entwicklung eines verschiebbaren rückenprofils für auto-und ruthesitze, *Ergonomics*, 1969, vol. 12, no. 2, pp. 226–233.

21 Squires, P. C.: *The shape of the normal work area*, Navy Department, Bureau of Medicine and Surgery, Medical Research Laboratory, New London, Conn., Report 275, July 23, 1956.

22 Tichauer, E. R.: Industrial engineering in the rehabilitation of the handicapped, *Proceedings of the 18th Annual Institute Conference and Convention*, American Institute of Industrial Engineers, May, 1967, pp. 171–180.

23 Ward, J. S.: Ergonomic techniques in the determination of optimum work surface heights, *Applied Ergonomics*, 1971, vol. 2, no. 3, pp. 171–177.

24 Ward, J. S., and N. S. Kirk: The relation between some anthropometric dimensions and preferred working surface heights in the kitchen, *Ergonomics*, 1970, vol. 13, no. 6, pp. 783–797.

25 *Weight, height, and selected body dimensions of adults: United States, 1960–1962*. Data from National Health Survey, USPHS Publication 1000, series 11, no. 8, June, 1965.

Chapter 11

Physical Space and Arrangement

A large proportion of the lifetimes of most people is spent within physical space environments that are man-made, ranging from "local" situations in which people find themselves (such as at workplaces, in a kitchen, or in an automobile), through intermediate types of situations (such as office buildings, homes, and theaters), to general environments (such as communities). Our common experience points up the effects that the designs of such space and facilities can have on people, including their performance, their comfort, and even their physical well-being. Since we cannot deal intensively with the many human factors aspects of the arrangement of the space and facilities people use, we will touch on only certain aspects, with particular attention to work stations.

CONSIDERATIONS IN LOCATION OF COMPONENTS

In the case of many systems and facilities there are various "components" that need to be located within the system or facility. (We will use the term *component* in this discussion to refer to virtually any relevant feature, such as displays, controls, materials, machines, work areas, or rooms.) It is reasonable to hypothesize that any given component has a generally "optimum" location for serving its purpose. This optimum would be predicated on the human sensory, anthropometric, and biomechanical characteristics that are concerned (i.e., reading a visual display, activating a foot push button, etc.), or on the performance of some operational activity (such as reaching for parts, preparing food in a restaurant, or storing material in a warehouse). Preferably, of course, components should be placed in their optimum locations, but since this frequently is not possible, priorities sometimes must be established. These priorities, however, do not descend from heaven like manna but must be otherwise determined, usually on the basis of some factors such as those mentioned below.

Guiding Principles of Arrangement

Before touching on a few methods that are used in trying to figure out what should go where, however, let us set down a few general guides (in addition to the idea of optimum location) that may be helpful. Depending on the circumstance, these guidelines can deal with either or both of two separated, but interrelated, phases, as follows: that concerned with the general location of components (such as specific components within a fixed workspace or larger components that might be located in a more general work area such as an office), and that concerned with the specific arrangement of components.

   Importance Principle   This principle deals with operational importance, that is, the degree to which the performance of the activity with the component is vital to the achievement of the objectives of the system or some other consideration. The determination of importance is largely a matter of judgment.

   Frequency-of-Use Principle   As implied by the name, this concept applies to the frequency with which some component is used. An additional twist to this idea places it in the frame of reference of information theory [21, pt. B, chap. 4, p. B.4-12A], in which both frequency of use and the number of possible components from which a selection is to be made are taken into account. (This will be discussed later.)

   Functional Principle   The functional principle of arrangement provides for the grouping of components according to their function, such as the grouping together of displays, or controls, that are functionally related in the operation of the system.

   Sequence-of-Use Principle   In the use of certain items, there are sequences or patterns of relationship that frequently occur in the operation of the equipment. In applying this principle, the items would be so arranged as to take advantage of such patterns; thus, items used in sequence would be in close physical relationship with each other.

   Discussion   In putting together, as pieces of a jigsaw puzzle, the various components of a system, it is manifest that no single guideline can, or should, be applied consistently, across all situations. But, in a very general way, and in addition to the optimum premise, the notions of importance and frequency probably are particularly applicable to the more basic phase of *locating* components in a *general area* in the workspace; in turn, the sequence-of-use and functional principles tend to apply more to the *arrangement* of components *within* a general area.

The application of these various principles of arrangement of components generally has had to be predicated on rational, judgmental considerations, since there has been little empirical evidence available regarding the evaluation of these principles. There are, however, some data from at least one study that cast a bit of light on this matter. The study in question, by Fowler et al. [7], consisted of the evaluation of various control panel layouts in which the controls and displays had been arranged following each of the four principles described above. The panels included 126 standard military controls and displays. The arrangement of these following the four principles need not be described, but it should be added that, for each principle, three control panels were developed, varying in terms of three "levels" of application of the principle (based on a scoring scheme), these being high, medium, and low. The 200 male college student subjects used the various arrangements in a simulated task. Their performance was measured in terms of time and errors, the results for the time criterion being shown in Figure 11-1. This figure (and corre-

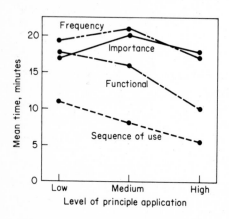

Figure 11-1 Time required to carry out a standard simulated task in the use of controls and displays arranged on the basis of four principles. In the case of each principle, three control panels were used, these varying in terms of the rated "level" or degree to which the principle has been applied in the panel design. [Adapted from Fowler et al., 7.]

sponding data regarding errors) showed a clear superiority for the sequence-of-use principle. Even the arrangements that were based on the "low" and "medium" application levels of this principle were better than, or equal to, the "high" levels for the other principles.

Although the sequence-of-use principle came out on top in this study, this principle obviously could be applied only in circumstances (as in this study) in which the operational requirements actually do involve the use of the components in question in rather consistent sequences. Where this is the case, this principle certainly should be followed.

METHODS OF ANALYSIS

As in the case of other aspects of human factors, relevant data can be useful in the design process. In considering physical space, perhaps the most important types of data relate to the human activities or operations that are to be carried out (such as physical movements, psychomotor activities, or visual activities). The relevant criteria of such activities include frequency, their sequence and interrelationships, their judged importance, and the time devoted to them. In some circumstances subjective criteria of comfort, convenience, or preference might also be used.

Methods of Activity Analysis

When a modification of an existing system or facility is to be developed, it is possible to carry out activity analyses of persons in an existing model with the expectation of using the data so collected to design an improved model. In this connection, a number of the techniques used in the field of industrial engineering can be used, such as: film analysis; observation; and the use of eye-movement recordings (with eye cameras or other related devices). The data so generated typically are categorized in terms of the various types of activities or operations. Other methods of data collection include the use of interviews or questionnaires with personnel experienced in the use of the system or facility, in order to elicit their opinions (such as about the frequency or importance of certain activities or the ease of use of some device) or expressions of preference for certain design features.

In the case of new systems or facilities (without current counterparts), the frequency, sequence, and other activity parameters must be inferred from whatever tentative drawings, plans, or concepts are available. In some cases (especially during early design stages) the designer may have to rely on some rather gross estimates; but as plans become further crystallized, it may be possible to use a more systematic method in the analysis of the activities that would be involved in the use of the system or facility. Such analyses may, in turn, aid in rearranging the features until, through an iterative process, a satisfactory arrangement is developed. Such a tentative arrangement sometimes can be further tested by the use of prototypes or physical models (if these are developed).

SUMMARIZING AND USING ACTIVITY DATA

The activity data collected by whatever method—whether of the hand movements of an assembler, of the sequence of activities of a housewife baking a cake, or of the movement flow of people in an airport—usually need to be summarized in order to aid in their interpretation and use. Frequently it is sufficient to derive simple sums (such as frequency of use of a brake pedal) or simple averages (such as the average time spent by nurses walking to and from patients' rooms). In other circumstances, however, some special manipulations in summarizing such data may be handy. A couple of such examples will be given below—in particular, those relating to the use of components in a system.

Indexes of Priorities of System Components

In the case of some systems the relative priorities of the various components may be clearly obvious from the summarization of relevant data or even from casual knowledge about the system and the operation. In other cases, however, it may be desirable to derive some index of priorities to provide guidance in locating components. In such instances the designer has to decide what the appropriate basis for the priorities (such as rated importance, frequency of use, etc.) should be. When two or more such factors seem relevant, however, it is possible to combine them into a single index, either by adding the ratings on, say, importance and frequency, or by multiplying them. The table below shows, for each of five displays (A, B, C, D, and E) the average ratings on these two factors (a rating of 3 being high and a rating of 1 being low); in addition, the sums and the products of the two ratings are given for each display.

	Display				
	A	B	C	D	E
1. Av. rating, importance	3.0	1.7	1.6	2.5	1.1
2. Av. rating, frequency	3.0	1.2	1.7	1.0	2.7
3. Sum of 1 and 2	6.0	2.9	3.3	3.5	3.8
4. Product of 1 and 2	9.0	2.0	3.7	2.5	3.0

The sums or products reflect the relative priorities of the displays, for consideration in deciding where to place them. Since the *relative* values of the sums and the products can be different (as in the above example), the manner in which the ratings are combined (including any weighting system) should be given careful thought.

Links as Indexes of Interrelationships

The operational relationships between people and between people and physical components usually can be expressed in terms of *link* values. Link values can be developed for a wide range of such relationships, although they fall generally into three classes, namely, communication links, control links, and movement links. Communication and control links can be considered as functional. Movement links generally reflect sequential movements from one component to another. Some versions of the three types follow:

> **I** Communication links
>   **A** Visual (man to man or equipment to man)
>   **B** Auditory, voice (man to man, man to equipment, or equipment to man)

    C   Auditory, nonvoice (equipment to man)
    D   Touch (man to man or man to equipment)
II Control links
    E   Control (man to equipment)
III Movement link (movements from one location to another)
    F   Eye movements
    G   Manual and/or foot movements
    H   Body movements

Link indexes can be used as aids in connection with the general location of components or with their relative arrangements. In some circumstances they can be used as the bases for assignment of priorities.

Derivation of Link Values In the case of existing systems or facilities, certain link values can be derived from empirical observation or films, especially link values of frequency (such as frequency of control actions or of movements from one component to another). In the case of systems or facilities being developed, frequency usually would have to be estimated. Link values of importance almost of necessity need to be based on judgment. When link values are derived for both importance and frequency, it is usually the practice to compute a composite link value by multiplying or adding the importance and frequency values of the individual links, in much the same manner as illustrated above in the case of priority indexes.

Link values in operational procedures sometimes can be derived by a graphic approach in which the sequential steps in an operation are recorded. Subsequently the functional links and the sequential links (showing relations in operation between all pairs of components) can be tallied in the manner presented by Haygood et al. [10] and illustrated in Table 11-1. This table actu-

Table 11-1 Link Analysis of Certain Panels of Flight System Checkout Console

Panel	Functional links		Sequential links among panels				
	Visual	Control	2	3	4	5	6
1. Master selector	175	884	348	97	44	7	1
2. Programmer	31	637		32	11	10	3
3. Recorder	51	81			0	1	2
4. Meter panel	63	0				8	0
5. Power supply	5	49					0
6. Oscilloscope	12	12					

Source: Adapted from Haygood et al. [10], tables 1 and 2.

ally presents data for only a few of the "panels" that were considered as components, but these will at least illustrate the nature of the results. Incidentally, the purpose of this investigation was that of comparing a computerized approach to the development of link values with a more conventional graphic approach. It was found that both methods produced substantially the same results in link values, but that the computerized approach was more economical of time and cost.

Use of Activity Data in Arranging Components

In the use of activity data for developing a reasonably optimal arrangement of components, the typical approach is through some form of physical simulation. However, in some circumstances, more systematic, quantitative methods can be used.

Arrangement by Physical Simulation   In the use of a physical simulation approach, the various components are juggled around on paper (in graphic form) or in the form of models or mock-ups, until an arrangement is achieved that is judged to be reasonably optimum for whatever considerations are relevant (e.g., the optimum location of components, their priorities, or their functional or sequential link values). An example of this is shown in Figure 11-2.

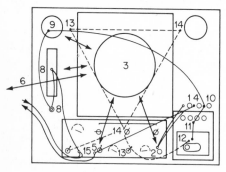
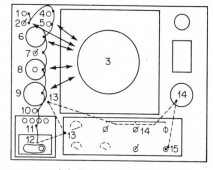

(*a*) Original layout   (*b*) Redesigned layout

Figure 11-2   Sequence of operations required by a prototype panel layout (*a*), and the redesigned layout after analysis of the operating sequence (*b*). [22, figs. 5.2*a* and 5.2*b*.]

Part *a* shows an initial panel layout that was developed as a prototype, the sequence of operations being shown by serial numbers and lines. After the sequence was analyzed on the basis of the tryout, a revised design, shown as part *b*, was developed for operational use. The greater simplicity of the redesigned layout is obvious.

Another example of an arrangement developed by essentially graphic methods comes from a study of the layout of aircraft instruments [Jones, Milton, and Fitts, 12]. The basic data for the study came from the use of an eye camera that records the reflection of the light from the cornea of the eye. Recordings were made of the eye movements of 36 pilots during various maneuvers. Figure 11-3 shows the sequential link values between the various instruments (expressed as percentages of eye shifts between instruments) for one particular maneuver. Data such as shown in these figures were used to develop a standard instrument arrangement (basically the one shown) which has been accepted by the U.S. Air Force, the U.S. Navy, U.S. airlines, the Royal Canadian Air Force, and the Royal Air Force (U.K.).

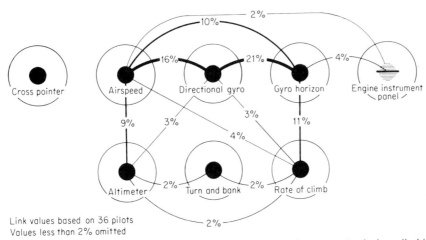

Link values based on 36 pilots
Values less than 2% omitted

Figure 11-3 Eye-movement link values between aircraft instruments during climbing maneuver with constant heading. [Courtesy of USAF, AMRL, Behavioral Sciences Laboratory.]

Quantitative Solutions to Arrangement Problems Especially in simple problems of arranging components, it would be gilding the lily to apply sophisticated quantitative methods. But with complex systems that have many components, some quantitative attack may well be justified. One such method is that of linear programming. This is a statistical method that results in the optimizing of some criterion or dependent variable by manipulation of various independent variables. The optimum in some cases would be the minimum, and in other cases the maximum—whichever is the desired value in terms of the criterion.

An example of this technique draws upon data from the following two sources: (1) data on frequencies with which a pilot made task responses, with eight controls, in flying simulated cargo missions in a C-131 aircraft, over 139 one-minute periods [Deininger, 3]; and (2) data on the accuracy of manual blind-positioning responses in various areas, based on the study by Fitts [6]; for this particular purpose accuracy data for 8 of the 20 target areas were used, as shown in Figure 11-4. The accuracy scores are average errors in inches in reaching to targets in the 8 areas. Linear programming, when used in the analysis of the data by Freund and Sadosky [8], involved the derivation of a *utility cost* rating for each of eight controls in each of the eight areas shown in Figure 11-4. Each one was computed by multiplying the *frequency* of responses involving each control (from Deininger) by the *accuracy* of responses in each area (from Fitts). For example, the values for the C control (cross-pointer set) for the eight areas range from 42.8 to 69.9, each such value being the product of the *frequency* value for the control (in this case 20) and the mean *accuracy* scores for the areas as shown in Figure 11-4. For any possible arrangement of controls one can derive a *total cost*, this being the sum of the utility costs of the several controls in their locations specified by that arrangement. By linear programming it was possible to identify the particular arrangement for which

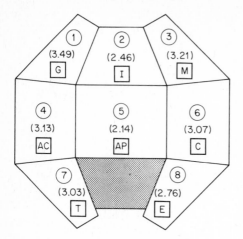

Figure 11-4 Perspective drawing of eight areas used in application of linear programming to the arrangement of eight aircraft controls [Freund and Sadosky, 8]. Mean accuracy scores of blind-positioning movements are given in parentheses for the eight target areas [based on data from Fitts, 6]. The letters in boxes are symbols for the eight controls, their locations shown here representing the optimum linear programming solution.

the total cost was minimum. The resulting optimum arrangement is that shown in Figure 11-4, the individual controls simply being identified by a letter code. Note that each control is not in its *own* optimum location, but that *collectively* the derived arrangement is optimum in terms of the criterion of total cost. (Control C, mentioned above, was thus placed in space 6, even though its utility cost in that location was less.)

Although this particular exercise in linear programming deals with the location of controls, this and other quantitative techniques can also be applied to the arrangement of displays, of items of equipment, or of people.

GENERAL LOCATION OF COMPONENTS

As indicated above, it is reasonable to assume that any given component in a system or facility would have some reasonably optimum location, as predicated on whatever sensory, anthropometric, biomechanical, or other considerations are relevant. Although the optimum locations of some specific components probably would depend upon situational factors, some generalizations can be made about certain classes of components. Certain of these will be discussed below.

Visual Displays

The normal line of sight is usually considered to be about 15° below the horizon. Visual sensitivity accompanied by moderate eye and head movements permits fairly convenient visual scanning of an area around the normal line of sight. The area for most convenient visual regard (and therefore generally preferred for visual displays) has generally been considered to be defined by a circle roughly 10 to 15° in radius around the normal line of sight.

However, there are indications that the area of most effective visual regard is not a circle around the line of sight but rather is more oval. Such an indication is found in the research of Haines and Gilliland [9], who measured subjects'

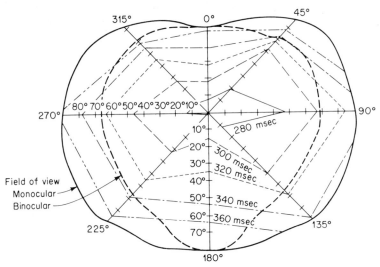

Figure 11-5 Isoresponse times within the visual field. Each line depicts an area within which the mean response time to lights is about the same. [Adapted from Haines and Gilliland, 9, fig. 2. Copyright 1973 by the American Psychological Association and reproduced by permission.]

times of response to a small light that flashed on in various locations of the visual field. Figure 11-5 shows the mean response times to the lights in the various regions of the visual field. Each boundary on that figure indicates the region within which mean response time can be expected to be about the same (the "isoresponse" time region). The generally concentric lines tend to form ovals—but slightly lopsided ovals, being flatter above the line of sight than below.

The subjects in this study detected the lights even toward the outer fringes of this area, but the primary implications of this (and some other) research are that critical visual displays should be placed within a reasonably moderate oval around the normal line of sight.

Hand Controls

The optimum location of hand control devices is, of course, a function of the type of control, the mode of operation, and the appropriate criterion of performance (accuracy, speed, force, etc.). Certain preceding chapters have dealt with some tangents to this matter, such as the discussion of the work-space envelope in Chapter 10.

Controls That Require Force Many controls are easily activated, so a major consideration in their location is essentially one of ease of reach. Controls that require at least a moderate force to apply (such as certain control levers and hand brakes in some tractors), however, bring in another factor, that of force that can be exerted in a given direction with, say, the hand in a given position. Investigations by Dupuis [4] and Dupuis, Preuschen, and Schulte

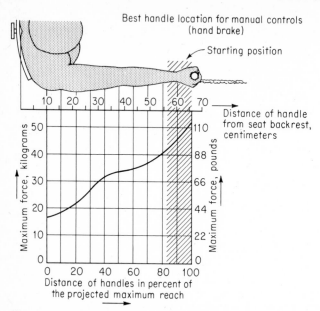

Figure 11-6 Relationship between maximum pulling force (such as on a hand brake) and location of control handle. [From Dupuis, Preuschen, and Schulte, 5.]

[5] dealt with this question, specifically the pulling force that can be exerted, when seated, when the hand is at various distances from the body (actually, from a seat reference point). Figure 11-6, which illustrates the results, shows the serious reduction in effective force as the arm is flexed when pulling toward the body. The maximum force that can be exerted by pulling is about 57 to 66 cm forward from the seat reference point, and this span, of course, defines the optimum location of a lever control (such as a hand brake) if the pulling force is to be reasonably high.

According to Kroemer [13] the best location for cranks and levers (especially those to be operated continuously to and fro) is in front of the sitting or standing operator so that the handle travels at about waist height in the sagittal plane (i.e., the vertical plane from front to back) passing through the shoulder.

Controls on Panels   Many controls are positioned on panels or in areas forward of the person who is to use them. Because of the anthropometric and biomechanical characteristics of people, controls in certain locations can be operated more effectively than those in other locations. Figure 11-7 shows one set of preferred areas for four classes of control devices as based on relative priorities [23]. Although this proposed arrangement is based on data for military personnel, it would of course have more general applicability.

When there are many controls and displays to arrange in a console or panel, the use of angled side panels may place more of the controls within convenient access. The advantage of this was demonstrated empirically by Siegel and Brown [17] in a study in which subjects, using a 48-in front panel with side

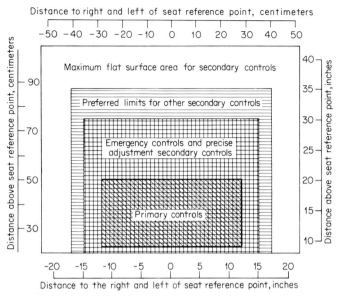

Figure 11-7 Preferred vertical surface areas and limits for different classes of manual controls. [Adapted from *Personnel subsystems,* 23, chap. 2, sec. 2D6, p. 7.]

panels at 35, 45, 55, and 65°, followed a sequence of verbal instructions to use the controls on the panels. A number of criteria were obtained, including objective criteria of average number of seat movements, average seat displacement, average body movements (number and extent), average number of arm extensions (part and full), subjective criteria based on the subjects' responses of degree of ease or difficulty, judgments that the panels should be wider apart or closer together, and preference ranking for the four angles.

Only some of the data will be presented, but they characterize the results generally. Figure 11-8 shows data for four of the criteria for the four angles. The criterion scales have been converted here to fairly arbitrary values, and only the "desirable" and "undesirable" directions are indicated. It can be seen, however, that all four criteria were best for the 65°-angle side panels. The consistency across all criteria (these and the others) was quite evident.

Although Figures 11-7 and 11-8 provide general guidelines for locating controls, including consideration of relative priorities, it should be added that the operational requirements of certain specific types of controls sometimes impose impossible constraints on their location. This was illustrated, for example, by the results of a study by Sharp and Hornseth [16] in which seated subjects operated each of 3 types of controls (knobs, toggle switches, and push buttons) at each of 12 locations in each of 3 consoles (far, middle, and close). The controls of the close console were positioned for convenience of reach of individuals of small build (about the 5th percentile of males), and only data for this console will be given here.

Figure 11-9, part *a,* shows the time in seconds to activate the three types of

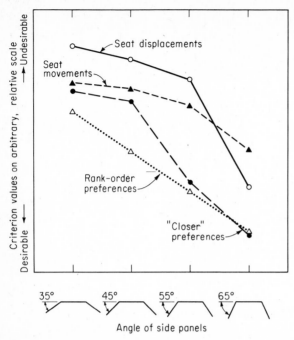

Figure 11-8 Representation of criteria from study relating to angles of side panels of console. The criterion values are all converted to an arbitrary scale for comparative purposes, but they all indicate the desirability of the 65° angle panels over the others. [Adapted from Siegel and Brown, 17, courtesy of Applied Psychological Services, Wayne, Pa.]

controls when located at various angles from the center position, that part indicating that activation time was minimum for all three controls at about 25° from center. Part *b* in turn shows the areas of the shortest performance times for each type of control (e.g., the area within which times were *within 5 percent* of *minimum* times for the control in question); the smaller area for the toggle

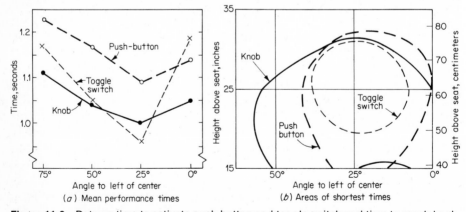

Figure 11-9 Data on time to activate push button and toggle switch and time to reach knob, when in various positions. Data are for *left* hand, and for close console. Part *a* gives mean times for controls positioned 25 in above seat reference level. Part *b* gives contours of the area for which times were *shortest* for the particular control (e.g., times within 5 percent of the minimum time for the control). [Adapted from Sharp and Hornseth, 16.]

switch suggests that the selection of a location for such devices may be more critical than for the other devices if time is of the essence. However, as suggested by the investigators, a well-designed system preferably should not impose response-time requirements on operators in which differences of 0.1, 0.2, or 0.3 s are crucial.

Foot Controls

Since only the most loose-jointed among us can put their feet behind their heads, the location of foot controls generally needs to be in the fairly conventional areas, such as those depicted in Figure 11-10. These areas, differentiated as optimal and maximal, for toe-operated and heel-operated controls, have been delineated on the basis of dynamic anthropometric data. The maximum areas indicated require a fair amount of thigh or leg movement or both, and preferably should be avoided as locations for frequent or continual pedal use. Incidentally, that figure is predicated on the use of a horizontal seat pan; with an angular seat pan (and more angled back rest) the pedal locations need to be manipulated accordingly (although such adjustments have been published, they will not be given here).

The areas given in Figure 11-10 generally apply to foot controls that do not

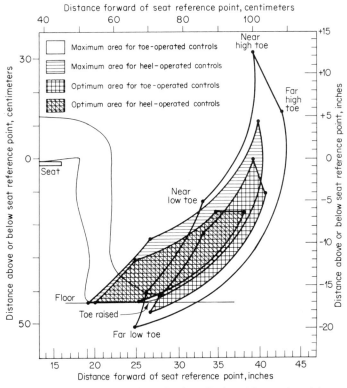

Figure 11-10 Optimal and maximal vertical and forward pedal space for seated operators [Adapted from *Personnel subsystems,* 23, chap. 2, sec. 2D7, p. 3.]

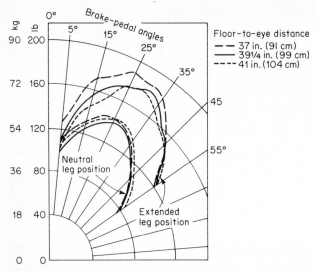

Figure 11-11 Mean maximum brake-pedal forces exerted by 100 Air Force pilots in two leg positions and for various brake pedal angles and three "floor-to-eye" distances. (The seat level was adjusted for three conditions so the eye level was at the three specified distances above the floor.) [Adapted from Hertzberg and Burke, 11, fig. 6.]

require substantial force. For applying considerable force, a pedal preferably should be fairly well forward. This point is illustrated by Figure 11-11, which shows the mean maximum brake-pedal forces for 100 Air Force pilots when the leg was in a "normal" position and in an extended position (with the leg essentially forward from the seat). The figure shows that the maximum brake-pedal forces were clearly greater for the extended position for three seat levels that were used in the experiment. (The seat level was adjusted so the "floor-to-eye" distances were standardized.) As an aside, the figure also shows that the forces for both leg positions were greatest when the foot angle on the pedal was between about 15 and 35° from the vertical.

SPECIFIC ARRANGEMENT OF COMPONENTS

Once a determination has been made of the general locations of components (such as controls and displays), the next process is that of arranging those which are within the same general area. There are actually two aspects of the process, namely, the basic layout of components relative to each other, and the spacing between and among them (especially of controls).

Layout of Components

As indicated before, the arrangement of groups of components can be based on the principles of sequence or function. Where there are common sequences, or at least frequent relationships, in the use of displays, controls, or other components, the layout usually should be such as to facilitate the sequential process—as in hand movements, eye movements, etc. This is essentially a matter of minimizing link values. Figures 11-2 and 11-3 shown earlier in this chapter are examples.

When there are no fixed or common sequences, the components should be grouped on the basis of function. In such instances the various groups should be clearly indicated by borders, color, or shading or otherwise. An example of such grouping is shown in Figure 11-12.

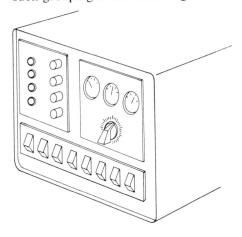

Figure 11-12 Example of controls and displays grouped by function, in which the different groups are clearly indicated.

SPACING OF CONTROL DEVICES

Although we have talked about minimizing the distances between components, such as the sequential links between controls, there are obvious lower-bound constraints that need to be respected, such as the physical space required in the operation of individual controls to avoid touching other controls. Whatever lower-bound constraints there might be would be predicated on the combination of anthropometric factors (such as of the fingers and hands) and on the precision of normal psychomotor movements made in the use of control devices.

An illustration of the effects of such factors is given in Figure 11-13, this showing inadvertent "touching errors" in the use of knobs of various diameters as a function of the distances between their edges. In this instance the figure shows that errors dropped sharply with increasing distances between knobs up to about 1 inch (2.5 cm), while beyond that distance performance improved at a much slower rate. When separate comparisons were made between knob centers (rather than edges), however, performance was more nearly error-free for $1/2$-in-diameter knobs than for the larger knobs. This suggests that when panel space is at a high premium, the smaller-diameter knobs are to be preferred. By referring again to Figure 11-13, it can be seen that touching errors were greatest for the knob to the right of the one to be operated, and were minimal for the one on the left.

On the basis of various studies of control devices Chapanis in Van Cott and Kinkade [2] have set forth certain recommended distances (preferred and minimal) between pairs of similar devices, as given in Figure 11-14.

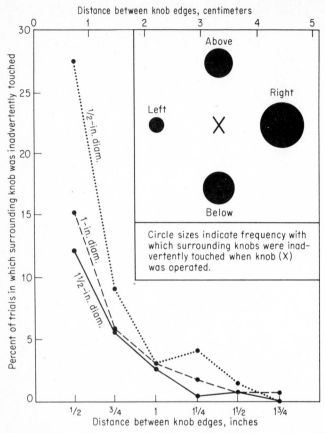

Figure 11-13 Frequency of inadvertent touching errors for knobs of various diameters as a function of the distance between knob edges. Areas of circles in inset indicate relatively the frequency with which the four surrounding knobs were inadvertently touched when knob *x* was being operated. [From Bradley, 1, fig. 2.]

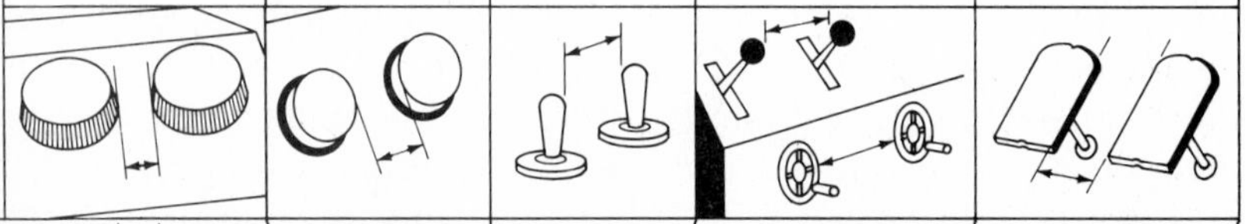

Number of body members and → type of use			Knobs	Push buttons	Toggle switches	Cranks, levers	Pedals
1, randomly	in	2(1)		2(½)	2(¾)	4(2)	6(4)
	cm	5(2.5)		5(1.3)	5(1.8)	10(5)	15(10)
1, sequentially	in			1(¼)	1(1½)		4(2)
	cm			2.5(.6)	2.5(1.3)		10(5)
2, simultaneously	in	5(3)				5(3)	
	cm	12.7(7.6)				12.7(7.6)	
2, randomly, sequentially	in	½(½)		¾(⅝)			
	cm	1.3(1.3)		1.8(1.6)			

Figure 11-14 Recommended separation in inches and centimeters between adjacent controls. Preferred separations are given for certain types of use with corresponding minimum separation in parentheses. [Adapted from Chapanis, 2.]

GENERAL GUIDELINES IN DESIGNING INDIVIDUAL WORKPLACES

In designing workplaces some compromises are almost inevitable because of competing priorities. In this regard, however, appropriate link values can aid in the trade-off process. Some general guidelines for designing workplaces that involve displays and controls are given below [adapted from Van Cott and Kinkade, 20, chap. 9]:

- *First priority:* Primary visual tasks
- *Second priority:* Primary controls that interact with primary visual tasks
- *Third priority:* Control/display relationships (put controls near associated displays; compatible movement relationships, etc.)
- *Fourth priority:* Arrangement of elements to be used in sequence
- *Fifth priority:* Convenient location of elements that are used frequently
- *Sixth priority:* Consistency with other layouts within the system or in other systems

EXAMPLES OF INDIVIDUAL WORKPLACES

Although it is not feasible to illustrate the physical arrangement of very many types of workplaces, a few examples will be given below—in particular, certain ones for which some design guidelines have been developed or that represent special design problems.

Consoles

Consoles of one sort or another are used for operators of various types of systems, the consoles usually including both displays and controls. Figure 11-15 illustrates the design features of consoles that are recommended on the basis of anthropometric characteristics of people along with consideration to their visual and psychomotor skills.

Vehicle Cabs

A similar set of recommended design features of vehicle cabs is shown in Figure 11-16, this also representing something of a general composite. One particular feature should be given special attention, namely, the visual field that can be viewed from the driver's position. (In Figure 11-16, with a vertically adjustable seat, the eye position of most drivers would be within 2 inches above or below the position shown in the figure.) Other important features are: the seat height, depth, and back angle; leg and knee clearance; and hand and foot reach requirements for control action.

BART Rapid Transit System

The Bay Area Rapid Transit District, in the development of the BART car for use in the San Francisco Bay area, set forth certain human factors design cri-

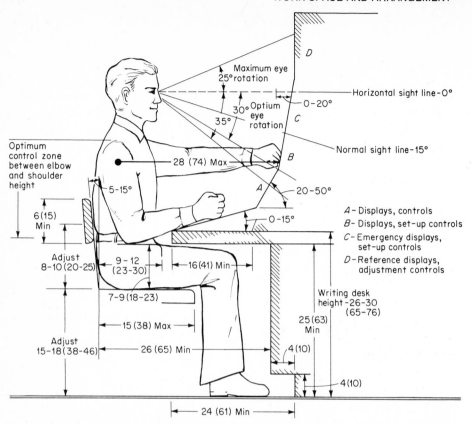

Figure 11-15 Recommended design features of consoles for seated operators. These features are designed to be suitable for persons ranging from the 5th to the 95th percentiles of an Air Force population.) [Adapted from Van Cott and Kinkade, 20, fig. 9-6.]

teria for the system, some of these relating to the comfort and convenience of the passengers, and others relating to the operation of the car. In the latter connection the system provides for electronic control by a central computer system, but with an attendant on each train for various purposes, especially for overriding the central computer system in case of emergency. Actually the attendant is in a detachable control *pod* that can be positioned at the front of the first car. In the design of this pod, a prime requirement was that of ensuring adequate forward visibility to observe signals, the track ahead, and around curves, and backward visibility to observe the interior of the car. Figure 11-17 shows the control console of the prototype pod.

Crane Controls and Cabs

In the iron and steel industries, as well as certain others, traveling overhead cranes are used to move materials from one place to another. The operational requirements of this moving control would suggest the need for unobstructed vision, easily operated controls that respond with acceptable fidelity to the operator's responses, and a relatively comfortable posture. The British Iron and Steel Research Association has interested itself in the design and arrangement

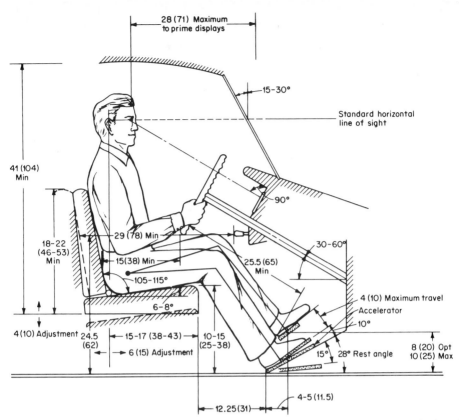

Figure 11-16 Recommended design features of vehicle cabs. These features are designed to be suitable for persons between the 5th and 95th percentile of operators. [Adapted from Van Cott and Kinkade, 20, fig. 9-11.]

Figure 11-17 Attendant's control console for the prototype of the pod of the BART rapid-transit car described by Sundberg and Ferar [19]. [Photograph courtesy of Sundberg-Ferar, Southfield, Mich.]

of equipment for use in iron and steel operations, including this one. On the basis of some research, plus the application of sound human factors principles, a modification was made of an original crane cab along with its associated controls. The before-and-after designs are given in Figure 11-18. The improved

(*a*) Original crane cab   (*b*) Original crane controls

(*c*) Improved crane cab   (*d*) Improved crane controls

Figure 11-18 Illustration of original and improved cab and controls for overhead traveling crane used in steel mill. [From Laner, 14, and Sell, 15. Courtesy The British Iron and Steel Research Association.]

visibility, better arrangement of the controls, and posture of the operator can be seen from these photographs.

DISCUSSION

The variety of physical situations in which people find themselves is of course almost infinite, and we have here been able to touch on only a few aspects of the design of such situations. The basic point that is intended is that the design of such situations can have an effect upon any of a number of relevant criteria, such as the performance of intended functions (such as operation of a processing operation), safety (as in the operation of vehicles), and comfort. Thus, in the design process the designers need to utilize whatever relevant anthropometric and performance data are available (plus good judgment) in designing the system or facility for suitable human use.

REFERENCES

1 Bradley, J. V.: Optimum knob crowding, *Human Factors*, 1969, vol. 11, no. 3, pp. 227–238.

2 Chapanis, A.: "Design of Controls," in H. P. Van Cott and R. G. Kinkade (eds.), *Human engineering guide to equipment design*, rev. ed., chap. 8. U.S. Government Printing Office, Washington, 1972, chap. 8.

3 Deininger, R. L.: *Process sampling, workplace arrangements, and operator activity levels*, unpublished report, Engineering Psychology Branch, USAF, WADD, 1958.

4 Dupuis, H.: *Farm tractor operation and human stresses*, paper presented at the meeting of the American Society of Agricultural Engineers, Chicago, Dec. 15–18, 1957.

5 Dupuis, H., R. Preuschen, and B. Schulte: *Zweckmäbige gestaltung des schlepperführerstandes*, Max Planck Institutes für Arbeitsphysiologie, Dortmund, Germany, 1955.

6 Fitts, P. M.: "A study of location discrimination ability," in P. M. Fitts (ed.), *Psychological research on equipment design*, Army Air Force, Aviation Psychology Program, Research Report 19, 1947.

7 Fowler, R. L., W. E. Williams, M. G. Fowler, and D. D. Young: An investigation of the relationship between operator performance and operator panel layout for continuous tasks, USAF AMRL-TR-68–170, December, 1968 (AD-692 126).

8 Freund, L. E., and T. L. Sadosky: Linear programming applied to optimization of instrument panel and work-place layout, *Human Factors*, 1967, vol. 9, no. 4, pp. 295–300.

9 Haines, R. F., and K. Gilliland: Response time in the full visual field, *Journal of Applied Psychology*, 1973, vol. 58, no. 3, pp. 289–295.

10 Haygood, R. C., K. S. Teel, and C. P. Greening: Link analysis by computer, *Human Factors*, 1964, vol. 6, no. 1, pp. 63–78.

11 Hertzberg, H. T. E., and F. E. Burke: Foot forces exerted at various aircraft brake-pedal angles, *Human Factors*, 1971, vol. 13, no. 5, pp. 445–456.

12 Jones, R. E., J. L. Milton, and P. M. Fitts: *Eye fixations of aircraft pilots: IV. Frequency, duration, and sequence of fixations during routine instrument flight*, USAF, AF TR 5975, 1949.

13 Kroemer, K. H. E.: Human strength: terminology, measurement, and interpretation of data, *Human Factors*, 1970, vol. 12, no. 3, pp. 297–313.

14 Laner, S.: *Ergonomics in the steel industry*, The British Iron and Steel Research Association, Report 19/61, List 120, November–December, 1961.

15 Sell, R. G.: The ergonomic aspects of the design of cranes, *Journal of the Iron and Steel Institute*, 1958, vol. 190, pp. 171–177.

16 Sharp, E. and J. P. Hornseth: *The effects of control location upon performance time for knob, toggle switch, and push button*, AMRL, TR 65–41, October, 1965.

17 Siegel, A. I., and F. R. Brown: An experimental study of control console design, *Ergonomics*, 1958, vol. 1, pp. 251-257.

18 Stevens, P. H., D. O. Chase, and A. W. Brownlie: Industrial design of a narrow aisle sit-down lift truck, *Human Factors*, 1966, vol. 8, no. 4, pp. 317-325.

19 Sundberg, C. W., and M. Ferar: Design of rapid transit equipment for the San Francisco Bay Area rapid transit system, *Human Factors*, 1966, vol. 8, no. 4, pp. 339-346.

20 Van Cott, H. P., and R. G. Kinkade: "Design of individual workplaces," in H. P. Van Cott and R. G. Kinkade (eds.), *Human engineering guide to equipment design*, rev. ed., U.S. Government Printing Office, Washington, D.C., 1972, chap. 9.

21 *Handbook of instructions for aerospace personnel subsystem design* (HIAPSD), USAF, AFSC Manual, 80-3, 1967.

22 "Layout of panels and machines," *Applied ergonomics handbook*, chap. 5, in *Applied Ergonomics*, 1970, vol. 1, no. 2, pp. 107-112.

23 *Personnel subsystems*, USAF, AFSC design handbook, Series 1-0, General, AFSC DH 1-3, Jan. 1, 1972, Headquarters, AFSC, 2d ed.

Part Five

Environment

Chapter 12

Illumination

In many aspects of life we depend upon the sun as our source of illumination, as in driving in the daylight, playing golf, or picking tomatoes. When human activities are carried on indoors, however, it is usually necessary to provide some artificial illumination. Within those circumstances in which artificial illumination is used (or at least could be used), it is of course possible to design the illumination system so the illumination provided is satisfactory for the "users" of the environment. This chapter will deal with some aspects of illumination from the human factors point of view. First, however, we should discuss the nature and measurement of light.

THE NATURE AND MEASUREMENT OF LIGHT

Light is *visually evaluated radiant energy*. The entire radiant energy (electromagnetic) spectrum consists of waves of radiant energy that vary from about 1/1 billion of a millionth to about 100 million meters (m) in length. This tremendous range includes cosmic rays; gamma rays; x-rays; ultraviolet rays; the visible spectrum; infrared rays; radar; FM, TV, and radio broadcast waves; and power transmission—as illustrated in Figure 12-1.

The visible spectrum ranges from about 380 to 780 nanometers (nm). The

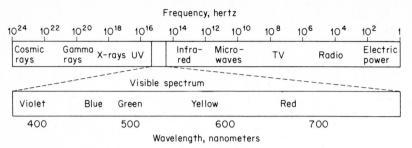

Figure 12-1 The radiant energy (electromagnetic) spectrum, showing the visible spectrum. [Adapted from *Light and color,* 26, p. 5.]

nanometer (formerly referred to as a "millimicron") is a unit of wavelength equal to 10^{-9} (one-billionth) m. Light can be thought of as the aspect of radiant energy that is visible; it is then basically psychophysical in nature rather than purely physical or purely psychological. Variations in wavelength within the visible spectrum give rise to the perception of color, the violets being around 400 nm, blending into the blues (around 450 nm), the greens (around 500 nm), the yellow-oranges (around 600 nm), and the reds (around 700 nm and above).

Light comes to us from two sources: incandescent bodies ("hot" sources, such as the sun, luminaries, or a flame) and luminescent bodies ("cold" sources, i.e., the objects we see in our environment, which reflect light to us). A hot light source that includes all wavelengths in about equal proportions is called *white light.* Most light sources, as luminaires, have spectrums that include most wavelengths but that tend to have more energy in certain areas of the spectrum than in others. These differences make the lights appear yellowish, greenish, bluish, etc.

As light from a hot source falls upon an object, some specific combination of wavelengths is absorbed by the object. The light that is so reflected is the effect of the interaction of the spectral characteristics of the light source with the spectral absorption characteristics of the object. If a colored object is viewed under white light, it will be seen in its "natural" color. If it is viewed under a light that has a concentration of energy in a limited segment of the spectrum, the reflected light may alter the apparent color of the object, such as a blue necktie appearing green when viewed under a yellow light. This effect is illustrated in Figure 12-2, this figure showing how the apparent complexion of a person is influenced by the nature of the light source.

The Measurement of Light: Photometry

There are many concepts and terms that relate to the measurement of light. A few of these will be defined here. The *candela* (cd), formerly called the *candle,* is a measure of luminous intensity.[1] *Candlepower* (cp) is a measure of the luminous intensity of a light source as expressed in candelas. *Luminous flux* is the

---

[1]The candela is defined as the luminous intensity of $^1/_{600,000}$ square meter of projected area of a blackbody radiator operating at the temperature of the solidification of platinum (2047°K).

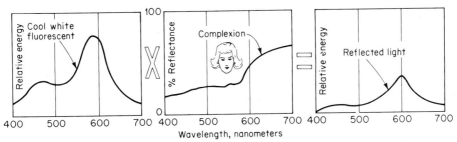

Figure 12-2 Illustration of the effect of the spectrum of a light source on the spectrum of reflected light, in this instance the light reflected from the complexion of a person's face. [Adapted from *Light and color*, 26, p. 25.]

time rate of flow of light measured in lumens (lm). The light from a point source of one candela (12.57 lumens) falling on an area of 1 ft² at a distance of 1 ft would be 1 lumen. *Illuminance* (or luminous intensity) is the light falling on a surface; one measure of illuminance is the *footcandle* (fc).

The *footcandle* (fc) is the unit of illumination on a surface 1 ft² in area on which there is a uniformly distributed flux of 1 lumen, or the illumination produced on a surface all points of which are at a distance of 1 ft from a directionally uniform point source of 1 candela. The distribution of illumination follows the *inverse-square law*, as follows:

$$\text{fc} = \frac{\text{cp}}{D^2} \text{ in which } D \text{ is the distance in feet}$$

At 2 ft a 1-cp source would produce $^1/_4$ fc, and at 3 ft it would produce $^1/_9$ fc. The inverse-square law is illustrated in Figure 12-3. *Dekalux* is also a measure of luminance, and is equal to 1.076 fc. One dekalux equals 10 lux. *Luminance* is the amount of light per unit area reflected from or emitted by a surface. It is usually measured in units of *millilamberts* (mL) or *footlamberts* (fL). The *footlambert* (fL) is a measure of luminance, and is equal to $1/\pi$ candela per square foot. The *millilambert* (mL) is equal to 0.929 fL, and for practical purposes can be considered as an equivalent measure. The light that is reflected from the surface of objects that strike the eye (specifically the retina) causes us to "see" objects—their configuration and color.

Color

The light that is reflected from objects—and that produces our sensations of color—can be described in terms of three characteristics: the *dominant* wavelength; *saturation* (i.e., the predominance of a narrow range of wavelengths, as contrasted with an admixture of various wavelengths); and *luminance*. These physical characteristics of the light, in turn, influence our perceptions of color in terms of three corresponding attributes, respectively: hue; saturation (the attribute that determines the degree of difference of a color from a gray with the same lightness); and lightness (the attribute associated with the relative amount

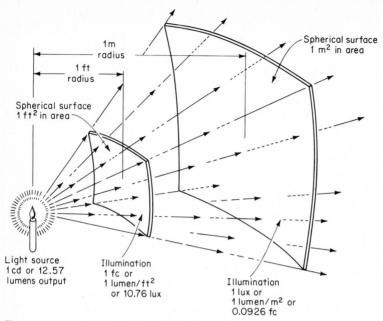

1m
radius

1 ft
radius

Spherical surface
1 m² in area

Spherical surface
1 ft² in area

Light source
1 cd or 12.57
lumens output

Illumination
1 fc or
1 lumen/ft²
or 10.76 lux

Illumination
1 lux or
1 lumen/m² or
0.0926 fc

Figure 12-3 Illustration of the distribution of light from a light source following the inverse-square law. [From *Light measurement and control,* 27, p. 5.]

of the incident light). These three attributes are depicted in the color cone shown in Figure 12-4. In the color cone, hue is indicated by position around the circumference. Saturation (sometimes called *purity* or *chroma*) is shown in the color cone as the radius. A saturated color consists of a single hue and would be positioned on the circumference of the cone. Colors toward the center are mix-

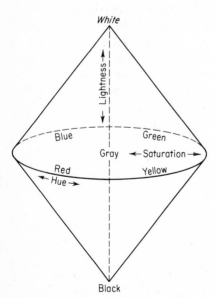

White

Lightness

Blue

Green

Gray   ← Saturation →

Red

Yellow

← Hue →

Black

Figure 12-4 The color cone. Hue is shown on the circumference, lightness (from light to dark) on the vertical, and saturation on the radius from circumference to the center.

tures of various hues, and while they may have a dominant hue, they do not appear to be pure. Lightness (sometimes referred to as *value* or *brightness*) is shown on the vertical dimension, the center of which ranges from white through varying levels of gray, to black. Any color of a given hue and saturation can be varied in its lightness. Although there is a general relationship between the luminance of light and the subjective response of lightness, all colors that reflect equal total amounts of light energy are not necessarily perceived as equal in lightness. This is due to the fact that the eye is differentially sensitive to various wavelengths.

Color Systems   Two types of color systems are used as standards of colors. First are those which consist of color plates or color chips for use as standards in characterizing colorants. The Munsell color system [28], the Ostwald color system [24], and the Maerz and Paul *Dictionary of color* [10] are of this type. Some of these (for example, the Munsell and Ostwald systems) correspond substantially to the color cone and provide standard nomenclature to identify selected colors of certain hues, saturations, and lightness levels. The Ostwald system, for example, identifies 680 color samples (28 color chips for each of 24 hues, plus 8 representing lightness levels on the white-to-black continuum).

The CIE color system, developed by the Commission Internationale de l'Eclairage [29], provides for designating colors in terms of their relative percentages of each of the three primary colors of light, namely, red (X), green (Y), and blue (Z). All possible colors can be designated with the CIE system on a chromaticity diagram, whether they are emitted, transmitted, or reflected.

AMOUNT OF ILLUMINATION

The problem of determining the level of illumination that should be provided for various visual tasks has occupied the attention of illuminating engineers, psychologists, and others for many years. Over a period of a number of years, however, Blackwell [1, 2, 3, 4, 5] has carried out an extensive research program dealing with illumination that has served as the basis for the establishment of recommended illumination levels by the IES for various tasks and activities [25].

Visibility Research of Blackwell

Because Blackwell's research has been accepted by the IES as the primary base for specifying interior illumination levels for tasks and activities, a summary of some of his work will be in order.[2]

Laboratory Procedures and Results   Blackwell's research was concerned primarily with the study of basic parameters of visibility. His laboratory setup

[2]The details of his research and of the procedures for specifying illumination levels for various tasks and activities are available in other sources [Blackwell, 1, 2; Crouch, 9; *IES lighting handbook*, 25].

consisted of a large cubicle painted white and illuminated by concealed lighting from the side. The subjects viewed one side of this that consisted of a translucent screen, the luminance of which could be varied from 0.001 to 800 fL. The target to be viewed consisted of a disk of light which could be projected onto that screen. The disk could be varied in size (1 minute of arc to 64), ratio of contrast with background (0.01 to 300), and time of exposure (0.001 to 1.0 s). A given disk was projected during any one of four time intervals, these being delineated by a buzzer sound at the beginning of each interval. As a disk was presented on the screen (during one of the four intervals delineated by the buzzer), the subject was to identify the interval and press a button that identified the interval in question. Some of the results are summarized in Figure 12-5. This shows, for various target sizes (1, 2, 4, and 10

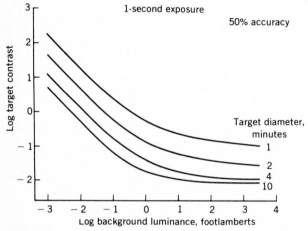

Figure 12-5 Relationship between luminance of the background and contrast required for 50 percent accurate discriminability of targets of different sizes (minutes of arc). [From Blackwell, as presented by Crouch, 9.]

minutes of arc), the relationship between luminance of the background and the contrast between the background and the target disk required for 50 percent accuracy of identification of the disk. For estimating values that would be required for 99 percent accuracy, the contrast value should be multiplied by a factor of about 2[Crouch, 9]. One can here see the beginnings of a scheme for utilizing such data for determining illumination requirements for various tasks.

Using a research procedure that involved the identification of defects in moving targets (a procedure we need not describe here), it was possible to equate the visibility requirements of moving versus static targets.

Visibility Level (VL) as Visual Performance Criterion The use of Blackwell's research in establishing recommended illumination levels is based on the concepts of visibility level (VL) and Visibility Reference Function. Let us start with the Visibility Reference Function, which is shown as the bottom curve of Figure 12-6. That curve—derived from Blackwell's research—repre-

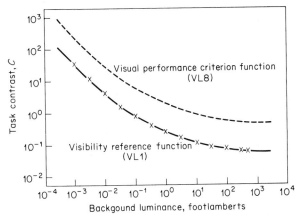

Background luminance, footlamberts

Figure 12-6 Two visibility level (VL) curves, each representing combinations of contrast and luminance required for equal visibility. VL1 (Visibility Reference Function) is a basic curve relating to the luminance needed to achieve threshold visibility of a static task of discriminating a 4-minutes-of-arc luminous disk exposed for $^1/_5$ s. In turn, VL8 (Visual Performance Criterion Function) shows the level adjusted for three factors: dynamic versus static target; not knowing versus knowing where and when the target will appear; and requiring a 99 versus 50 percent probability of detection. [From *IES lighting handbook*, 25, fig. 3-24, as based on work of Blackwell.]

sents the luminance contrast (on the vertical scale) that is required at different levels of task background luminance (on the horizontal scale) to achieve visibility threshold for a standard visual task. This task consisted of the detection of a luminous disc, 4 minutes of visual arc in size, exposed for $^1/_5$ s. (The observers used in this phase adjusted the contrast of the disc relative to its background to the borderline between "visibility" and "invisibility.") That curve shows in effect the trade-off between contrast and luminance required for equal visibility of a given visual target. That basic Visibility Reference Function is one of a number of possible visibility levels (VL), this particular one being designated VL1. One could envision various VL curves above this, each of which would reflect equal visibility levels greater than the threshold curve VL1.

In practical situations one would of course want to use a level of luminance that provided for more than "threshold" discrimination of such a task as that used in Blackwell's experiment. In the development of a "common-sense" level that might be suitable for many actual tasks, three factors were taken into account, and for each one a compensating adjustment was made to the VL1. The adjustments were for the difference in visibility levels required to compensate for the differences in the following factors: (1) a static versus dynamic presentation of the same visual task (i.e., target); (2) the difference between knowing and not knowing where and when a visual task (i.e., target) might appear in the visual field; and (3) the differences between a 50 percent probability of detection of the target and the more practical standard of a 99 percent probability. The statistical multipliers for these three factors are 2.78, 1.5, and 1.9, respectively, which, multiplied together, give a product of about 8. The upper curve of Figure 12-6 reflects the combination of these and is labeled VL8. This curve, then, can be considered as characterizing the combinations of contrast and luminance required for equal visibility of a standard visual task under the conditions of the three factors mentioned above.

Establishing Illumination Level Recommendations

The research of Blackwell provides a basis for determining the illumination levels for various tasks. (The specific procedures for doing this are discussed in the *IES lighting handbook*, 25.) We will here discuss only a couple of factors that are basic to the procedures, namely, luminous contrast of task detail and task background luminance.

A visual task typically requires the visual detection of certain detailed features of an object or whatever one sees in the visual field. "Luminous contrast of task detail" refers to the difference in luminance between those features and the background, such as between letters on a road sign and their background. (This contrast, in combination with the size of the detailed features, affects the difficulty of the task.) "Task background luminance" refers to the amount of light that is reflected from the task background (such as from a road sign).

In determining the illumination requirements for any given task, a value of "equivalent contrast" with the standard task (represented as VL8 in Figure 12-6) is determined by a method that need not be described. This value, in turn, can be used to determine the required luminance in fL, either by reading from Figure 12-6 or by another method. In turn, the required luminance in fL serves as the basis for determining the illumination in fc required to produce the required luminance. Since this depends on the reflectance of the task background, an additional adjustment is made to account for this in deriving the required fc for the task, based on the following relationship:

$$\text{Required illumination, fc} = \frac{\text{required luminance, fL}}{\% \text{ reflectance}}$$

The very important influence of reflectance on illumination requirements in fc is illustrated below for three background reflectances (80, 50, and 20 percent) for two levels of required luminance (40 and 100 fL):

Required luminance, fL	Required illumination, fc, for three reflectance levels		
	80%	50%	20%
40 fL	50	80	200
100 fL	125	200	500

The procedures provided by the IES can be used for deriving illumination requirements for any individual task. However, such procedures have already been applied to many types of tasks and are published [*IES lighting handbook*, 25, chap. 9]. Thus, for at least many types of tasks, these recommendations can be used. Some examples are given in Table 12-1.

Table 12-1 Illumination Standards Recommended by the IES for Several Selected Types of Situations and Tasks

Situation or task	Recommended illumination, fc	Dekalux
Assembly:		
Rough easy seeing	30	32
Rough difficult seeing	50	54
Medium	100	110
Fine	500	540
Extra fine	1000	1080
Machine shops:		
Rough bench and machine work	50	54
Medium bench and machine work	100	110
Fine bench and machine work	500	540
Extra-fine bench and machine work, grinding—fine work	1000	1080
Storage rooms or warehouses: Inactive	5	5.4
Offices:		
Cartography, designing, detailed drafting	200	220
Accounting, bookkeeping, etc.	150	160
Mail sorting	100	110
Corridor, elevators, stairways	20	22
Residences:		
Kitchen, food preparation	150	160
Reading, writing	70	75

Source: Examples selected from *IES lighting handbook* [25, pp. 9–81 to 9–95].

As most persons of advancing age can testify, visual performance tends to go to pot with increasing age. Thus, illumination levels that are adequate for young persons may not be adequate for older persons. In this regard Blackwell and Blackwell [7], on the basis of visual performance by people of various ages, have derived "contrast multipliers" for different age groups, using the median performance of a 20–25-yr group as a base of 1.00. The multipliers for the median values of certain other age groups are: 40 yr, 1.17; 50 yr, 1.58; and 65 yr, 2.66. To provide for higher proportions of the age-group populations (such as 90 or 95 percent instead of 50 percent), higher multipliers would be required, up to 6.92.

Although there have been many laboratory studies of visual performance under varying illumination conditions, oddly enough there have been relatively few documented studies of actual work performance under various levels of illumination. In one summary of such studies [McCormick, 19] data are given on the "original" illumination levels (some of which were very low) for 19 different jobs, the "new" levels (which in some instances were still markedly below current recommended practice), and the percent change in work output of workers. There were reported increases in work output in all instances, ranging from 4 to 35 percent. For various reasons one needs to be a bit circumspect in evaluating such results, but they do tend to indicate that actual work performance is influenced somewhat by illumination level.

Discussion

In considering the illumination levels for various tasks, it should be noted that there has been an ongoing controversy over the years that has not yet abated, the central issue relating to whether visual performance does continue to improve with increasing levels of luminance [Hopkinson and Collins, 1970, Faulkner and Murphy, 11]. The luminance levels of the IES recommendations as based on Blackwell's research tend to level off with decreasing task contrast, as shown in Figure 12-6, particularly in the region of task brightness above 10 fL. This indicates that a slight decrease in contrast would require a large increase in illumination level to maintain adequate task visibility.

And although high levels of illumination might be warranted in some circumstances, there are definite indications that for certain tasks high levels of illumination can weaken information cues [Logan and Berger, 17], by suppressing the "visual gradients" of the pattern density of the objects being viewed. This effect is illustrated in Figure 12-7, which shows the effects of general

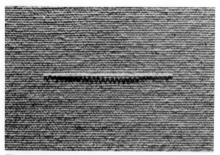

Figure 12-7 Illustration of the effects of general versus surface-grazing illumination on the visibility of a loose thread on cloth. [From Faulkner and Murphy, 11, Fig. 3 and 4.]

versus "surface-grazing" illumination. Thus, in some circumstances lower levels of illumination (albeit of the appropriate type) would be better than high levels. Thus, one probably needs to recognize that the illumination requirements of at least some visual tasks are not covered by the present recommendations.

GLARE

Glare is produced by brightness within the field of vision that is sufficiently greater than the luminance to which the eyes are adapted to cause annoyance, discomfort, or loss in visual performance and visibility. *Direct* glare is caused by light sources in the field of view, and *reflected* or *specular* glare is caused by reflections of high brightness from polished or glossy surfaces that are reflected toward an individual.

Effects of Glare on Visual Performance

The effects of direct glare on the visual performance of people are illustrated by the results of a study in which the subjects viewed test targets with a glare source of a 100-W inside-frosted tungsten-filament lamp in various positions in the field of vision [18]. The test targets consisted of parallel bars of different sizes and contrasts with their background. The glare source was varied in position in relation to the direct line of vision, these positions being at 5, 10, 20, and 40° with the direct line of vision, as indicated in Figure 12-8. The effect of the

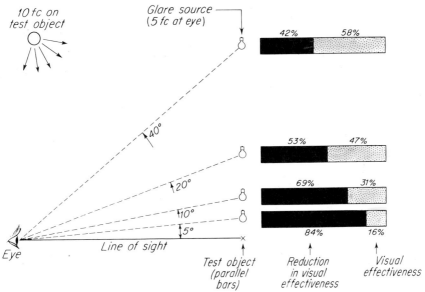

Figure 12-8 Effects of direct glare on visual effectiveness. The effects of glare become worse as the glare source gets closer to the line of sight. [After Luckiesh and Moss, 18.]

glare on visual performance is shown as a percentage of the visual effectiveness that would be possible without the glare source. It will be seen that with the glare source at a 40° angle, the visual effectiveness is 58 percent, this being reduced to 16 percent at an angle of 5°.

Glare and Visual Comfort

Visual discomfort from glare is, unfortunately, a common experience, and a major concern in the design of luminaires and lighting installations should be to minimize such discomfort. Toward this end, there has been considerable research relating to glare and its effects on the subjective sensations of visual comfort and discomfort. As a direct result of such research, the IES has adopted a standard procedure for computing *discomfort glare ratings* (DGR) for luminaires and for tentatively planned interior lighting situations [31].

Derivation of Discomfort Glare Ratings (DGR)

The calculation of DGR ratings for specific lighting layouts takes into account most of the situational factors that affect visual comfort, as follows: (1) room size and shape; (2) room surface reflectances; (3) illumination level; (4) luminaire type, size, and light distribution; (5) number and location of luminaires; (6) luminance of entire field of view; (7) observer location and line of sight; and (8) equipment and furniture. In addition, the procedures can take into account a ninth variable, if desired, namely, differences in individual glare sensitivity.

The scheme for estimating the DGR for any specific lighting layout is complex and will not be described here [see *IES lighting handbook*, 25, pp. 324–328]. The derived DGR value for any given circumstance, however, can be converted into a visual comfort rating, which is an estimate of the visual comfort probabilities (VCP), that is, the percent of observers who would be expected to judge a lighting condition to be at, or more comfortable than, the borderline (i.e., the threshold) between comfort and discomfort. The relationships between DGR and VCP are shown in Figure 12-9.

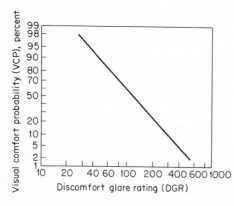

Figure 12-9 Relationship between discomfort glare ratings (DGR) of specific illumination conditions to visual comfort probabilities (VCP), that is, the percent of observers who could be expected to judge a lighting condition to be at or more comfortable than the borderline between comfort and discomfort. [Adapted from *IES lighting handbook*, 25, fig. 3-27.]

Following the procedures for deriving the DGR values, one can derive, for any given type of luminaire (as mounted lengthwise or crosswise for a room), the VCP values for any given room size (width, length, and ceiling height). In the application of the procedure, it has been stated by the IES Subcommittee on Direct Glare [31, p. 643] that direct glare will not be a problem in a lighting installation if three specific conditions are fulfilled. One of these is that the VCP be 70 or more.[3]

Although most of the interest in direct glare has been with respect to glare from luminaires, Hopkinson [12] has carried out some research relating to glare from daylight (through windows) and presents a modified formula for computing a glare index for the daylighting of interiors. Windows, of course,

---

[3]The others are that the ratio of maximum-to-average luminaire luminance should not exceed 5:1 (preferably 3:1) at 45, 55, 65, 75, and 85° from the nadir crosswise and lengthwise, and maximum luminances of luminaires should not exceed values of 2250, 1605, 1125, 750, and 495 fL, respectively, at these angles.

can provide a useful source of illumination, so a basic problem is that of designing windows in order to provide daylight illumination (along with other desirable features, such as ventilation, possibly a view, etc.) with minimum glare.

The Reduction of Glare

• *To reduce direct glare from luminaires:* (1) Select luminaires with low DGR; (2) reduce the luminance of the light sources (for example, by using several low-intensity luminaires instead of a few very bright ones); (3) position luminaires as far from line of sight as feasible; (4) increase luminance of area around any glare source, so the luminance (brightness) ratio is less; and (5) use light shields, hoods, and visors where glare source cannot be reduced.

• *To reduce direct glare from windows*: (1) Use windows that are set some distance above the floor; (2) construct an outdoor overhang above the window; (3) construct vertical fins by the window extending into the room (to restrict direct line of sight to the windows); (4) use light surrounds to window (to minimize contrast with light from window); and (5) use shades, blinds, or louvers [Hopkinson, 12].

• *To reduce reflected glare:* (1) Keep luminance level of luminaires as low as feasible; (2) provide good level of general illumination (such as with many small light sources and use of indirect lights); (3) use diffuse light, indirect light, baffles, window shades, etc.; (4) position light source or work area so reflected light will not be directed toward the eyes; and (5) use surfaces that diffuse light, such as flat paint, nonglossy paper, and crinkled finish on office machines; avoid bright metal, glass, glossy paper, etc.

DISTRIBUTION OF LIGHT

There are bits and pieces of evidence that visual performance is generally (but not always) better if there is a reasonable level of general illumination at the work area.

Luminance Ratio

The *luminance ratio* is the ratio of the luminance of a given area (as the work area) to the surrounding area. The ratios recommended by the IES for various areas relative to the visual task, for both office and industrial situations, are given in Table 12-2. In this, the ratio for the task and its adjacent surroundings is given as 3:1 (except if the surroundings are very light in color).

Reflectance

The distribution of light within a room is not only a function of the amount of light and the location of the luminaires, but is also influenced by the reflectance of the walls, ceilings, and other room surfaces. Tied in with reflection is the concept of the *utilization coefficient*, which is the percentage of light that is reflected, collectively, by the surfaces in the room or area. The influence of

Table 12-2 Recommended Luminance Ratios for Offices and Industrial Situations

Areas	Recommended maximum luminance ratio	
	Office	Industrial
Task and adjacent surroundings	3:1	
Task and adjacent darker surroundings		3:1
Task and adjacent lighter surroundings		1:3
Task and more remote darker surfaces	5:1	10:1
Task and more remote lighter surfaces	1:5	1:10
Luminaires (or windows, etc.) and surfaces adjacent to them		20:1
Anywhere within normal field of view		40:1

Source: From *IES lighting handbook,* 25, fig. 11-2, p. 11-3, and fig. 14-2, p. 14-3.

reflection on the utilization coefficient is illustrated by the following example [*IES lighting handbook,* 25, fig. 5-17, p. 5-17]:

Reflection of surface, %				Utilization
Ceiling	Walls	Floor	Furniture	coefficient %
65	40	12	28	29
85	72	85	50	57

In order to contribute to the effective distribution and utilization of light in a room, it is generally desirable to use rather light walls, ceilings, and other surfaces. However, areas of high reflectance in the visual field can become sources of reflected glare. For this and other reasons (including practical considerations), the reflectances of surfaces in a room (such as an office) generally increase from the floor to the ceiling. Figure 12-10 illustrates the IES recom-

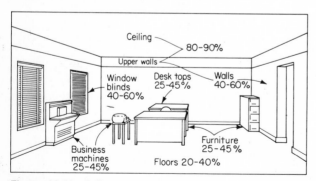

Figure 12-10 Reflectances recommended for room and furniture surfaces in an office. [*Journal of the IES,* 23, fig. 11.]

mendations on this score, indicating for each type of surface the range of acceptable reflectance levels. Although that figure applies specifically to offices, essentially the same reflectance values can be applied to other work situations, as in industry.

Discussion

Although illumination recommendations usually provide for at least a moderate level of general illumination (i.e., a low luminance ratio), there are certain factors that argue against the use of high levels of general illumination. One such circumstance is when a high level of illumination suppresses the "visual gradient" as shown before in Figure 12-7. Another is when it is desirable to draw the visual attention to a particular area. There is a phenomenon called *prototropism*—that is, the tendency of the eyes to turn toward the light. By proper positioning of the light (and minimization of general illumination) one can "light up" the area of most important visual attention. This is illustrated in Figure 12-11 with the effects of change in location of luminaires on a wood planer [Hopkinson and Longmore, 14].

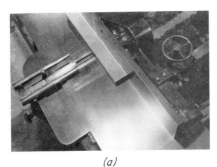 

(a)   *(b)*

Figure 12-11 Effects of change of location of luminaires on wood planer. Original position (left) caused reflection from polished surface, which distracted attention from cutters. Relighting (right) gives lower brightness reflection, which makes cutter the center of attraction. [From Hopkinson and Longmore, 14, by permission of the Controller of H. M. Stationery Office.]

ILLUMINATION FOR SPECIAL SITUATIONS

There are, of course, many circumstances that require special illumination installations in order to make the necessary visual discriminations possible. Two examples will be given to illustrate the human factors questions that bob up in illumination problems.

Roadway Illumination

Among the postulated advantages of roadway illumination is that of safety. The before-and-after accident records of streets and highways which have been illuminated provide extremely persuasive evidence that this pays off. Following is a summary of the data compiled by the Street and Highway Lighting Bureau of Cleveland for 31 thoroughfare locations throughout the country showing traffic deaths for the year before they were illuminated and for the year after [30, pp. 585–602]:

| | | Reduction | |
Year before illumination	Year after	No.	Percent
556	202	354	64

Such evidence is fairly persuasive, and argues for the expansion of highway illumination programs.

Roadway Features That Influence Visibility   Several aspects of roadways presumably have some effect upon visibility in night driving, and therefore possibly are related to accidents. One of these is transition lighting. Where roadway luminaires are to be used (such as at an intersection), it is desirable to provide some "transition" on the approaches and exits by graduating the size of lamps used. This helps to facilitate visual adaptation to and from the highly illuminated area. Another factor is what is referred to as *system geometry*, which refers particularly to the mounting height of luminaires; in some installations the mounting height is 35 ft or more, and in some European systems installations at heights of 40 ft or more are used. Such heights (while requiring larger lamps) help to increase the *cutoff* distance, that is, the distance from the light source at which the top of the windshield cuts off the view of the luminaire from the driver's eyes. Under average conditions this cutoff is about 3.5 times the mounting height (MH) of the luminaire.

Still another factor is that of luminaire design, which can control the "cutoff" of luminaire candlepower at greater approach distances, thus improving comfort and visibility. The effects of the "cutoff" type of luminaire are shown below in terms of glare ratings and visibility ratings of 121 drivers [Christie, 8]. The rating scales ranged from −2 to +4.

| | Type of luminaire | |
	Cutoff	Noncutoff
Glare rating, mean	1.8	− 0.9
Visibility rating, mean	1.3	0.4

While on the subject of roadway luminaires, it might be relevant to illustrate the differences in the amount of illumination produced by luminaires of different types. Figure 12-12 shows the minimum and average illumination (in footcandles) and pavement illumination (in footlamberts) produced by high-intensity sodium, iodine mercury, and clear mercury lamps (400-watt) at two mounting heights [Rex, 21]. Translating such data into economic terms, it is proposed by Rex that the high-intensity sodium lamp is the best investment now in units of illumination per mile.

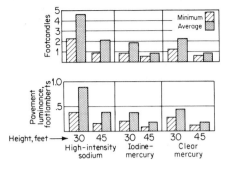

Figure 12-12 Minimum and average illumination (footcandles) and pavement luminance (footlamberts) produced by 400-watt luminaires of three types of roadway lights at two mounting heights (MH) above ground. [Adapted from Rex, 21.]

In addition, the reflectivity of roadway surfaces varies substantially, creating a factor which in turn affects visibility. For example, asphalt pavement surface (about 6 years old) reflects approximately 8 percent of the light, whereas concrete pavement has a surface reflectance of about 20 percent. It has been estimated that the amount of illumination (in footlamberts) required for equal brightness of asphalt pavements, compared with concrete, is of approximately the ratio 1.92:1 [Blackwell, Prichard, and Schwab, 6]; this is roughly a 2:1 ratio.

Vehicle Headlights

Another aspect of illumination for drivers is the type of headlight used in vehicles, which can be assessed in terms of visibility and visual comfort. An interesting approach for eliciting evaluation of comfort with three headlight systems was used in a study by Schwab et al. [22], the systems being: (1) a low-beam (LB) system; (2) a high-beam (HB) system; and (3) a high-intensity polarized system (HIP). At 15-minute intervals during the first two hours drivers could choose between the LB and the HIP, and during the next two hours between HB and HIP, the opposing traffic in any case using the same as the method the subject chose. In the case of each pair of options, one was alternatively "free" while the other was available for a "price" (to be deducted from the $2.50 per hour which the subjects were being paid). The "price" a person was willing to pay for one system over the other could be thought of as an indication of his preference. The average amount the subjects paid for the HIP and for the LB over the HB are given below for two age groups:

	Young (18–28)	Old (47–64)
Pay for HIP over HB	40¢/h	23¢/h
Pay for LB over HB	16¢/h	13¢/h

Although there were age differences, the preference was greatest for the polarized system (HIP) and next greatest for the low-beam system (LB). The high-beam system (HB) was clearly less acceptable.

Inspection Processes

Inspection processes frequently call for some type of special-purpose lighting. In fact, Faulkner and Murphy [11] indicate that, on the basis of experience gained through the design of 50 inspection stations, general illumination is seldom, if ever, the best choice for inspection lighting. They report that the Eastman Kodak Human Factors Group has developed a "lighting library" which contains examples of about 20 different types of special-purpose lighting, including: use of polarized light; spotlighting; edge lighting; stroboscopic lighting; black light; and "surface grazing" or "shadowing." A visual illustration of the comparative effects of general versus surface-grazing illumination was shown above in Figure 12-7.

As an example of how visual inspection performance can be affected by the nature of the illumination, Lion [15] compared performance on six tasks using tungsten filament and fluorescent lighting. The results are given below:

Task and criterion	Tungsten	Fluorescent
1. Grading ball bearings, no. sorted	306.2	322.1*
1. Grading ball bearings, no. errors	30.7	34.1
2. Needles, no. threaded	149.7	156.2*
3. Number reading, no. read	91.6	91.7
4. Rod measuring, time, s	718.3	695.8*
4. Rod measuring, no. errors	13.4	13.3

*Statistically significant.

On the three manipulative tasks (1, 2, 4) the subjects worked more quickly under fluorescent light without any detrimental effect on accuracy of performance; it was suggested that this might be the consequence of lower glare with fluorescent than with incandescent tungsten lamps. On the basis of a subsequent study with other items [Lion et al., 16], the investigators inferred that the advantage of fluorescent lamps in certain inspection tasks may be due to the greater "diffusion" of the illumination (as contrasted with incandescent lamps, which are "point" sources).

DISCUSSION

The human visual apparatus makes possible the sensing of a tremendous amount of visual detail, but only under two conditions, namely, that the eye is not seriously impaired and that the illumination available is adequate. Even under conditions of partial impairment and somewhat inadequate illumination, the eye is capable of remarkable sensitivity. But in the long-range interests of helping to maintain the visual skills of people and of enhancing the visual "performance" of people, it is important that the illumination provided be appropriate for the visual tasks in question. This is of course achieved by consideration of the visual task at hand and the application of the data and principles of illuminating engineering that have been based on research, such as those discussed in this chapter.

REFERENCES

1 Blackwell, H. R.: Development and use of a quantitative method for specification of interior illumination levels on the basis of performance data, *Illuminating Engineering*, 1959, vol. 54, pp. 317–353.

2 Blackwell, H. R.: Development of visual task evaluators for use in specifying recommended illumination levels, *Illuminating Engineering*, 1961, vol. 56, pp. 543–544.

3 Blackwell, H. R.: Further validation studies of visual task evaluation, *Illuminating Engineering*, 1964, vol. 59, no. 9, pp. 627–641.

4 Blackwell, H. R.: A human factors approach to lighting recommendations and standards, in *Proceedings of the Sixteenth Annual Meeting of the Human Factors Society*, Santa Monica, Calif., 1972, pp. 441–449.

5 Blackwell, H. R., and O. M. Blackwell: The effect of illumination quantity upon the performance of different visual tasks, *Illuminating Engineering*, 1968, vol. 63, no. 3, pp. 143–152.

6 Blackwell, H. R., B. S. Prichard, and R. N. Schwab: *Illumination requirements for roadway visual tasks*, Highway Research Board Bulletin 255, Publication 764, 1960.

7 Blackwell, O. M., and H. R. Blackwell: Visual performance data for 156 normal observers of various ages, *Journal of Illuminating Engineering Society*, October, 1971, vol. 1, no. 1, pp. 3–13.

8 Christie, A. W.: Visibility in lighted streets and the effect of the arrangement and light distribution of the lanterns, *Ergonomics*, 1963, vol. 6, no. 4, pp. 385–391.

9 Crouch, C. L.: New method of determining illumination required for tasks, *Illuminating Engineering*, 1958, vol. 53, pp. 416–422.

10 Faulkner, T. W., and T. J. Murphy: *Illumination: A human factors viewpoint*, Paper presented at 15th annual meeting of Human Factors Society, 1971.

11 Faulkner, T. W., and T. J. Murphy: Lighting for difficult visual tasks, *Human Factors*, 1973, vol. 15, no. 2, pp. 149–162.

12 Hopkinson, R. G.: Glare from daylighting in buildings, *Applied Ergonomics*, 1972, vol. 3, no. 4, pp. 206–215.

13 Hopkinson, R. G., and J. Collins: *The ergonomics of lighting*, McDonald, London, 1970.

14 Hopkinson, R. G., and J. Longmore: Attention and distraction in the lighting of work-places, *Ergonomics*, 1959, vol. 2, pp. 321–334.

15 Lion, J. S.: The performance of manipulative and inspection tasks under tungsten and fluorescent lighting, *Ergonomics*, 1964, vol. 7, no. 1, pp. 51–61.

16 Lion, J. S., E. Richardson, and R. C. Browne: A study of the performance of industrial inspectors under two kinds of lighting, *Ergonomics*, 1968, vol. 11, no. 1, pp. 23–34.

17 Logan, H. L., and E. Berger: Measurement of visual information cues, *Illuminating Engineering*, 1961, vol. 56, pp. 393–403.

18 Luckiesh, M., and F. K. Moss: "The new science of seeing," in *Interpreting the science of seeing into lighting practice*, vol. 1, 1927–1932, General Electric Co., Cleveland.

19 McCormick, E. J.: *Human factors engineering*, 3d ed., McGraw-Hill Book Company, New York, 1970.

20 Maerz, A., and M. R. Paul: *Dictionary of color*, 2d ed., McGraw-Hill Book Company, New York, 1950.

21 Rex, C. H.: Roadway lighting for the motorist, *Illuminating Engineering*, 1967, vol. 62, no. 2, pp. 98–110.

22 Schwab, R. N., D. Solomon, and J. F. Lyons: Monetary values drivers place on comfort in night driving, *Journal of the Illuminating Engineering Society*, January, 1973, vol. 2, no. 2, pp. 104–112.

23 American national standard practice for office lighting, *Journal of the Illuminating Engineering Society*, 1973, vol. 3, no. 1, pp. 3–27.

24 *Color harmony manual*, Container Corporation of America, Chicago, 1942.

25 *IES lighting handbook*, 5th ed., IES, New York, 1972.

26 *Light and color:* TP-119, Large Lamp Department, General Electric Company, Nela Park, Cleveland, Ohio, August, 1968.

27 *Light measurement and control:* TP-118, Large Lamp Department, General Electric Company, Nela Park, Cleveland, Ohio, March, 1965.

28 *Munsell book of color*, Munsell Color Co., Baltimore, 1929.

29 Optical Society of America, Committee on Colorimetry: *The science of color*, Thomas Y. Crowell Company, New York, 1953.

30 Public lighting needs, *Illuminating Engineering*, 1966, vol. 61, no. 9, pp. 585–602.

31 Visual comfort ratings for interior lighting: Report 2 (prepared by Subcommittee on Direct Glare, Committee on Recommendations for Quality and Quantity of Illumination, IES), *Illuminating Engineering*, 1966, vol. 61, no. 10, pp. 643–666.

Chapter 13

Atmospheric Conditions

Since the current model of the human organism is the result of evolutionary processes over millions of years, it has developed substantial adaptability to the environmental variables within the world in which we live, including its atmosphere. There are, however, limits to man's range of adaptability. In addition, science and technology are busy developing new kinds of environments for the human being, involving space capsules, cold-storage warehouses, and underwater caissons, and causing changes in our natural environment as the unintentional by-product of civilization—such as smog and air pollution. Since we cannot cover all aspects of the ambient environment, we will cover those that are of most general concern.

THE HEAT-EXCHANGE PROCESS

The human body is continually generating heat as the consequence of metabolic activity. In a state of rest, an adult male generates a little over 1 kcal/min; some sedentary activities cause expenditures of from 1.5 to 2.0 kcal/min, and physical activities range up to about 5.0 kcal/min for moderate activity and up to 10 or even 20 kcal/min for extremely heavy work. Since the metabolic activity is continuous, the body is continually in the process of trying to maintain thermal equilibrium with its environment.

Body Changes during Thermal Adjustment

When the body changes from one thermal environment to another, certain physical adjustments are made by the body, especially the following:

• *Changes from optimum environment to cold one*: (1) The skin becomes cool; (2) the blood is routed away from the skin and more to the central part of the body, where it is warmed before flowing back to the skin area; (3) rectal temperature rises slightly; and (4) shivering and "goose flesh" may occur. The body may stabilize with large areas of skin receiving little blood.
• *Changes from cool environment to warm one*: (1) more blood is routed to the surface of the body, resulting in higher skin temperature; (2) rectal temperature falls; and (3) sweating may begin. The body may stabilize with increased sweating and increased blood flow to the surface of the body.

Acclimatization to Heat and Cold   Acclimatization consists of a series of physiological adjustments that occur when an individual is habitually exposed to extreme thermal conditions, hot or cold, as the case may be. An illustration of the changes that occur during acclimatization to heat is given in Figure 13-1.

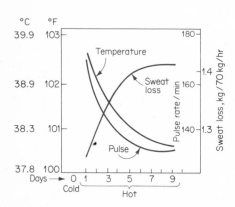

Figure 13-1 Changes in rectal temperature, pulse rate, and sweat loss of a group of men during 9 days of exposure to a hot climate (dry-bulb and wet-bulb temperatures of 120 and 80°F), during which they worked for 100 min at an energy expenditure of 300 kcal/h. Comparative temperature and pulse values are given for a preceding control day (day 0) during which the men worked in a cool climate. [Adapted from Lind and Bass, 21.]

The men involved in the study worked each of 9 days for 100 min at an energy expenditure of 300 kcal/h in a hot climate [Lind and Bass, 21]. The figure shows changes in rectal temperature, pulse rate, and sweat loss. The changes in these physiological indices are very distinct. Although there is some question about the time required to become fully acclimatized, it is evident that much acclimatization to heat occurs within 4 to 7 days, and reasonably complete acclimatization occurs usually in 12 to 14 successive days of heat exposure [Leithead and Lind, 18, p. 21]. Quite a bit of acclimatization to cold occurs within a week of exposure, but full acclimatization may take months or even years [McBlair et al., 25]. Even complete acclimatization, however, does not fully protect an individual from extreme heat or cold, although such an individual can tolerate extremes better than his unacclimatized brother.

Methods of Heat Exchange

Given the heat produced by the metabolic process, primary sources of heat gain and heat loss to and from the body are (1) convection (heat gain or loss by contact with the air), (2) radiation (heat gain or loss depending on skin temperature and the temperature of surrounding areas), and (3) evaporation (under most circumstances this results in heat loss), these being further influenced by the work activity of the individual. The thermodynamic process in which these are involved in the heat exchange between man and his environment may be described by the following general heat balance equation, describing the heat gained from, and lost to, the environment [Gagge et al., 11]:

$$S \text{ (storage)} = M \text{ (metabolism)} - E \text{ (evaporation)} \pm R \text{ (radiation)} \pm C \text{ (convection)} - W \text{ (work accomplished)}$$

The S factor is the amount of heat gained or lost; if the body is in a state of thermal balance, S becomes zero. It should be added that heat may also be exchanged by conduction or by ingestion of hot or cold fluids, but their influence typically is very nominal.

Factors That Affect Heat Exchange

The discussion above has given hints of the environmental variables that affect the heat-exchange process, but it may be useful here to pin these down specifically. These factors are: air temperature; air humidity; air movement; and radiant temperature (temperature of walls, ceilings, and other surfaces in the area).

Heat Exchange under Various Conditions   The relative importance of convection, radiation, and evaporation in maintaining thermal equilibrium depends very much upon the four factors listed above, these interactions being illustrated in Figure 13-2. This figure shows, for each of five conditions of air and wall temperature, the percentage of heat loss by evaporation, radiation, and convection. In particular, conditions d and e illustrate the fact that, with high air and wall temperatures, convection and radiation cannot dissipate much body heat, and the burden of heat dissipation is thrown on the evaporative process. But, as we all know, evaporative heat loss is limited by the humidity; there is indeed truth to the old statement that "it isn't the heat—it's the humidity." To see the limiting effect of humidity on evaporation, let us look at Figure 13-3. This shows the upper limits of tolerance in relation to temperature and relative humidity for working and for resting nude subjects. For any one of the three curves in Figure 13-3, combinations of temperature and humidity to the right of the curve represent conditions which, if accentuated enough and long

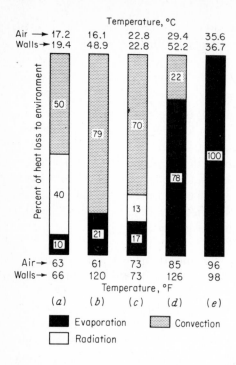

Figure 13-2 Percentage of heat loss to environment by evaporation, radiation, and convection under different conditions of air and wall temperature. [Adapted from Winslow and Herrington, 32.]

enough, could result in some physiological consequences ranging up to heat-stroke or death. A comparison of the two curves at the right shows that air movement usually helps to make conditions more tolerable by exposing the surface of the body to more air.

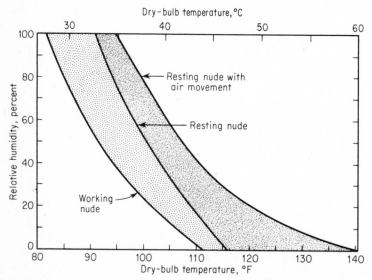

Figure 13-3 Approximate upper limits of tolerance for heat loss by evaporation. For any of the three conditions, temperature and humidity combinations to the right of the curve prevent evaporation. [Derived from data by Winslow et al., 42.]

Clo Unit: A Measure of Insulation The basic heat-exchange process is of course influenced by the insulating effects of the clothing worn, varying from a bikini bathing suit to a parka as worn in the arctic. Such insulation is measured by the *clo* unit. The clo unit is a measure of the thermal insulation necessary to maintain in comfort a sitting, resting subject in a normally ventilated room at 70°F temperature and 50 percent relative humidity [Gagge et al., 11][1] Since the typical individual in the nude is comfortable at about 86°F, one clo unit would be required to produce an equal sensation at about 70°F; a clo unit has very roughly the amount of insulation required to compensate for a drop of about 16°F, and is approximately equivalent to the insulation of the clothing normally worn by men. To lend support to the old adage that there is nothing new under the sun, the Chinese have for years described the weather in terms of the number of suits required to keep warm, such as a "one-suit" day (reasonably comfortable), a "two-suit" day (a bit chilly), up to a limit of a "twelve-suit" day (which would be really bitter weather).

Composite Indexes of Environmental Factors

Since there are different environmental factors that affect the heat-exchange process, it would be desirable to have some index of the stress imposed by combinations of environmental factors, and of the strain induced in the individual by exposure to the environment. As Macpherson points out [31], many different indexes of heat stress have been developed over the years, but the fact that many have been devised implies that none is entirely satisfactory. Such physiological variables as body temperature, heart rate, and amount of sweat can serve as indexes of strain. A few indexes of environmental factors are described below.

Effective Temperature There are two indexes of *effective temperature*, both of which were developed under the sponsorship of the American Society of Heating, Refrigerating, and Air-Conditioning Engineers (ASHRAE). The original effective temperature (ET) scale was developed many years ago as an empirical sensory index, combining into a single value the effect of temperature and humidity on thermal sensations, with an adjustment for the effects of air movement. It was developed through a series of studies in which subjects compared the relative thermal sensations of various air conditions in adjoining rooms by passing back and forth from one room to the other. Any given ET— such as 70°—was operationally characterized as the thermal sensation produced by a dry-bulb temperature of that same temperature (e.g., 70°) in combination with relative humidity (rh) of 100 percent. However, other combinations of dry-bulb temperature and rh that produced the same sensation under the experimental condition used have that same ET (e.g., 70°). In general terms, the "other" combinations for any given ET were characterized in terms

---

[1]A clo unit is defined as follows:

$$\text{Clo unit} = \frac{°F}{\text{Btu/(h)(ft}^2 \text{ of body area)}}$$

of *higher* dry-bulb temperatures and *lower* rh values. A simplified version of the original effective temperature scale (ET) is shown in Figure 13-4.

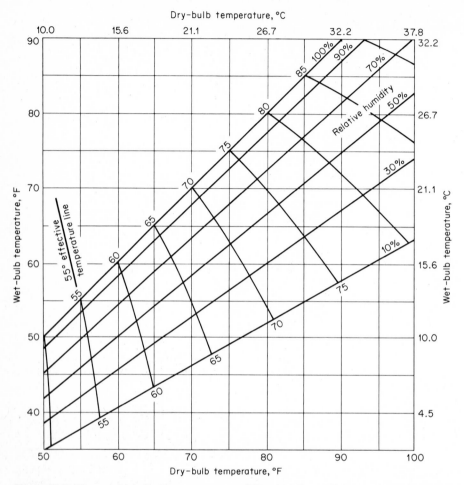

Figure 13-4 The original effective temperature (ET) scale. Each ET line (e.g., 70°) reflects the combinations of dry-bulb temperature and relative humidity that produced the same thermal sensation under the experimental conditions. [Adapted from ASHRAE handbook of fundamentals, 46.]

The original ET scale has been demonstrated to overemphasize the effect of humidity in cool conditions and to underemphasize its effect in warm conditions, and it does not account fully for air movement in hot-humid conditions. Further, its use was limited to sedentary conditions. Therefore, the ASHRAE has sponsored the development of a new effective temperature scale (ET^x) that is illustrated in Figure 13-5. This scale is based in part on consideration of the physiology of heat regulation as it applies to comfort, temperature sensation, and health [Gagge et al., 12], especially as heat regulation depends upon evaporative heat loss. Evaporative heat loss consists of three parts. That heat which is lost from the lungs through respiration, and the heat lost through vaporized water diffusing through the skin layer, are both called "insensible"

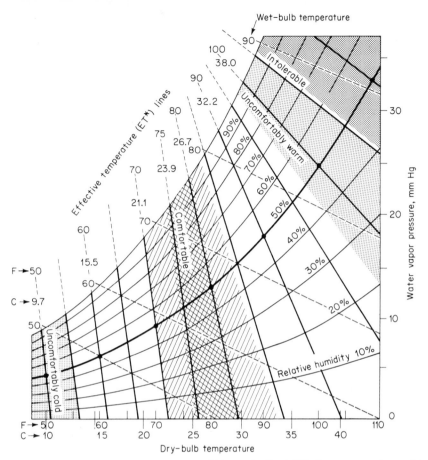

Figure 13-5 New effective temperature (ET^x) scale. Each ET^x line represents combinations of dry-bulb temperature and relative humidity that generally produce the same level of skin "wettedness" as caused by regulatory sweating. The figure also shows areas of various thermal sensations. [Adapted from ASHRAE handbook of fundamentals, 47, chap. 7, fig. 16.]

heat loss. The heat of vaporized sweat necessary for regulation of body temperature is called "sensible" heat loss. The body generally is in a state of thermal neutrality with respect to "regulatory" (i.e., sensible) heat loss when the dry-bulb temperature is about 77°F (25°C) and the rh is 50 percent. Holding rh constant (at 50 percent), higher or lower temperatures would alter the evaporative process, thus affecting the ratios of skin "wettedness." The ET^x scale is based essentially on these ratios, Figure 13-5 representing these at the 5 percent rh line. However, any given level of skin "wettedness" can be produced by *other* combinations of dry-bulb temperature and rh, and these are represented by the various ET^x lines on the figure.

However, the ET^x lines also generally reflect varying levels of thermal sensation, these being shown in Figure 13-5. The *comfortable* range (from ET^x values of about 74 to 81) is bordered by *slightly cool* and *slightly warm* ranges, with *uncomfortably cold*, *uncomfortably warm*, and *intolerable* ranges also being shown.

It should be noted that the original ET scale had the loci of the ET lines at the intersection of corresponding dry-bulb temperatures in combination with 100 percent rh. Since the loci of the new ET^x lines are at the 50 rh lines, the new ET^x values would tend to be numerically higher than the older ET values for corresponding conditions. (Most of the later references to effective temperature are to the old ET scale rather than to the new ET^x scale.)

Operative Temperature   Since the ET scale does not account for radiation to or from wall surfaces, Winslow, Herrington, and Gagge [44] developed a scale of *operative temperature*.[2] This scale takes into account air temperature and wall temperature, but not humidity or air flow.

Heat-Stress Index   This index, originally developed by Belding and Hatch [4], has been modified by Hatch [14] and Hertig and Belding [15]. The index expresses the heat load in terms of the amount of perspiration that must be evaporated to maintain heat balance; this is referred to as E_{req} (the *required evaporation* heat loss). In turn, it is possible to determine the maximum heat that can be lost through evaporation E_{max}, from assumptions of body size, weight, and temperature and by taking into account water-vapor pressure of the environment and air velocity. While the details of these derivations will not be repeated, the *difference* between these two values (expressed in Btu per hour) indicates the load that must be reduced or dissipated otherwise. The *otherwise* can take various forms, such as further reduction of convection or radiation sources, reduction of task demands by reducing physical requirements or by rest pauses, and by proper clothing.

To illustrate this general approach, Hertig and Belding [15] present a hypothetical example of a task as follows:

Source of heat load	Btu/hr
Metabolism (based on type of activity)	800
Radiation	2800
Convection	60
E_{reg}	3660
E_{max} (computed by formula)	2530
Difference	1130

[2]Operative temperature is the sum of a radiation constant multiplied by the mean wall temperature, and a convection constant multiplied by the mean air temperature, divided by the sum of the two constants, as derived from this formula.

$$T_o = \frac{K_R T_W + K_C T_A}{K_R + K_C}$$

in which T_o = operative temperature
K_R = radiation constant
T_W = temperature of walls
K_C = convection constant
T_A = temperature of air

In this particular case, heat by radiation is the major source of the heat load and would be the primary aspect of the situation to do something about.

In connection with the heat-stress index, McKarnes and Brief [26] have worked up a set of nomographs based on the modifications reported by Hertig and Belding, and have also elaborated on the theme by setting up a formula for estimation of allowable exposure time (AET) and minimum recovery time (MRT).

Oxford Index The Oxford index, or WD index, is a simple weighting of wet-bulb and dry-bulb temperatures, as follows [Leithead and Lind, 18, p.82]:

$$WD = 0.85w \text{ (wet-bulb temperature)} + 0.15d \text{ (dry-bulb temperature)}$$

It has been found to be a reasonably satisfactory index to equate climates with similar tolerance limits.

Predicted 4-h Sweat Rate (P4SR) In a series of investigations McArdle et al. [24] found that sweat loss was the physiological measurement that correlated best with the severity of experimental environments. They then developed a scale based on sweat loss, the basic reference value being the amount of sweat produced in 4 hours by acclimatized young men performing a prescribed amount of work. The index takes into account globe or dry-bulb temperature, wet-bulb temperature, and air speed, with adjustments for energy expenditure and clothing worn, and in effect predicts the sweat rate for the reference group from these environmental factors. Macpherson points out [31] that it is based on the assumption that the level of environmental heat stress can be expressed as a function of the amount of sweat produced by the individual. Leithead and Lind [18, pp. 62–68] express the opinion that it is the most accurate scale of heat stress now available. It would be applicable, however, only in those circumstances in which sweating would occur.

THERMAL SENSATIONS

Granting that there are individual differences in the sensations of people under various thermal conditions, Figure 13-5 showed, for various ranges of effective temperatures (ET^x, as based on the new scale), the general sensations associated with various ranges. On the basis of a study with 1600 subjects, Rohles and Nevins [33] identified a range of conditions called a "comfort envelope" in which resting subjects, lightly clothed (about 0.6 to 0.8 clo units), tended to report the greatest comfort. This envelope falls between the 75 and 80 ET^x lines, forming something of a vertical diamond shape. However, with more normal clothing (0.8 to 1.0 clo units) the ASHRAE [47, chap. 7, fig. 17] presents an area that is somewhat lower, ranging from an ET^x of about 72 to 77 with rh values of about 20 to 60 percent.

Our sensations about various environmental conditions, however, are in part a function of the physical activities involved and the apparel worn. Some

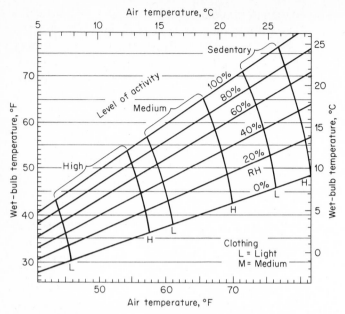

Figure 13-6 Comfort lines for persons engaged in three levels of work activity with light clothing (0.5 clo) and medium clothing (1.0 clo). (These data are for a low level of air velocity.) [Adapted from Fanger, 9, as presented in *ASHRAE handbook of fundamentals,* 47, chap. 7, fig. 18.]

indication of the influence of these factors is reflected in Figure 13-6, which shows "comfort lines" for each of three levels of activity (sedentary, medium, and high) for persons with light and with medium clothing. The differences in these lines (they are actually ET^x lines) reflect the very definite effects on comfort of work activity and clothing.

HEAT STRESS

Having discussed the heat-exchange process and the various indexes of environmental conditions, let us now turn to the effects of heat stress.

Physiological Effects

One of the most direct effects of heat stress is on the temperature of the body. Body temperature measurements include measures of *core* temperature (oral, rectal, etc.) and of *shell* temperature (usually skin temperature). In discussing skin and rectal temperatures Welch et al. [40] call our attention to the fact that the relationship between them is relatively limited except under conditions of high temperature and humidity. They add the comment that in their experience rectal temperature is the best physiological index of "fatigue," pointing out that rectal temperatures of 38.8°C (101.8°F) in most cases coincide with the onset of exhaustion.

Recognizing that skin and rectal temperatures cannot be used interchangeably, let us first see the relationship between ET and skin temperature as summarized from different studies by Leithead and Lind [18] as shown in Figure 13-7. The upswing of the curves probably reflects the consequence of

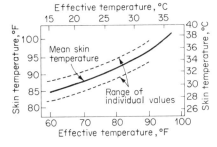

Figure 13-7 Relationship between effective temperature (ET) and skin temperature as consolidated from various studies. [Adapted from Leithead and Lind, 18, fig. 6.]

the flow of blood toward the surface of the skin to increase the dissipation of surplus heat with increasing effective temperatures. As a follow-up, Figure 13-8 represents the relation between ET and rectal temperature for one individ-

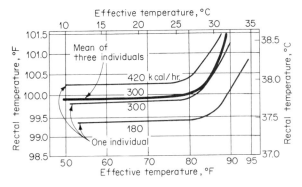

Figure 13-8 Rectal temperature as related to effective temperature (ET) for one individual working at three levels of work activity and for three individuals working at the same level (300 kcal/h). [Adapted from Lind, 19.]

ual engaged in three levels of physical activity and for the average of three individuals at a single level of activity. In each instance the long, flat sections of the curves represent what Lind [19] refers to as the *prescriptive zone* for whatever activity is involved. As a prelude to a later discussion of the effects of work load, we can see the differential upswings for the three levels of work activity represented, the upswing starting at successively lower points with increasing work load. The middle, light line (that represents a single individual working at a level of 300 kcal) parallels the heavy line (that represents the average for three individuals), a fact which indicates considerable stability of the patterns of the curves.

As shown in Figure 13-8, the level of work activity contributes to the strain. This is illustrated further in Figure 13-9, which shows the metabolic cost of carrying various loads on the level and on a slope.

Levels of work in combination with environmental conditions that bring

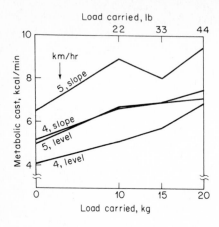

Figure 13-9 Metabolic cost of carrying various loads on the level and on a 4 percent slope at 4 and 5 km/h. [Adapted from Kamon and Belding, 17, table 3.]

about the rise in core temperature (as shown in Figure 13-8) also induce other corresponding physiological changes, which, if continued, can cause hypothermia, a condition that renders normal heat loss more difficult. Dehydration, such as from sweat, is another possible consequence of heat stress. In the case of 51 operating engineers working during mild summer weather in California's Central Valley, the average weight loss from water depletion per day was about 5 lb 2 oz. Such a deficit, of course, needs to be replaced during the evening and night.

Heat Stress and Performance

Most of the studies on the relationship between heat stress and performance have been carried out in controlled, pristine laboratory situations. As an aside, the results of these represent a mixed bag, indicating that the way in which environmental variables affect such performance is indeed complicated, the effects being related in part to the type of work activity. Let us review briefly a few relevant studies.

Physical Work The effects of heat stress on performance of physical work are rather well documented, such as by one study reported by Mackworth [30]. In this experiment the subjects performed on a pull test that required raising and lowering a 15-lb weight by bending and straightening the arm to a metronome that beat every second, until the men could not lift the weight again, the task being performed under different ETs. The results, shown in Figure 13-10, indicate that performance (the amount of work done) deteriorated with higher ETs, both for good and for average subjects, under both low- and high-incentive conditions. A similar pattern of deterioration occurred in performance in a heavy pursuit meter task [Mackworth, 30] and in the case of gold miners filling mine cars in South Africa [Wyndham et al., 45]; and Pepler [in Leithead and Lind, 18, chap. 12] summarizes the results of several industrial surveys that further nail down the point that heat stress takes its toll in performance in physical work activities.

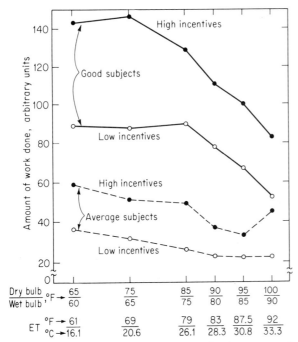

Figure 13-10 Relationship between effective temperature (ET) and performance on a weight-lifting test for good and average subjects under high and low incentives. [From Mackworth, 30.]

But, for most practical purposes, one would like to have some inkling of the tolerance and comfort of people under different *levels* of stress, for different *levels* of energy expenditure, and for different periods of *time*. Although there is probably no simple scheme or set of data that completely fulfills this order, some indication of these relationships is shown in Figure 13-11. That figure shows the tolerance times of men working at three levels of energy expenditure in relation to variations in the Oxford index described above.

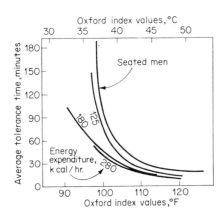

Figure 13-11 Average tolerance times of men seated, and of men working at three levels of energy expenditure, in relationship to Oxford index values. Safe tolerance values should be taken to be no more than about 75 percent of the times shown. [Adapted from Lind, 20, in Hardy, 13, as based on data from other studies.]

Tracking Tasks   An example of a study of the effects of heat stress on tracking tasks is that reported by Azer et al. [1]. Different groups of subjects performed a tracking task under each of the following conditions, as well as under a control condition (°C are shown in parentheses):

	Temperature	Humidity	ET
1	95 (35.0)	50	84.9 (29.4)
2	95 (35.0)	75	89.8 (32.1)
3	100 (37.8)	50	88.4 (31.3)
Control	76 (24.4)	50	71.0 (21.7)

All of the experimental conditions resulted in significant increases (as compared with the control condition) in average skin temperature, weight loss ratio, heart rate, and thermal sensation reported by the subjects. However, only condition 2 resulted in a significant degradation in performance, and this was the only condition that resulted in a significant increase in core (rectal) temperature. (Condition 2 had the highest ET, which was determined by its higher humidity.) This leads to the inference that core temperature is a manifestation of the strain that induces performance degradation.

Vigilance Tasks   Interestingly enough, in a vigilance task it was also found that core temperature was the one factor that was associated with performance decrement [Colquhoun and Goldman, 7]. The investigators attribute the effect to an "arousal" hypothesis, namely, that the increased body temperature increased the general level of arousal. In turn, they suggest that it is only when the level of arousal exceeds the "optimum" value for performance on the task that the increase in body temperature affects task performance. (It should be added, however, that "arousal" is not exclusively a function of body temperature, as demonstrated by Provins et al. [32], but is also influenced by the ambient conditions.)

Mental Activities   The effects of heat stress on performance of mental activities are intertwined with the environmental conditions (such as ET) and duration of the work. A generalized pattern of the temperature-duration function related to mental performance has been developed by Wing [41] on the basis of a very thorough analysis of the results of 15 previous studies. In this analysis he identified, from the several studies, the lowest temperature at which a statistically significant performance decrement occurred. These points, plotted at their respective exposure durations, were then used in drawing the curve shown in Figure 13-12. In addition, that figure shows, for comparative purposes, curves of tolerable and marginal physiological limits based on earlier studies by other investigators. Wing suggests that the thresholds for at least some mental tasks might be between the lower curve and the tolerable physiological-limit curve in the case of fully acclimatized or highly practiced individuals.

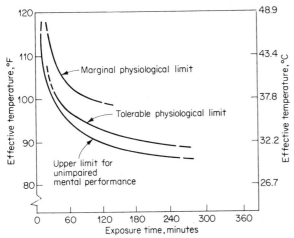

Figure 13-12 Tentative upper limit of effective temperature (ET) for unimpaired mental performance as related to exposure time; data are based on an analysis of 15 studies. Comparative curves of tolerable and marginal physiological limits are also given. [Adapted from Wing, 41, fig. 9. See Wing's report for original sources.]

Performance on Industrial Jobs

Although most of the studies of heat stress on performance have been carried out in laboratories, there have been a few studies in actual job situations, which, in general, support the indications of laboratory studies that heat stress does take its toll. An example is one reported by Tichauer [37] dealing with the performance of the job of "picking" in cotton weaving. Figure 13-13 shows an

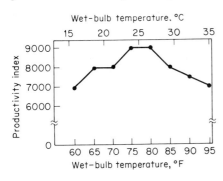

Figure 13-13 Relationship between temperature and performance of "pickers" in cotton weaving. Cotton fiber can be processed best in a warm, humid atmosphere, which of course is bad for the efficiency of man. In this instance, a wet-bulb temperature of 80°F (27°C) represents a satisfactory compromise. [Adapted from Tichauer, 37.]

index of the number of "picks" per day under various temperature conditions, the results showing the highest production levels within the range of 75 to 80°F (24 to 27°C) wet-bulb temperatures.

Discussion

The several factors associated with heat stress preclude the setting of any single set of environmental conditions that could be specified as being acceptable on an across-the-board basis. In this connection Mackworth [30] does propose the concept of a *critical region* above which most acclimatized men dressed in shorts will not work effectively, this being between ETs of 83 and 87.5°F (28.3 and 31.0°C). Such a range, however, may be on the high side for heavy work in terms of the prescriptive zones shown in Figure 13-8 [Lind, 19] for three levels of work activity, these ETs being 80.4, 81.3, and 86.1°F (26.4, 27.4, and 30.0°C). And on the other hand, in the case of at least a tracking task, performance was not affected by an ET as high as 88.4.

Aside from the possible effects of heat stress on work performance, however, one should also be concerned about the comfort of people carrying out different types of work. In general terms, comfort levels are generally below those at which performance begins to go to pot.

COLD

While civilization is generally reducing the requirement for many people to work in cold environments, there still are some circumstances where people must work, and live, in cold environments. These situations include outdoor work in winter, arctic locations (especially in military and exploration activities), cold-storage warehouses, and food lockers. As in the case of heat exposure, there are a number of interlaced factors that affect the tolerance, comfort, and performance of people in cold environments; these include the level of activity, degree of acclimatization, duration, and insulation.

Indexes Related to Effects of Cold

First, let us mention one of the indexes that is related to the effects of cold.

Wind Chill One of these is a wind-chill index [Siple and Passel, 35]. It provides a means for making a quantitative comparison of combinations of temperature and wind speed. The quantitative value corresponds to a calorie scale (actually kilocalories per square meter per hour) but is converted into a sensation scale ranging from hot (about 80), to pleasant (about 200), cool (400), cold (800), bitterly cold (1200), and even colder values (including that at which exposed flesh freezes in 1 min or less). Although the wind-chill index itself is not given here, Table 13-1 shows the cooling effects of combinations of certain temperatures and wind speeds derived from the scale, these being expressed as "equivalent temperatures." For example, an air temperature of 10°F (−12°C) with a 20-mph wind produces the same cooling effect as a temperature of −25°F (−33°C) under calm wind conditions.

Physiological Effects of Cold

With inadequate protection, exposure to the cold brings about a reduction of both core and shell temperatures. Continued exposure, of course, can bring about frostbite and other effects, and ultimately death. As a case in point, the mean body and skin temperatures for four subjects exposed to conditions of

Table 13-1 Cooling Effects of Temperature and Wind Speed

Wind speed		Air temperature, °F						Air temperature, °C					
mi/h	km/h	40	20	10	0	−10	−30	4	−7	−12	−18	−23	−29
		Equivalent temperature											
calm		40	20	10	0	−10	−20	4	−7	−12	−18	−23	−29
5	9	37	16	6	−5	−15	−26	3	−9	−14	−20	−26	−32
10	16	28	4	−9	−21	−33	−46	−2	−16	−23	−29	−36	−43
20	32	18	−10	−25	−39	−53	−67	−8	−23	−33	−37	−47	−55
30	49	13	−18	−33	−48	−63	−79	−11	−28	−36	−43	−52	−62
40	64	10	−21	−37	−53	−69	−85	−12	−29	−38	−47	−56	−65

Source: Adapted from Siple and Passel [35].

−26°F after about 75 min are given below, for the conditions indicated [Veghte and Clogston, 38].

Condition	Temperature, °F	
	Body	Skin
Walking, 1 mph (160 kcal/h)	95.6	89
Resting, 4.2 clo units	93.5	84
Resting, 2.3 clo units	91.8	79

Normal skin temperature is about 92°F, and the critical mean skin temperature—excluding the hands and feet—is about 76°F [Barnett, 3]. This is the level at which extreme discomfort generally occurs.

Performance in Cold

The primary interest in the effects of cold on performance relates to manual tasks of one sort or another. In this regard Fox [10] pulls together a fairly persuasive body of evidence to the effect that hand-skin temperature (HST) is a particularly critical factor in performance of manual tasks in cold temperatures. The critical region is apparently between about 55 and 60°F, as illustrated in Figure 13-14. Fine finger dexterity is more susceptible to adverse temperature

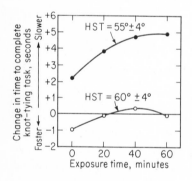

Figure 13-14 Changes in manual performance on knot-tying test as function of hand-skin temperature (HST) and duration of cold exposure. The differences for the 60°F HST are not significant. Those for the 55°F temperature are. [From Clark, 6.]

conditions than grosser types of manual activities [Dusek, 8]. There have been some inklings that manual performance can be maintained if hand-skin temperature is above the critical level of about 60°F [Clark, 6]; more recent evidence, however, indicates that this is not universally the case, and that, for certain types of tasks, HST must be maintained at higher, more normal levels to preclude work decrement [Lockhart, 22].

Because of the importance of maintaining a reasonable HST when working in cold areas, some schemes have been used experimentally to keep the hands warm. In one such investigation, for example, the hands were warmed with radiant heat from infrared heaters [Lockhart and Kiess, 23]. Five different tasks were carried out by subjects under various ambient temperature conditions, including two dexterity tests, a block-stringing task, a knot-tying task, and a

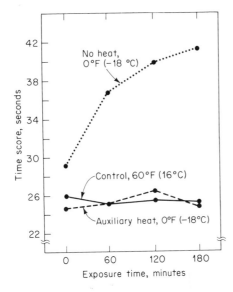

Figure 13-15 Time to perform the assembly part of the Purdue Pegboard Test under air conditions of 0°F (− 18°C) with and without auxiliary warming of the hands, and for a control condition. [Adapted from Lockhart and Keiss, 23, fig. 1.]

screw-tightening task. Figure 13-15 shows the results with one of the dexterity tests (the assembly part of the Purdue pegboard), the figure showing the time required to complete the test at four points in time under three conditions. It can be seen that, with air temperature of 0°F, the auxiliary hand warming permitted the test to be performed about as well as under a control condition of 60°F, whereas without the hand warming performance deteriorated over time.

In outdoor activities, wind chill is a factor to be contended with. Among its effects are increased numbness [Mackworth, 29] and increased reaction time [Teichner, 36]. In this connection acclimatization somewhat increases tolerance to wind chill but does not eliminate the effects of cold on manual performance. This was shown, for example, with a group of 22 men who lived in a room with temperatures of a brisk −20°F for a period of 8 to 14 days [Horvath and Freedman, 16]. It was found that such continuous exposure resulted in deteriorating performance on both a gear-assembly test and a code test with the use of a pencil. Two other interesting points also were discovered, namely, that neither mental performance (as measured by the code test) nor visual performance (as measured by speed or precision in responding) was particularly affected by the cold.

THE MANAGEMENT OF TEMPERATURE PROBLEMS

It is neither feasible nor appropriate here to try to set forth the engineering and other specifications for the resolution of all kinds of existing or anticipated temperature problems. However, let us touch very briefly on the major aspects of the management of such problems.

In the case of indoor situations, atmospheric control can be effected through heating, air conditioning, circulation of air, humidity control, insulation, and shielding against radiation, as well as by other techniques. For persons who are exposed to extreme conditions, appropriate clothing and protective gear may be useful. In the case of cold temperature, for example, the use of warm clothing can extend the tolerance level of people, as illustrated in Figure 13-16. That shows the combination of exposure time and temperature that was

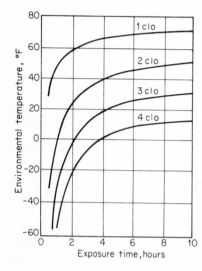

Figure 13-16 Exposure time and temperature that are tolerable for different levels of insulation (clo units). [Adapted from Burton and Edholm, 5, as presented in Webb, 39, p. 125.]

tolerable for each of four insulation levels; the difference in tolerance between 1 and 4 clo units is upward of 60°F or more. This figure demonstrates the trade-off effects, in terms of maintaining heat balance, of exposure time and insulation. At the hot end of the scale, clothing needs to be light and loose to permit evaporation and convection heat loss, and to minimize absorption of radiant energy from the sun. The rather common use in the tropics of loose-fitting, white apparel probably is not happenstance, but rather is the result of experience which has proved such apparel more suitable than other types; white fabric, for example, would absorb less energy from the sun than dark fabric.

Where it is not possible or practical to modify extreme environmental conditions to bring them into the bounds of normal, continuous human tolerance, there are certain actions which can be taken in the management of personnel who are to work or live in the environment that may make it tolerable. Some such actions are as follows [adapted in part from Machle, 27]: (1) selection of personnel who can tolerate the condition (sometimes by tryout for 4 or 5 days); (2) permitting people to become gradually acclimatized; (3) establishing appropriate work and rest schedules; (4) rotating personnel; (5) modifying the work (such as by reducing energy requirements); and (6) maintaining hydration by seeing that people drink enough water to replace that lost (thirst is not an adequate indication of water requirements).

AIR PRESSURE AND ALTITUDE

In the mundane lives of most mortals, the atmospheric variables that we complain about most are temperature and humidity. In less normal environments, however, other variables may play first fiddle. For people in high altitudes (e.g., mountainous areas and aircraft) and for people below sea level (e.g., diving and underwater construction work) air pressure and associated problems can be of paramount importance to well-being and to human performance.

The Atmosphere around Us

The atmosphere of the earth consists primarily of oxygen (21 percent) and nitrogen (78 percent), but also includes a bit of carbon dioxide (0.03 percent), plus other odds and ends. The density of the atmosphere, however, is reduced at higher altitudes; thus, at an altitude of, say, 20,000 ft, there is less air in a given volume that at sea level. Further, because of the weight of the atmosphere, the pressure decreases with altitude. Air pressure is frequently measured in pounds per square inch (psi) or in millimeters of mercury (mm Hg). At sea level the air pressure is 14.71 psi (760 mm Hg). Figure 13-17 shows the

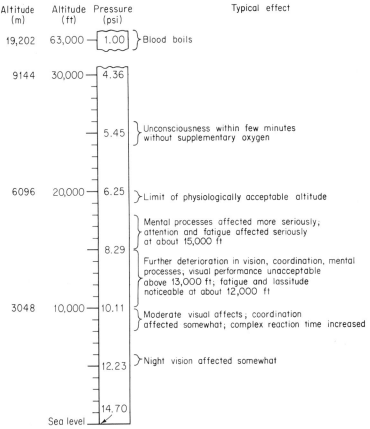

Figure 13-17 General effects of hypoxia at various altitude levels and equivalent pressure levels. [Adapted in part from Roth, 34, vol. 3, sec. 10; *Your body in flight,* 52; *Flight surgeon's manual,* 49; *Handbook of human engineering data,* 50, part 7, chap. 2, sec. 14, tables 4-1 to 4-4; and Balke, 2.]

pressures at other altitudes. A reduction of pressure by some ratio would mean that the volume occupied by a given amount of air would be increased inversely (such as doubling the volume if the pressure is reduced by half).

Air Pressure and Oxygen Supply

A primary function of the respiratory system is transporting oxygen from the lungs to the body tissue and picking up carbon dioxide on the return trip and carting it back to the lungs, where it is exhaled. Under normal circumstances (including near-sea-level pressure) the blood (actually the red blood cells) carries oxygen up to about 95 percent of the red blood cells' capacity. As air pressure is reduced, however, the amount of oxygen that the blood will absorb is reduced. For example, at approximately 10 psi (equivalent to about a 10,000-ft altitude) the blood will hold about 90 percent of its potential capacity; at about 7.3 psi (18,000 ft) the percentage drops to about 70. At 1.0 psi (63,000 ft) the pressure is so low that the blood actually boils, like water in a tea kettle.

Hypoxia

If the oxygen supply is reduced, a condition of hypoxia (also called *anoxia*) can occur, the effects varying with the degree of reduction. Some indication of these is given in Figure 13-18 as related to the amounts of oxygen that would be

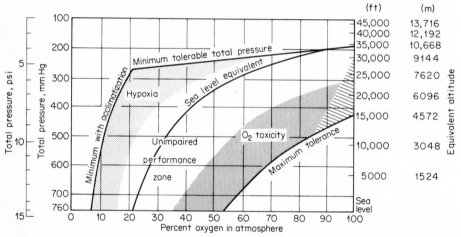

Figure 13-18 Physiological relationships between the percentage of oxygen in the atmosphere and pressure (and equivalent altitude). The white band under the "sea-level equivalent" curve shows the conditions in which human performance normally is unimpaired. [Adapted from Roth, 34, fig. 1-2, based on data compiled by U. C. Luft and originally drawn by E. H. Green of the Garrett Corporation.]

available at various altitudes (or their equivalents). Generally speaking, the effects below about 8000 ft are fairly nominal, but above that level, or at least above about 10,000 ft, the effects become progressively more serious, as shown in the figure. In connection with the effects of hypoxia, however, two qualifications are in order: (1) There are marked individual differences and (2) acclimatization does increase tolerance somewhat.

Use of Oxygen   At altitudes where the hypoxia effects would normally be of some consequence, the use of oxygen masks can stave off the onset of hypoxia or minimize its degree. This is illustrated, for example, in Figure 13-19, which shows the relationship between altitude and percentage of oxygen

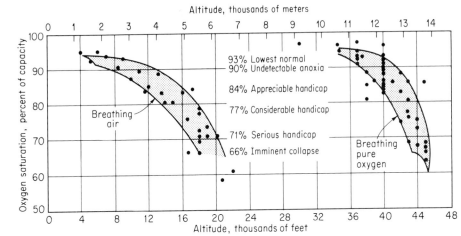

Figure 13-19   Relationship between altitude and oxygen saturation (percentage of capacity) for subjects breathing air and those breathing pure oxygen. [From *Flight surgeon's manual*, 49.]

capacity for subjects breathing air versus pure oxygen. This also shows the approximate degree of handicap for various percentage values. It can be seen, for example, that while breathing pure air reduces the percentage of capacity to about 70 at around 18,000 ft, the use of oxygen delays this amount of reduction up to about 42,000 ft.

Pressurization   The ideal scheme for avoiding hypoxia at high altitudes is the use of a pressurized cabin that maintains the atmospheric conditions of some lower altitude. This is done in high-altitude civilian planes and in some military planes. The Air Force generally requires the use of oxygen equipment for aircraft that will operate at altitudes of 10,000 ft or above, or at 8000 ft on flights of 4 h or longer [*System safety design handbook*, 51, chap. 3, sec. 3P], and at higher altitudes prescribes minimum differentials in pressure for different types of planes and flights [52]. Pressure suits usually are prescribed for high-altitude aircraft, in part as protection against decompression.

Decompression

It has been indicated that the volume of gas expands or contracts in proportion to the pressure applied to it (this is Boyle's law). The atmosphere within the human body is not immune from this law. Thus, as an individual changes from one pressure to another the atmosphere in body tissues and cavities expands or contracts. In this connection, as the external air pressure is increased, that within the body follows suit fairly closely, and (aside from some discomfort here and there) there are no serious consequences. But when the external air pressure is reduced suddenly, there can be some unhappy consequences, generally referred to as *decompression sickness*. Actually, there are different kinds of physiological reactions. Among these is the formation of nitrogen bubbles in the blood, body tissue, and around the joints; the effect is somewhat like taking the top off a bottle of soda water. The manifestations of the physiological effects include various symptoms, such as those referred to familiarly as the *bends* and the *chokes*, and various skin manifestations. The bends consist of generalized pains in the joints and muscles. The chokes are characterized by breathing (choking) difficulty, coughing, respiratory distress, and accompanying chest pains; the skin sensations include hot and cold sensations and itching of the skin, and a mottling of the skin surface sometimes occurs. In extreme cases, the above symptoms become more severe, and in some cases shock, delirium, and coma occur; fatalities may occur in such cases.

Protection from Decompression Decompression sickness occurs primarily in underwater construction work (in sealed caissons), in underwater diving operations, in submarines, and in aircraft (as in the case of rapid ascent to, say, 30,000 ft or more, or in rapid decompression from a rupture in a sealed cabin). Where rapid decompression has occurred (say, by accident), it is usually the practice to subject the person to a higher pressure (near that to which he was originally exposed) and then gradually bring the pressure back to that of the earth.

AIR POLLUTION

We humans have inherited from our natural environment a number of environmental hazards to contend with. But our science and technology have resulted in the creation of a whole host of new environmental hazards, especially pollutants, including smoke, exhaust fumes, toxic vapors and gases, insecticides, herbicides, and ionizing radiation. Some of these contaminants are the by-products of industrial processes and tend to be confined to their industrial environments, whereas others escape into the general atmosphere; and our automobiles, household furnaces, and other nonindustrial sources add their bit to the ever-increasing level of pollution in the air we breathe. The constraint of pages precludes a thorough discussion of this topic.[3] However, Figure 13-20 is presented for the purpose of illustrating the effects of atmospheric pollutants; this particular example deals with the effects of carbon monoxide on man as a function of the degree of concentration and exposure time. For some other types of pollutants, exposure limits are set in terms of the amount or concentration, and in some instances the exposure times that should not be exceeded.

---

[3]The interested reader is referred to such sources as Roth [34, sec. 13, and the *Bioastronautics data book*, 48, chap. 3].

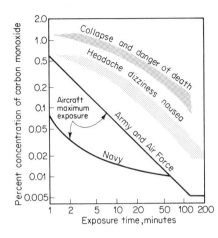

Figure 13-20 Effects of carbon monoxide on man as functions of concentration and exposure time. Shaded areas show conditions that are dangerous and lethal (heavy shading) and that are milder (light shading). The solid lines represent exposure limits set by the military services for aircraft. [Adapted from *Bioastronautics data book*, 48, fig. 10-2, p. 468, as based on various sources; for original sources, see that reference.]

REFERENCES

1 Azer, N. Z., P. E. McNall, and H. C. Leung: Effects of heat stress on performance, *Ergonomics*, 1972, vol. 15, no. 6. pp. 681–691.

2 Balke, B.: *Human tolerances*, Civil Aeromedical Research Institute, Federal Aviation Agency, Aeronautical Center, Oklahoma City, Report 62–6, April, 1962.

3 Barnett, P.: *Field tests of two anti-exposure assemblies*, Arctic Aeromedical Laboratory, Fort Wainwright, Alaska, AAL TDR 61–56, 1962.

4 Belding, H. S., and T. F. Hatch: Index for evaluating heat stress in terms of resulting physiological strains, *Heating, Piping and Air Conditioning*, August, 1955, pp. 129–136.

5 Burton, A. C., and O. G. Edholm: *Man in a cold environment: physiological and pathological effects of exposure to low temperatures*, Edward Arnold (Publishers), Ltd., London, 1955.

6 Clark, R. E.: *The limiting hand skin temperature for unaffected manual performance in the cold*, USA Quartermaster Research and Engineering Center, TR EP-147, February, 1961.

7 Colquhoun, W. P., and R. F. Goldman: Vigilance under induced hyperthermia, *Ergonomics*, 1972, vol. 15, no. 6, pp. 621–632.

8 Dusek, E. R.: *Manual performance and finger temperature as a function of skin temperature*, USA Quartermaster Research and Engineering Center, TR EP-68, October, 1957.

9 Fanger, P. O.: *Thermal comfort, analysis and applications in environmental engineering*, Danish Technical Press, Copenhagen, 1970.

10 Fox, W. F.: Human performance in the cold, *Human Factors*, vol. 9, no. 3, pp. 203–220.

11 Gagge, A. P., A. C. Burton, and H. C. Bazett: A practical system of units for the description of the heat exchange of man with his environment, *Science*, 1941, vol. 94, pp. 428–430.

12 Gagge, A. P., J. A. J. Stolwijk, and Y. Nishi: An effective temperature scale based on a simple model of human physiological regulatory response, *ASHRAE Transactions*, 1971, vol. 77, pt I., pp. 247–257.

13   Hardy, J. D. (ed.): *Temperature—its measurement and control in science and industry*, Reinhold Publishing Corporation, New York, 1963.

14   Hatch, T. F.: "Assessment of heat stress," in J. D. Hardy (ed.), *Temperature—its measurement and control in science and industry*, vol. 3, pt. 3, pp. 307–318, Reinhold Publishing Corporation, New York, 1963.

15   Hertig, B. A., and H. S. Belding: "Evaluation and control of heat hazards," in J. D. Hardy (ed.), *Temperature—its measurement and control in science and industry*, vol. 3, pt. 3, pp. 347–355, Reinhold Publishing Corporation, New York, 1963.

16   Horvath, S. M., and A. Freedman: The influence of cold upon the efficiency of man, *Journal of Aviation Medicine*, 1947, vol. 18, pp. 158–164.

17   Kamon, E. and H. S. Belding: The physiological cost of carrying loads in temperate and hot environments, *Human Factors*, 1971, vol. 13, no. 2, pp. 153–161.

18   Leithead, C. S., and A. R. Lind: *Heat stress and heat disorders*, Cassell & Co., Ltd., London, 1964.

19   Lind, A. R.: A physiological criterion for setting thermal environmental limits for everyday work, *Journal of Applied Physiology*, 1963, vol. 18, pp. 51–56.

20   Lind, A. R.: "Tolerable limits for prolonged and intermittent exposures to heat," in J. D. Hardy (ed.), *Temperature—its measurement and control in science and industry*, vol. 3, pt. 3, pp. 337–345, Reinhold Publishing Corporation, New York, 1963.

21   Lind, A. R., and D. E. Bass: The optimal exposure time for the development of acclimatization to heat, *Federal Proceedings*, 1963, vol. 22, no. 3, pp. 704–708.

22   Lockhart, J. M.: Extreme body cooling and psychomotor performance, *Ergonomics*, 1968, vol. 11, no. 3, pp. 249–260.

23   Lockhart, J. M., and H. O. Keiss: Auxiliary heating of the hands during cold exposure and manual performance, *Human Factors*, 1971, vol. 13, no. 6, pp. 457–465.

24   McArdle, B., et al.: *The prediction of the physiological effects of warm and hot environments*, Medical Research Council (Great Britain), H.S. 194 Royal Naval Personnel Research Committee, Report R.N.P. 47/391, October, 1947.

25   McBlair, W., D. Rumbaugh, and J. Fozard: *Ventilation, temperature, humidity*, San Diego State College Foundation, Contract Nonr-1268 (01), December, 1955.

26   McKarnes, J. S., and R. S. Brief: Nomographs give refined estimate of heat stress, *Heating, Piping and Air Conditioning*, January, 1966, vol. 38, no. 1, pp. 113–116.

27   Machle, W.: Control of heat in industry, *Occupational Medicine*, 1946, vol. 2, pp. 350–359.

28   Mackworth, N. H.: Effects of heat on wireless telegraphy operators hearing and recording Morse messages, *British Journal of Industrial Medicine*, 1946, vol. 3, pp. 143–158.

29   Mackworth, N. H.: Cold acclimatization and finger numbness, *Proceedings of the Royal Society*, 1955, vol. B, 143, pp. 392–407.

30   Mackworth, N. H.: *Researches on the measurement of human performance*, Medical Research Council (Great Britain), Special Report Series 268, 1950. Reprinted in H. W. Sinaiko (ed.), *Selected papers on human factors in the design and use of control systems*, Dover Publications, Inc., New York, 1961.

31   Macpherson, R. K.: Thermal stress and thermal comfort, *Ergonomics*, 1973, vol. 16, no. 5, pp. 611–623.

32   Provins, K. A., D. J. Glencross, and C. J. Cooper: Thermal stress and arousal, *Ergonomics*, 1973, vol. 16, no. 5, pp. 623–631.

33 Rohles, R. H., Jr., and R. G. Nevins: The nature of thermal comfort for sedentary man, *ASHRAE Transactions*, 1971, vol. 77, pt. I, pp. 239–246.

34 Roth, E. M. (ed.): *Compendium of human responses to the aerospace environment*, NASA CR-1205, vols. 1–4, November, 1968.

35 Siple, P. A., and C. F. Passel: Measurement of dry atmospheric cooling in subfreezing temperatures, *Proceedings of the American Philosophical Society*, 1945, vol. 89, pp. 177–199.

36 Teichner, W. H.: Reaction time in the cold, *Journal of Applied Psychology*, 1958, vol. 42, pp. 54–59.

37 Tichauer, E. R.: The effects of climate on working efficiency, *Impetus*, Australia, July 1962, vol. 1, no. 5, pp. 24–31.

38 Veghte, J. A., and J. R. Clogston: *A new heavy winter flying clothing assembly*, Arctic Aeromedical Laboratory, Fort Wainwright, Alaska, AAL TN 61–4, 1961.

39 Webb, P. (ed.): *Bioastronautics data book*, NASA SP–3006, 1964.

40 Welch, R. B., E. O. Longley, and O. Lomeav: The measurement of fatigue in hot working conditions, *Ergonomics*, 1971, vol. 14, no. 1, pp. 84–90.

41 Wing, J. F.: *A review of the effects of high ambient temperature on mental performance*, USAF, AMRL, TR 65–102, September, 1965.

42 Winslow, C. E. A., et al.: Physiological influence of atmospheric humidity: second report of the ASHVE Technical Advisory Committee on Physiological Reactions, *Transactions of the ASHVE*, 1942, vol. 48, pp. 317–326.

43 Winslow, C. E. A., and L. P. Herrington: *Temperature and human life*, Princeton University Press, Princeton, N.J., 1949.

44 Winslow, C. E. A., L. P. Herrington, and A. P. Gagge: Physiological reactions of the human body to varying environmental temperatures, *American Journal of Physiology*, 1937, vol. 120, pp. 1–22.

45 Wyndham, C. H., N. B. Strydom, H. M. Cook, and J. S. Mavitz: *Studies on the effects of heat on performance of work*, Applied Physiology Laboratory Reports, 1-3/59, 1959, Transvaal and Orange Free State Chamber of Mines, Johannesburg, S. Africa.

46 *ASHRAE handbook of fundamentals*, 1967, ASHRAE, New York.

47 *ASHRAE handbook of fundamentals*, 1972, ASHRAE, New York.

48 *Bioastronautics data book*, 2d ed., NASA SP-3006, U.S. Government Printing Office–Washington, D.C., 1973.

49 *Flight surgeon's manual*, USAF Manual 160–5, July, 1954.

50 *Handbook of human engineering data*, 2d ed., Tufts University, Medford, Mass., 1952.

51 *System safety design handbook*, USAF, AFSC DH 1–6, 2d ed., July 20, 1968.

52 *Your body in flight*, USAF Pamphlet 160–10–3, Jan. 1, 1960.

Chapter 14

Noise

Before the days of machines and mechanical transportation equipment, mankind's noise environment consisted of noises such as those of household activities, domestic animals (and maybe a few blood-curdling wild ones), horse-drawn vehicles, hand tools, and weather. But man's ingenuity changed all this through his creation of machines, motor vehicles, subways, radios, guns, bombs, fire sirens, jet aircraft, and New Year's Eve horns. Noise has become such a pervasive aspect of working situations and community life as to be referred to as "noise pollution" and to be considered a health hazard.

Although noise has commonly been referred to as unwanted sound, a somewhat more definitive concept is the one proposed by Burrows [3], in which noise is considered in an information-theory context, as follows: Noise is "that auditory stimulus or stimuli bearing no informational relationship to the presence or completion of the immediate task." This concept applies equally well to attributes of task-related sounds that are informationally useless, as well as to sounds that are not task-related.

NOISE AND LOSS OF HEARING

Of the different possible effects of noise, one of the most important is hearing loss. There are really two primary types of deafness. One is called *nerve* deafness and most frequently is caused by a condition of the nerve cells of the inner ear that reduces sensitivity. The other is *conduction* deafness and is caused by some condition of the outer or middle ear that affects the transmission of sound waves to the inner ear.

The hearing loss in nerve deafness is typically uneven; usually the hearing loss is greater in the higher frequencies than in the lower ones. Normal deterioration of hearing through aging is usually of the nerve type, and continuous exposure to high noise levels also typically results in nerve deafness. Once nerve degeneration has occurred, it can rarely be corrected. Conduction deafness is only partial, since airborne sound waves strike the skull and may be transmitted to the inner ear by conduction through the bone. It may be caused by different conditions, such as adhesions in the middle ear that prevent the vibration of the ossicles, infection of the middle ear, wax or other substance in the outer ear, or scars from a perforated eardrum.

People with this type of damage sometimes are able to hear reasonably well, even in noisy places, if the sounds to which they are listening (for example, conversation) are at intensities above the background noise. This type of deafness can sometimes be arrested, or even improved. Hearing aids are more frequently useful in this type of deafness than they are in nerve deafness.

Measuring Hearing

In order to review the effects of noise on hearing, we should first see how hearing (or hearing loss) is measured. There are two basic methods of measurement, namely, the use of simple tests of hearing and the use of an audiometer.

Simple Hearing Tests   For some purposes simple hearing tests are used. These include a voice test, a whisper test, a coin-click test, and a watch-tick test. In the voice and whisper tests, the tester (out of sight) speaks or whispers to the testee, and the testee is asked to repeat what was said. This may be done at different distances and with different voice intensities. The primary shortcoming of such tests is that they usually lack standardization. If reasonable standardization can be achieved (such as using a particular person's voice or a particular watch), such tests may serve certain rough hearing-test purposes.

Audiometer Tests   Audiometers are of two types, the most common being an instrument that is used to measure hearing at various frequencies. It reproduces, through earphones, pure tones of different frequencies and intensities. As the intensity is increased or decreased, the testee is asked to indicate when he can hear the tone or when it ceases to be audible. It is then possible to determine for each frequency tested the lowest intensity that can just barely be heard; this is the *threshold* for the frequency.

Another type of audiometer is a speech audiometer. Direct speech, or a recording of speech, is reproduced to earphones or to a loudspeaker, and intensity is controlled. Various types of speech intelligibility tests were discussed in Chapter 6.

Normal Hearing and Hearing Loss

Before we see what effect noise has on hearing, we should first see what normal hearing is like.

Surveys of Hearing Loss   Surveys have been made to determine the hearing abilities of people. In such surveys individuals are tested at various frequencies to determine their loss of hearing at each of the tested frequencies. The average hearing loss at each frequency is then determined. An analysis of such data by Spoor [23] has given the results shown in Figure 14-1. The two sets of

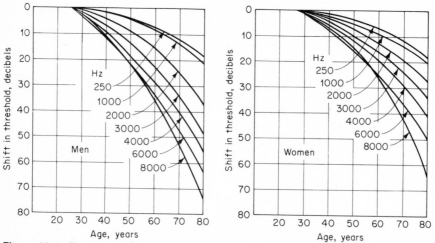

Figure 14-1   The average shifts with age of the threshold of hearing for pure tones of persons with "normal" hearing, using a 25-year-old group as a reference group. [From Spoor, 23, as presented by Peterson and Gross, 20.]

curves show, for men and women, the average shifts in the threshold of hearing for pure tones that occur with age. It is clear that hearing loss typically becomes increasingly severe at the higher frequencies, the average loss being a bit greater for men than for women in those frequencies. The hearing loss shown in that figure generally represents the effects of two factors, one being aging (presbycosis), and the other the normal stresses and nonoccupational noises of our current civilization.

Hearing Loss from Continuous Noise Exposure

The typical pattern of hearing loss for individuals is one which starts out with some temporary hearing loss (a temporary threshold shift, TS) which usually is recovered a few hours or days after exposure. However, with additional exposure the amount of recovery gradually becomes less and less, with some residual permanent loss (what is sometimes called *noise-induced permanent threshold shift*, NIPTS). This NIPTS, in turn tends to become progressively greater with subsequent exposure, as illustrated in Figure 14-2. That figure shows the NIPTS at three stages. Although the audiograms for different indi-

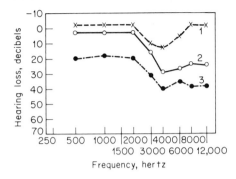

Figure 14-2 Audiograms for an individual showing three stages in the development of noise-induced permanent threshold shift (NIPTS). [From Tomlinson, 26, fig. 1.]

viduals vary in terms of the amount of hearing loss at different frequencies, there is a dip at about 4000 Hz that is characteristic of the pattern for many individuals.

There are of course individual differences in susceptibility to hearing loss, but the relationship between temporary and permanent hearing loss is such that temporary hearing loss can be used as a reasonably valid yardstick for predicting permanent hearing loss due to continuous exposure to noise [Kryter, 14].

Evidence of Hearing Loss from Noise There have been a number of specific studies, and summaries thereof, that provide persuasive testimony about the effects of noise on hearing [Cohen, 5; Kryter, 13; and LaBenz et al., 17]. In this connection, a committee under the aegis of the American Industrial Hygiene Association has teased out and condensed, from many sources, data that show the incidence of hearing impairment, in a consolidated figure, for each of several age groups of individuals who have been exposed (in their work) to noise intensities of different levels. This consolidation is shown in Figure 14-3. Actually this figure shows, for any group, the probabilities of individuals having hearing impairment (specifically, an average hearing threshold in excess of 15 dB at 500, 1000, and 2000 Hz). Although the probabilities of impairment (so defined) are not much above those of the general population for individuals exposed to 85 dB, the curves shoot up sharply at higher levels, except for the youngest group (but their time will come!).

Although Figure 14-3 shows us something about the *odds* of having impairment, it does not depict the nature of the impairment in terms of frequencies. For this purpose, let us refer to the very thorough longitudinal survey carried out by a subcommittee of the ANSI (formerly USASI) [37]. Audiograms were prepared on a selected group of 200 workers (1) who had been exposed continuously (during the working day) to a single type of noise for a period of time—the time ranging from 2 to 44 years; (2) who had no history of previous exposure to intense noise; and (3) whose noise environment in the plant was known (sound spectrums were developed for 30 different plant locations, with an octave-band analyzer).

By the use of an ingenious procedure that need not be described here, the investigators developed estimates of the hearing loss brought about by varying

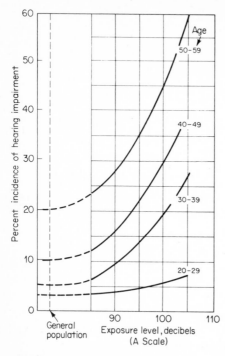

Figure 14-3 Incidence of hearing impairment in the general population and in selected populations by age group and occupational noise exposure; *impairment* is defined as a hearing threshold level in excess of an average of 15 dB at 500, 1000, and 2000 Hz. [From *Industrial noise manual,* 33, fig. 1, p. 420.]

lengths of exposure to noises of various levels. These estimates were made of hearing loss at each of three frequency levels, namely, 1000, 2000, and 4000 Hz. The estimates were corrected for normal hearing loss attributable to age. The results are shown in Figure 14-4. The three parts of this figure, *a*, *b*, and *c*, represent the estimated-average-trend curves for net hearing loss at 1000, 2000, and 4000 Hz, respectively. The decibel level for each contour is the noise level for a *specific octave, not* the *overall* level. Without belaboring the point, it can be said that the octave selected for use with any one of the three frequency levels is the octave for which the noise level was found to be most nearly correlated with hearing loss at the frequency in question. This does not necessarily mean, however, that the noise in this selected octave caused the hearing loss of the frequency in question.

Three implications can be drawn from this figure: (1) The amount of hearing loss is related to level of noise to which exposed; the greater the exposure intensity, the greater the hearing loss. (2) Hearing loss is greater in the 4000-Hz range than in the 1000- and 2000-Hz ranges; this has been found in numerous studies. (3) Hearing loss is associated with exposure time for higher exposure intensities, though to a limited extent, or not at all, for lower exposure intensities.

Noncontinuous Noise

The gamut of noncontinuous noise includes intermittent (but steady) noise (such as machines that operate for short, interrupted periods of time), impact

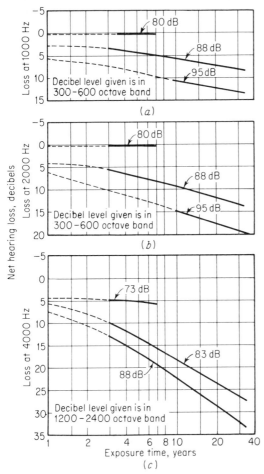

Figure 14-4 Estimated average trend curves for net hearing loss at 1000, 2000, and 4000 Hz, after continuous exposure to steady noise. Data are corrected for age but not for temporary threshold shift. Dotted portions of curves represent extrapolation from available data. [Adapted from ANSI, 37.]

noise (such as that from a drop forge), and impulsive noise (such as from gun fire). In heavy doses, such noise levies its toll in hearing loss, but the combinations and permutations of intensity, noise spectrum, frequency, duration of exposure, and other parameters preclude any simple, pat descriptions of the effects of such noise.[1] In the case of impact and impulsive noise, however, it might be noted that the toll sometimes is levied fairly promptly. For example, 35 drop-forge operators showed a noticeable increase in hearing threshold within as little as 2 years [37], and 45 gunnery instructors averaged 10 percent hearing loss over only 9 months, even though most of them had used hearing-protection devices [Machle, 18].

[1] A thorough treatment of exposure to intermittent noise is presented by Kryter et al. [16].

PHYSIOLOGICAL EFFECTS OF NOISE

Permanent hearing loss is of course the consequence of physiological damage to the mechanisms of the ear. Aside from possible damage to the ear itself, one would wonder whether continued exposure to noise might induce any other temporary or permanent physiological effects. To begin with, there are indications that exposure to reasonably intense noise can trigger certain "vegetative" reactions, such as vasoconstriction of the peripheral circulation system, and changes in pupil size, pulse amplitude, and EEG [Jansen, 10], and can induce a depression of the adreno-cortical function [Matsui and Sakamoto, 19]. It has been noted, however, that some initial physiological responses to noise (especially to sudden noise) sometimes settle back to normal or near-normal levels with continued exposure to the noise. The physiological reactions to noise, however, usually would not be regarded as being of a pathological nature if the noise occurs only a few times. However, there is an accumulating body of evidence that indicates that exposure to high noise levels (such as 95 dB or more) over a period of years can produce pathological side effects and thus can comprise a health hazard [Jansen, 10]. In this regard Cameron et al. [4] report the results of surveys in Los Angeles and Detroit in which the incidence of chronic and acute illness tended to be greater for persons who typically were exposed to a "great deal" of noise or sound as contrasted to those who were not so exposed, thus lending some evidence to the notion that noise may be a contributing factor to physical illness.

EFFECTS OF NOISE ON PERFORMANCE

By selection of relevant studies, one can "prove" that noise (1) produces a decrement in human performance, (2) has no effect on such performance, or (3) produces an increment in performance. This confusing state of affairs is reflected in a review by Teichner et al. [25], and probably somewhat supports the statements made in the *Industrial noise manual* [33] to the effect that the behavioral effects of noise are as complex and ill-defined as the noise itself and that a categorical statement that noise exposure has no ill effects on human behavior cannot be made at this time. In contemplating this unsatisfying state of affairs, Teichner et al. [25] chide those investigators who glibly attribute performance decrement under noise to "distraction," performance increment to motivational "compensation," and no effect to "lack of sensitivity" of the task. Further, they suggest that the results of various (and apparently conflicting) studies might be attributable to the way in which distraction, habituation, auditory adaptation, and bodily arousal combine to affect the performance in question. Although we probably can reject the notion that noise *generally* brings about degradation in human performance, there are accumulating indications from experimental studies that certain types of noise may adversely affect performance on certain types of tasks. For example, Broadbent [3] and Jerison [11] report degradation in vigilance tasks under certain noise conditions. In addition, it seems that performance on the following kinds of tasks is most likely to be affected by noise: certain types of complex mental tasks; tasks calling for skill and speed [Roth, 21]; tasks that demand a high level of perceptual capacity (such as in some time-shared tasks that press one's perceptual abilities) [Boggs and Simon, 1]; and complex psychomotor tasks [Eschenbrenner, 8].

A few other comments about the detrimental effects of noise are in order. For example, it has been suggested by Finkleman and Glass [9] that performance decrement would tend to occur when the task and concurrent environmental stress exceed the person's total information-handling capacity; in their own investigation they found support for this notion as reflected by degradation in a "subsidiary" task, this being interpreted as the effects of "overload" on the subjects. There are also inklings that unexpected or unpredictable noise tends to be more disruptive than continuous or periodic noise [Eschenbrenner, 8].

Although most discussion of the performance effects of noise relate to degradation, there are indications that noise can also serve to enhance performance in some circumstances. For example, Warner [28] reports decreases in errors in a target detection task with increasing levels of intermittent noise from 80 to 100 dB. Although Warner attributes his results to greater "flexibility of attention" caused by the increase in noise intensity, this is not a completely satisfying explanation for the facilitative effects of noise. Another explanation that seems more generally applicable is the "arousal" hypothesis. This hypothesis applies to relatively easy tasks, the argument for it [as discussed by Eschenbrenner, 8] being that extraneous stimulation tends to focus attention on the task and makes the person more alert with respect to the environment.

It should be stated that most of the evidence relating to either the degrading or enhancing effects of noise on performance is based on experimental studies and not on actual work situations. And extrapolation from such studies to actual work situations is probably a bit risky. In view of this and in view of the ambiguity relating to the effects of noise on performance, Roth's observations may be consoling [21]. He points out that the level of noise required to exert a measurably degrading effect on task performance (such as a sound-pressure level of 90 dB) is considerably higher than the highest levels that are acceptable by other criteria, such as hearing loss and effects on speech communications. Thus, if noise levels are kept within reasonable bounds in terms of, say, hearing loss considerations, the probabilities of serious effects on performance probably would be relatively nominal.

ANNOYANCE AND WELL-BEING

In discussing the annoying qualities of noise, Kryter [15] uses the term *perceived noisiness* as being somewhat synonymous with unwantedness, unacceptableness, annoyingness, objectionableness, and disturbingness. You and I know full well what he is talking about, and the noise characteristics that do bother us are intensity, bandwidth, spectral content, and duration.

Annoying Aspects of Community Noise

In a survey carried out in Boston, Los Angeles, and New York for the Federal Housing Administration [35], traffic was generally reported by people to be the most "bothersome" source of urban noise, with noise from planes, industry, children and neighbors, and sirens and horns coming in for frequent complaint, depending in part on location (as, for example, near airports) and distance (such as from streets). As an example, a 10-min sample of noise of heavy traffic in New York averaged 81 dB at 15 ft and 76 dB at 50 ft. As another illustration, a properly muffled diesel truck produces about 77 dB at 100 ft, and this value can increase by about 5 dB during its acceleration; the absence of a muffler can add 15 dB more. Although we are inclined to believe that community noise is growing worse (and it probably is in some locations), there are some indications that this is not universally the case. In fact, Soroka [22] reports an actual decline of the noise level at certain locations in New York over a period of seven years!

Noise around airports presents a special problem that has what Stockbridge and Lee [24] describe as having psychosocial consequences. As one indication of the annoyance of aircraft noise they report that 3000 complaints were received from the London Heathrow airport area during the summer months (April–September) of one year. Aside from being annoying, aircraft noise can also affect ongoing activities around airports. For example, Cohen [6] reports that in three schools around one airport there were between 40 and 60 interruptions per day, resulting in a loss of 10 to 20 min/day in each classroom.

In the case of those areas or communities where noise levels are high, the noise levels probably cannot be passed off simply as being "annoying"; they can approach the levels of a health hazard [Cohen, 6; Jones and Cohen, 12] and therefore deserve community attention.

Procedure for Rating Annoyance Some indication of the potential seriousness of the level of noise in a community can be derived by the use of some noise-rating procedure. One such procedure is that developed by the International Organization for Standardization (ISO) [36]. This procedure takes into account such factors (aside from the noise level itself) as the nature of the community, the time of day, the season of the year, and the initial effects (as in the case of a new noise source in the community). The procedures are outlined below:

1 Develop a spectrum of the noise (or the noise that is expected to be generated, in the case of a new facility).
2 Superimpose this spectrum over Figure 14-5, and determine the highest noise-rating curve N that the spectrum exceeds.
3 Apply the corrections shown in Table 14-1; add algebraically these corrections, and add or subtract from the noise rating N derived above. This is the *corrected noise-rating number*.
4 Refer to Table 14-2 to ascertain the estimated public reaction to the noise.

Such a procedure, although admittedly rough, probably would be particularly useful when the creation of a new facility that could be a potential source of noise is being considered for a community. If, for example, a noise rating for

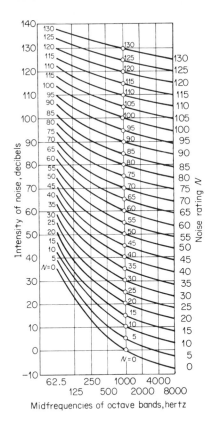

Intensity of noise, decibels

Noise rating N

Midfrequencies of octave bands, hertz

Figure 14-5 The International Organization for Standardization (ISO) noise-rating curves. Such curves can be used in the estimation of public reaction to noise sources (such as new sources) and for establishing noise ceilings for specific purposes (see text for discussion). [From 36.]

a facility being planned would turn out to be about 50, one could expect "widespread complaints" about it.

ACCEPTABLE LIMITS OF NOISE

In trying to figure out the upper ceiling of noise that would be acceptable in a given situation, the question of the criterion of acceptability immediately bobs up. Criteria in terms of speech communications were discussed in Chapter 6 [such as the speech interference level (SIL) and noise criteria (NC) curves], and the above discussion of residential noise levels dealt in part with the criterion of annoyance.

Hearing Loss Criteria

In typical work situations, hearing loss is perhaps the most important criterion by which to assess the acceptability of noise levels. In this connection, various damage risk criteria and other standards have been set forth by various organizations, these differentiating between continuous and noncontinuous noise.

Continuous Noise Actually, the ISO noise-rating curves given above in Figure 14-5 serve such purposes in connection with hearing loss that might occur from broad-band continuous noise during the working day (5 h or more).

Table 14-1   Correction to Noise Rating Number (Primarily Applicable to Residential Cases)

Influencing factor	Possible conditions	Corrections to uncorrected N
Noise spectrum character	Pure tone components	+5
	Wide-band noise	0
Peak factor	Impulsive	+5
	Nonimpulsive	0
Repetitive character	Continuous exposure to one-per-minute	0
(about $1/_2$ min noise	10–60 exposures per hour	−5
duration assumed)	1–10 exposures per hour	−10
	4–20 exposures per day	−15
	1–4 exposures per day	−20
	1 exposure per day	−25
Adjustment to exposure	No previous conditioning	0
	Considerable previous conditioning	−5
	Extreme conditioning	−10
Time of day and season	Only during daytime	−5
	At night	+5
	Winter	−5
	Summer	0
Allowance for local	Neighborhood:	
conditions	Rural	+5
	Suburban	0
	Residential, urban	−5
	Urban near light industry	−10
	Industrial area, heavy industry	−15

Source: ISO [36].

Table 14-2   Public Reaction to Noise in Residential Districts

Corrected noise rating number	Estimated public reaction
Below 40	No observed reaction
40–50	Sporadic complaints
45–55	Widespread complaints
50–60	Threats of community action
Above 65	Vigorous community action

Source: ISO [36].

In such continuous exposure, an N of 85 dB is suggested as an upper limit for conservation of hearing. In the United States the Occupational Safety and Health Administration (OSHA) of the Department of Labor has established a set of permissible noise exposures for persons working on jobs in industry [*Federal Register*, 30], the permissible levels depending on the duration of exposure, as shown (the dB levels given are based on the A scale):

Duration per day (hours)	Sound level (dB-A)
8	90
6	92
4	95
3	97
2	100
$1^1/_2$	102
1	105
$^1/_2$	110
$^1/_4$ or less	115

When the noise level is determined by octave-band analysis, the equivalent A-weighted sound level may be determined by the use of Figure 14-6. That figure shows several equivalent sound-level contours. In using this procedure, the spectrum of the noise is plotted over that figure, and the point of the highest

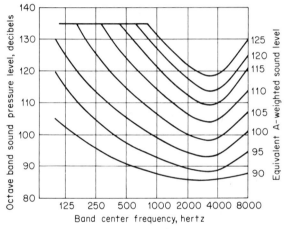

Figure 14-6 Equivalent sound-level contours used in determining the A-weighted sound level on the basis of an octave-band sound analysis. The curve at the point of the highest penetration of the noise spectrum reflects the A-weighted sound level. [From *Federal Register,* 30, p. 10518.]

penetration of that spectrum is the sound-level contour that is noted. That contour is the one that reflects the equivalent A-weighted sound level. That level, in turn, is used in determining the permissible duration of exposure as discussed above. However, this procedure offers only a rough approximation to dB-A, especially in instances of marked discontinuities in octave-band spectra.

The U.S. Air Force [29] has established a simple standard, using only the four octave-band pressure levels having center frequencies of 500, 1000, 2000, and 4000 Hz; this standard provides that the sound-pressure level should not exceed 85 dB for *any* of these four octave bands, for conventional daily exposure of 8 h (although there is provision for permitting 3 dB increases for each *halving* of the daily exposure duration).

Intermittent Noise In the case of intermittent noise, the tolerance limits depend on a trade-off between intensity and the relationship between the duration of exposure and the duration of subsequent nonexposure. An example of such trade-off is given in Figure 14-7.

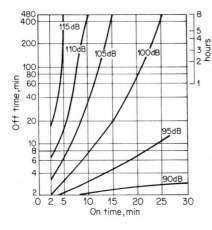

Figure 14-7 Guide to allowable exposure times for intermittent noise. (Each curve is labeled with the average value for octave bands with center-line frequencies of 500, 1000, and 2000 Hz.) The vertical scale shows the off time which must follow noise exposure of the duration shown on the horizontal scale (on time) to avoid temporary threshold shifts greater than 12 dB at 2000 Hz. [From *Guide for conservation of hearing in noise,* 32.]

THE HANDLING OF NOISE PROBLEMS

When a noise problem exists, or might be anticipated, there is no substitute for having good, solid information to bring to bear and for attacking the problem in a systematic manner. The manner of working out such problems is, of course, for the acoustical engineers, and we shall not go into this topic except in a very cursory way.

Defining the Noise Problem

The defining of a possible noise problem consists of essentially two phases. The first of these is the measurement of the noise itself (by the use of a sound-level meter and usually an octave-band analyzer). The second is to determine what noise level would be acceptable, in terms of hearing loss, annoyance, communications, etc. Such limits normally would be those adapted from relevant criteria such as discussed above. An example is shown in Figure 14-8. This figure shows the spectrum of the original noise of a foundry cleaning room, and a tentative-design base line that was derived from a set of relevant noise standards. Incidentally, part of the original spectrum was below this level, but the high frequencies were not; the difference represents the amount of reduction that should be achieved (in this case, the shaded area). The third line shows the noise level after abatement.

Noise Control

The general approaches to the control of noise include:

- Control at the source, such as proper design of machines, proper main-

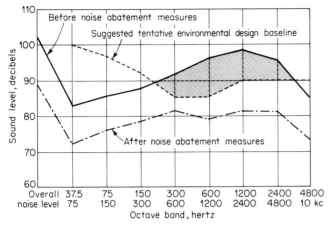

Figure 14-8 Spectrum of noise of foundry cleaning room before abatement, the base line that represents a desired upper ceiling, and the spectrum after abatement. The shaded area represents the desired reduction. The abatement consisted primarily in spraying a heavy coat of *deadener* on the tumbling barrels and surfaces of tote boxes. [From *Foundry noise manual*, 31, p. 52.]

tenance and lubrication, use of rubber mountings for machines, and use of vehicle mufflers

• Isolation of noise, as the use of enclosures, rooms, and other barriers; the closing of windows in a home typically reduces intensity by about 10 dB

    • Use of baffles and sound absorbers

    • Use of acoustical treatment

    • Proper layout

Actually, there are many ingenious variations of these and other means that have aided in noise reduction. Although we shall not discuss the effects of noise control, Figure 14-9 illustrates the possible effects of various noise-control

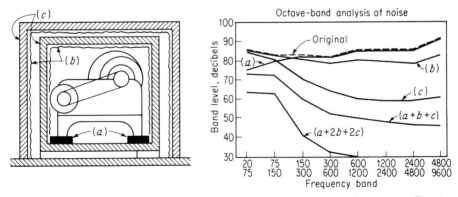

Figure 14-9 Illustrations of the possible effects of some noise-control measures. The lines on the graph show the possible reductions in noise (from the original level) that might be expected by vibration insulation *a*; an enclosure of acoustical absorbing material *b*; a rigid, sealed enclosure *c*; a single combined enclosure plus vibration insulation, *a + b + c*; and a double combined enclosure plus vibration insulation, *a + 2b + 2c*. [Adapted from Peterson and Gross, 20.]

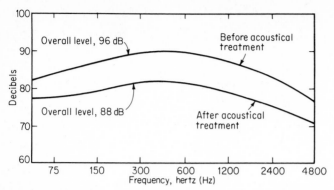

Figure 14-10 Spectrum of noise before and after acoustical treatment. [Adapted from Bon-vallet, in 34, p. 41.]

measures. As another example, Figure 14-10 shows the spectrum of noise in one situation before and after the use of acoustical treatment.

Ear Protection

Where the noise level cannot reasonably be reduced to "safe" limits, some form of ear protection should be considered for those people who are exposed to the noise. The two types of ear-protection devices that are reusable are earplugs and muffs. Disposable forms include dry cotton, waxed cotton, and glass down. Glass down is a form of glass wool in which the fibers are so fine that they form a material of downlike softness which has been reported to be quite harmless to the delicate skin of the ear canal [Coles, 7].

Effectiveness of Ear-Protection Devices The effectiveness of ear-protection devices is somewhat variable, depending on the nature of the noise, the duration of exposure, the fit of the device, its attenuating characteristics, and possibly other variables. By and large, however, such devices provide a worthwhile degree of protection from high noise levels. An example of such protection is shown in Figure 14-11 by the use of earplugs, earmuffs, and a combina-

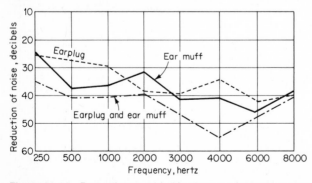

Figure 14-11 Protection provided by the use of earplugs, earmuffs, and a combination of the two [USAF, Wright-Patterson AFB].

tion. Coles [7] expresses the opinion that the fluid-seal type of earmuff comes as close to being the most efficient ear protectors that can be made. These have a plastic ring containing fluid (such as glycerin) that fits around the ear, thereby minimizing sound leakage. He also believes that the maximum amount by which an ear protector can reduce the sound reaching the ear is limited to about 35 dB at 250 Hz, up to about 60 dB at the higher frequencies. In practice, however, the typical attenuation is substantially lower.

REFERENCES

1 Boggs, D. H., and J. R. Simon: Differential effect of noise on tasks of varying complexity, *Journal of Applied Psychology*, 1968, vol. 52, no. 2, pp. 148–153.

2 Broadbent, D. E.: Effect of noise on an "intellectual" task, *Journal of the Acoustical Society of America*, 1958, vol. 30, pp. 824–827.

3 Burrows, A. A.: Acoustic noise, an informational definition, *Human Factors*, August, 1960, vol. 2, no. 3, pp. 163–168.

4 Cameron, P., D. Robertson, and J. Zaks: Sound pollution, noise pollution, and health: Community parameters, *Journal of Applied Psychology*, 1972, vol. 56, no. 1, pp. 67–74.

5 Cohen, A.: U.S. Public Health Service field work on the industrial noise hearing loss problem, *Occupational Health Review*, 1965, vol. 17, no. 3, pp. 3–10.

6 Cohen, A.: *Noise and psychological state*, USPHS, National Center for Urban and Industrial Health, RR-9, July, 1968.

7 Coles, R. R. A.: "Control of industrial noise through personal protection," in Ward and Frick [27].

8 Eschenbrenner, A. J., Jr.: Effects of intermittent noise on the performance of a complex psychomotor task, *Human Factors*, 1971, vol. 13, no. 1, pp. 59–63.

9 Finkelman, J. M., and D. C. Glass: Reappraisal of the relationship between noise and human performance by means of a subsidiary task measure, *Journal of Applied Psychology*, 1970, vol. 54, no. 3, pp. 211–213.

10 Jansen, G.: "Effects of noise on physiological state," in Ward and Frick [27].

11 Jerison, H. J.: Effects of noise on human performance, *Journal of Applied Psychology*, 1959, vol. 43, pp. 96–101.

12 Jones, H. H., and A. Cohen: Noise as a health hazard at work, in the community, and in the home, USPHS, *Public Health Reports*, July, 1968, vol. 83, no. 7, pp. 533–536.

13 Kryter, K. D.: The effects of noise on man, *Journal of Speech and Hearing Disorders, Monograph Supplement* 1, 1950, pp. 1–95.

14 Kryter, K. D.: Damage risk criterion and contours based on permanent and temporary hearing loss data, *American Industrial Hygiene Association Journal*, 1965, vol. 26, no. 1, pp. 34–44.

15 Kryter, K. D.: *The effects of noise on man*, Academic Press, Inc., New York, 1970.

16 Kryter, K. D., W. D. Ward, J. D. Miller, and D. H. Eldredge: Hazardous exposure to intermittent and steady-state noise, *Journal of the Acoustical Society of America*, 1966, vol. 39, pp. 451–463.

17 LaBenz, P., A. Cohen, and B. Pearson: A noise and hearing survey of earth-moving

equipment operators, *American Industrial Hygiene Association Journal*, March–April, 1967, vol. 28, pp. 117–128.

18   Machle, W.: The effect of gun blast on hearing, *Archives of Otolaryngology*, 1945, vol. 42, pp. 164–168.

19   Matsui, K., and H. Sakamoto: The understanding of complaints in a noisy workshop, *Ergonomics*, 1971, vol. 14, no. 1, pp. 95–102.

20   Peterson, A. P. G., and E. E. Gross, Jr.: *Handbook of noise measurement*, 7th ed., General Radio Co., New Concord, Mass., 1972.

21   Roth, E. M. (ed.): Compendium of human responses to the aerospace environment, NASA CR-1205 (5 vols.), November, 1968.

22   Soroka, W. W.: "Community noise surveys," in Ward and Frick [27].

23   Spoor, A.: Presbycusis values in relation to noise-induced hearing loss, *International Audiology*, July, 1967, vol. 6, no. 1, pp. 48-57.

24   Stockbridge, H. C. W., and M. Lee: The psycho-social consequences of aircraft noise, *Applied Ergonomics*, 1973, vol. 4, no. 1, pp. 44–45.

25   Teichner, W. H., E. Arees, and R. Reilly: Noise and human performance, a psychophysiological approach, *Ergonomics*, 1963, vol. 6, no. 1, pp. 83–97.

26   Tomlinson, R. W.: Estimation and reduction of risk of hearing: The background and a case study, *Applied Ergonomics*, 1971, vol. 2, no. 2, pp. 112–119.

27   Ward, W. D., and J. E. Frick (eds.): *Noise as a public health hazard*, Washington, D.C.: The American Speech and Hearing Association, ASHA Reports 4, February, 1969.

28   Warner, H. D.: Effects of intermittent noise on human target detection, *Human Factors*, 1969, vol. 11, no. 3, pp. 245–250.

29   AF Regulation 160–3, USAF, October, 1956.

30   *Federal Register*, May 29, 1971, vol. 36, no. 105.

31   *Foundry noise manual*, 2d ed., American Foundryman's Society, Des Plaines, Ill., 1966.

32   *Guide for conservation of hearing in noise*, prepared by Subcommittee on Noise, American Academy of Opthalmology and Otolaryngology, revised 1964.

33   *Industrial noise manual*, 2d ed., American Industrial Hygiene Association, Detroit, 1966.

34   *Noise*, lectures presented at the Inservice Training Course on the Acoustical Spectrum, Feb. 5–8, 1952, sponsored by the University of Michigan, School of Public Health and Institute of Industrial Health, University of Michigan Press, Ann Arbor.

35   *Noise environment of urban and suburban areas*, prepared by Bolt, Beranek, and Newman, Inc., Los Angeles, Federal Housing Administration, Superintendent of Documents, Washington, D.C., January, 1967.

36   *Rating noise with respect to hearing conservation, speech communication, and annoyance*, ISO, Technical Committee 43, Acoustics, Secretariat-139, August, 1961.

37   *The relation of hearing loss to noise exposure*, ANSI (formerly USASI), New York. 1954.

Chapter 15

MAN IN MOTION

Man's technological ingenuity in recent times has resulted in the creation of many methods of mobility that our ancestors never dreamed about. These include space capsules, aircraft, zero ground-pressure vehicles, rockets strapped to one's back, and of course various earthbound vehicles such as automobiles, buses, and trucks. Many of these make it possible for man to move at speeds and in environments he has never before experienced and to which he is not biologically adapted. The disparity between man's biological nature and that which is required for him to exist and perform in a different mobility environment specifies the domain within which human factors adaptations must be made.

The variables imposed by his increased mobility include vibration, acceleration and deceleration, weightlessness, and an assortment of more strictly psychological phenomena associated with his mobility, such as disorientation and other illusions. It is to some of these that we shall now turn our attention.

ACCELERATION AND VIBRATION TERMINOLOGY

Before discussing these, however, we should discuss certain of the concepts and terminology used in characterizing motion, and the sensory mechanisms that are related to motion and orientation.

Acceleration Terminology

Acceleration is a rate of change of motion of an object (a "mass"). The basic unit of measure of acceleration derives from the force of gravity in our earth-bound environment. The acceleration of a body in free fall is 32.24 feet per second, this rate being 1 *G*.

Acceleration forces applied to a mass can be either linear or angular. *Linear* acceleration is the rate of change of *velocity* of a mass, the direction of movement being kept constant. In turn, *angular* acceleration is the rate of change of *direction* of a mass, the velocity of which is kept constant. However, there are two forms of angular acceleration. One form, *radial* acceleration, is that in which the axis of rotation is external to the body (as in an aircraft turn). The other, commonly referred to as *angular* acceleration, is that in which the axis of rotation passes through the body (as when a ballet dancer is twirling around on her toe). In some circumstances people are subjected to an admixture of these.

Although angular acceleration is important in some circumstances (as in certain aircraft maneuvers), we will here discuss only linear acceleration. Such acceleration can of course occur in any direction with respect to the movement of the human body. In this regard, if the human body is caused forcibly to change velocity, that is, to accelerate or decelerate, as in a vehicle that is changing its velocity, the change in velocity causes a physiological reactive force that is opposite the direction of movement. This reactive force is manifested by a displacement of the heart and other organs, body tissue, and blood, since these body components are not rigid. Thus, as an automobile accelerates rapidly, the internal organs and the blood tend to "lag behind" the structure of the body as the body is propelled forward at increasing velocity.

The three possible directions of linear motion are depicted in Figure 15-1,

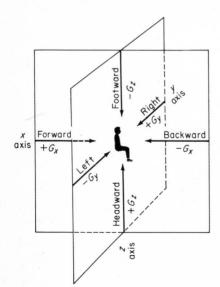

Figure 15-1 Illustration of three directions of linear acceleration. The direction of displacement of the heart and other organs is opposite that of the motion of the body.

these being called x (forward-backward), y (left and right), and z (upward-downward). Acceleration forces in these directions are expressed in terms of $\pm G$ units, as shown in column 4 of Table 15-1. Column 2 shows, for each direc-

Table 15-1 Acceleration and Vibration Terminology

Acceleration				Vibration		
1 Direction of motion	2 Heart motion toward	3 Motion of eyeballs	4 Symbol	5 Heart motion	6 Other description	7 Symbol
Forward	Spine	In	$+G_x$	Spine- sternum- spine	Fore-aft	$\pm g_x$
Backward	Ster- num	Out	$-G_x$			
To right	Left	Left	$+G_y$	Left- right- left	Side-to- side	$\pm g_y$
To left	Right	Right	$-G_y$			
Headward	Feet	Down	$+G_z$	Head- feet- head	Head-tail	$\pm g_z$
Footward	Head	Up	$-G_z$			

Source: Adapted from Hornick [21], table 7-1, p. 229.

tion of motion of the body, the (opposite) direction of motion of the heart and other internal organs. In turn, column 3 gives a vernacular label for each direction of motion as described by the sensation of movement of the eyeballs, such as "eyeballs in" or "eyeballs down."

Vibration Terminology

The earlier discussions of sound and noise have dealt with the physical parameters of vibrations that are audible. The same parameters, of course, apply to our present topic, except that here we are more concerned with vibrations of lower frequencies, generally below 100 Hz, most of which are not audible. The various physical parameters of vibration include on the one hand *frequency* (sinusoidal, combination of various frequencies, or random), and on the other hand some measure of *intensity*. Intensity can be measured in terms of maximum values of any of several parameters, such as: *displacement* (as in in or cm); *velocity*, the first derivative of displacement (as in in/s or cm/s); *acceleration*, the second derivative of displacement (as in in/s² or cm/s²), sometimes expressed as peak or maximum acceleration, and labeled in terms of lower-case g units as $\pm g_x$, $\pm g_y$, and $\pm g_z$, depending upon the direction of the oscillation as given in column 7 of Table 15-1; *rate of change of acceleration* or *jerk*, the third derivative of displacement (as in in/s^3 or cm/s^3); and power-spectral density (PSD) (used especially with random vibration, derived from the squares of the rms—root mean square—of the frequency bands).

When the body is caused to vibrate in a given direction (such as x, y, or z), the heart (and other body organs and members) oscillate to and fro in the plane of body movement, the directions of oscillation being given in columns 5 and 6 of Table 15-1.

MOTION AND ORIENTATION SENSES

The "five senses," in the Aristotelian tradition (vision, audition, smell, taste, and touch), basically deal with stimuli external to the body. The sensory receptors involved (the eyes, ears, etc.) are referred to as *exteroceptors*. There are, however, a number of other sensory receptors, and in this chapter we are concerned primarily with those that are related to motion and body orientation.

Proprioceptors

The *proprioceptors* are sensory receptors of various kinds that are imbedded within the subcutaneous tissues, such as in the muscles and tendons, in the coverings of the bones, and in the musculature surrounding certain of the internal organs. These receptors are stimulated primarily by the actions of the body itself. A special class of these are the kinesthetic receptors, which are concentrated around the joints and which are primarily associated with the discrimination of limb movements.

Semicircular Canals

The three semicircular canals in each ear are interconnected, doughnut-shaped tubes that form, roughly, a three-coordinate system, as shown in Figure 15-2

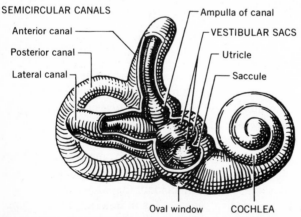

SEMICIRCULAR CANALS   Ampulla of canal

Anterior canal   VESTIBULAR SACS

Posterior canal   Utricle

Lateral canal   Saccule

Oval window   COCHLEA

Figure 15-2 The body orientation organs. The semicircular canals form roughly a three-coordinate system that provides information about body movement. The vestibular sacs (the utricle and saccule) respond to the forces of gravity and provide information about body position in relation to the vertical.

(along with the vestibular sacs). With changes in acceleration or deceleration, the fluid shifts its position in these tubes, which stimulates nerve endings that then transmit nerve impulses to the brain. It should be pointed out that movement of the body at a constant rate does not cause any stimulation of these canals. Rather, they are sensitive only to a *change* in rate (acceleration or deceleration).

Vestibular Sacs: Utricle and Saccule

The *vestibular sacs* (also called the *otolith organs*) are two organs with interior hair cells and containing a gelatinous substance. The *utricle* is generally positioned in a horizontal plane, and the *saccule* is more in a vertical plane. As the body changes position, the gelatinous substance is affected by gravity, setting up nerve impulses via the hair cells. The utricle apparently is the more important of the two organs. The primary function of these organs is that of sensing body posture in relation to the vertical and thus serving as something of a gyroscope that helps to keep us on an even keel. While their dominant role is that of aiding in sensing postural conditions of the body, they are also somewhat sensitive to acceleration and deceleration and presumably supplement the semicircular canals in sensing such changes.

Interdependence of Motion and Orientation Senses

In the maintenance of equilibrium and body orientation, or in reliably sensing motion and posture, all of these senses play a role, along with the skin senses, vision, and sometimes audition. The importance of vision in orientation was illustrated, for example, in experiments with people in a "tilting room," in some cases with a chair that could be tilted within the room [Witkin, 37]. Subjects seated in the room were tilted to various angles and asked to indicate what direction they considered vertical. When blindfolded, they were able to indicate the vertical more accurately than when they could see the inside of the room; when they could see, they tended more to indicate that the ceiling of the room was upright. The implication of such investigations is that misperceptions of the *true* upright direction may occur when there is a *conflict* between the sensations of gravity and visual perceptions; in such a case one's visual perceptions may dominate, even when they are erroneous.

WHOLE-BODY VIBRATION

Every object (or "mass") has a resonant frequency, in somewhat the same sense that a pendulum has a natural frequency. For example, rubber-tired earth-moving vehicles have resonant frequencies primarily in the range from 1 to 4 Hz. The predominant frequencies for farm tractors and highway trucks are from about 2 to 7 Hz, and for track types of vehicles from about 4 Hz up [Radke, 29]. In a similar manner, the body itself, and the individual body members and organs, have their own resonant frequencies.

As the human body is caused to vibrate (as in a vehicle), the effect upon the body and upon the body members and organs is the consequence of the interaction between the frequency of the vibrating source and the resonant frequencies of the individual "masses." Since the body members and organs have different resonant frequencies, and since they are not attached rigidly to the body structure, they tend to vibrate at different frequencies, rather than in unison. The resulting tensions and deformation can be the source of localized pain (such as abdominal, chest, and testicular pain), head symptoms, localized symptoms of other types, general discomfort, and anxiety [Magid, Coermann, and Ziegenruecker, 25]. There is, however, a fair amount of variation in the frequencies at which these various symptoms occur.

Amplification and Attenuation Effects

As vibration is transmitted to the body, it can be amplified, or attenuated, as the consequence of body posture (whether standing or sitting), the type of seating, and the frequency of the vibration. When the vibrating frequency of a seat is high, relative to the resonant frequency of the "mass" (e.g., the body or a part thereof), the displacement of the mass is high—which means that the vibration of the mass is amplified. This occurs when the ratio of those two frequencies is less than about 1.414 [Radke, 29]. When that ratio is greater than about 1.414, there is an attenuation, which means that the displacement amplitude of the mass is less than that of the seat. The amplification becomes greatest where this ratio is around 1.0 (where the seat is vibrating at, or near, the resonant frequency of the body). With particular reference to the head, as frequency increases above resonance, the displacement amplitude decreases, and at around 10 Hz the head and seat vibrate at about equal amplitudes [Cope, 14]. At higher vibrating frequencies, the transmission to the head is decreased, until at a vibrating frequency of about 70 Hz only about 10 percent of the amplitude of the seat is transmitted to the head [Coermann, 13, p. 36].

The amplification and attenuation that would result from the ratio mentioned above, however, would be expected with no "damping." In practice, some damping usually would be provided, either by intent or otherwise. Examples of the actual amplification and attenuation of the human body while standing and seated are illustrated in Figure 15-3. With the vibration of the vibrating table used as a base (expressed as 100 percent), the amplification, or attenuation, is measured by the use of accelerometers positioned at different locations, such as at the belt, the neck, and the head. For a person standing, there is a typical attenuation effect; this is because the legs serve to absorb the effects as the individual bends and straightens his legs in response to the movement. In the case of a seated individual, however, there is an amplification effect (at the various body locations) especially in the range of frequencies of about $3^1/_2$ to 4 Hz. At frequencies of 5 to 6 Hz the transmissibility is reduced to about the amplitude of the vibrating source (in experimental situations the vibrating table); the difference between these empirical results and possible theoretical (undamped) values probably can be attributed to some

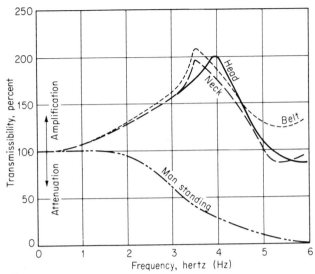

Figure 15-3 Mechanical response of man to vertical vibration, showing amplification and attenuation, by frequency, for seated and standing positions. [From Radke, 29.]

form of damping in the system, including the damping effects of the body itself, especially the buttocks.

Damping can be controlled to some degree by appropriate seat design, spring design, cushioning, etc. Figure 15-4 illustrates the amplification and attenuation effects of various seats. It can be seen that, in this particular comparison, the suspension seat resulted in substantial attenuation over the critical frequencies of 3 to 6 Hz.

Subjective Responses to Whole-Body Vibration

From the above discussion we can see that the physical effects of vibration are influenced by such factors as body posture and any materials interposed between the body and the source of vibration (such as the nature of the seat, padding, and springs). As one would expect, these and other factors also influence the subjective reaction of people to vibration. Since we cannot probe all of these as they relate to subjective responses (even if data were available on each), we shall select a couple of examples for illustrations.

As one example, Chaney [10] shows comparative data on subjective reactions of standing and seated subjects when subjected to vibration of various frequencies on a shake table. Three tolerance curves are shown in Figure 15-5, these being the mean levels that were reported by subjects to be mildly annoying, extremely annoying, and alarming. The greater tolerance of the standing subjects is undoubtedly the consequence of the damping effects of leg flexion, as illustrated above in Figure 15-3. The lowest tolerance level is in the range from about 4 to 6 or 8 Hz.

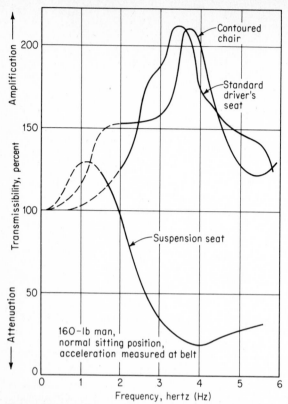

Figure 15-4 Mechanical response of a man's body to vibrations, when seated in different seats. [From Simons, Radke, and Oswald, 34.]

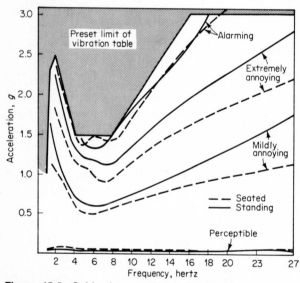

Figure 15-5 Subjective reactions of subjects, seated and standing, when subjected to whole-body vibration. Each curve represents the combinations of frequencies and amplitudes that were judged by the subjects to produce the level of sensation in question (such as mildly annoying). [From Chaney, 10.]

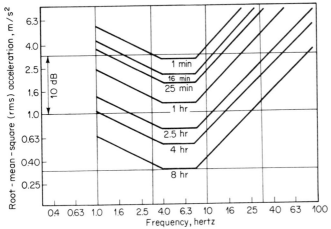

Figure 15-6 Tentative fatigue-decreased proficiency boundary for vertical vibration proposed by the International Organization for Standardization (ISO). The following weighting factors are proposed for the specified conditions: fore-and-aft or side-to-side vibration, subtract 3 dB from values; boundary of reduced comfort, subtract 10 dB; and safe exposure limits, add 6 dB. [From *Revised proposal to the secretariat: guide for the evaluation of human exposure to whole-body vibration,* 39.]

As you would expect, people can tolerate mild vibration longer than really rough jarring. Some indication of this is shown in Figure 15-6, each line representing an estimate of the upper limit of the amplitudes of vibrations of various frequencies that people generally can tolerate for the specified time limits before fatigue effects would catch up with them; amplitudes of about 10 dB less than those indicated tend to characterize the upper boundary of comfort, and the safe exposure limits are about 6 dB higher than the values shown. The particular set of curves is based on a draft proposal to the secretariat of the ISO, as prepared by a working group [39], and probably reflects the best available vibration-exposure criteria as a function of frequency.

Effects of Whole-Body Vibration on Performance

On the basis of a review of various vibration studies Grether [19] draws out the following generalizations concerning the effects of vibration on human performance:

1 Vibration causes an impairment of visual acuity that is proportional to vibration amplitude and is greatest at frequencies in the range from 10 to 25 Hz.

2 Vibration causes impairment of human tracking ability proportional to the vibration amplitude and is greatest at very low frequencies, i.e., below 5 Hz.

3 Other tasks that require steadiness or precision of muscular control are likely to show decrements from vibration.

4 Tasks that measure primarily central neural processes, such as reaction time, monitoring, and pattern recognition, appear to be highly resistant to degradation during vibration.

It should be noted that the motion amplitude is considered to be the primary factor in causing decrement in tracking performance, the decrement being greatest for low frequencies. This is due to the fact that, at low frequencies (for a given g value), the amplitude is greater than at high frequencies.

Although most vibration research has dealt with vertical sinusoidal vibration, the vibrations encountered in real life (as in vehicles) tend to be more or less random, and in some instances have lateral (i.e., sway) components. Some vehicles and other items of equipment have complex vibration characteristics, including a dominant resonant frequency. On the basis of a review of studies in which nonsinusoidal vertical vibration has been investigated, Grether [19] indicates that there seem to be no overall differences in effect on tracking performance among three types of vertical vibration (sinusoidal, sinusoidal with random amplitude, and random frequency and amplitude), except that tracking decrement was more closely related to vibration amplitude for sinusoidal vibration than for the random conditions. Although very few investigations have dealt with lateral vibration, Lovesey [24] points out that low-frequency vibration along the lateral axis has a significant effect upon tracking performance and on comfort. As an illustration of this effect, he presents the following data, which reflect tracking errors in an aircraft simulator during a "climb" maneuver:

Type of error	No vibration	Vertical	Vertical and lateral
Azimuth error	1.00	1.21	1.56
Pitch error	1.00	.93	1.37

ACCELERATION

Acceleration of the human body is most typically experienced in vehicles of various types. In moderate degree (as in most automobile driving) its effect is nominal, but in accentuated degree (as in high-speed aircraft and space capsules) its effects can be of consequence.

The physiological effects of sustained acceleration arise primarily from the effective increase in weight of the body—in particular, its fluid components. The effective weight of the body and body members is proportional to the number of G units. Thus, at $2G$ the body members have twice the effective weight of normal. In discussing this effect, Fraser [17] notes that at $+5G_z$, while maintaining mean arterial pressure at heart level of 120 mm Hg, the theoretical pressure at the base of the brain would be zero while at the feet it would be 370 mm Hg.

Some of the effects of $+G_z$ forces (headward acceleration) are as follows [Fraser, 17]: $2\frac{1}{2} + G_z$, difficult to raise oneself; 3 to $4 + G_z$, impossible to raise oneself, progressive dimming of vision; $4\frac{1}{2}$ to $6 + G_z$, diminution of vision progressing to blackout after 5 s, hearing and consciousness lost if exposure is continued, convulsions in some subjects, inspiration difficult.

In the case of $-G_z$ (footward acceleration) the following effects occur: -1 G_z, unpleasant but tolerable facial suffusion and congestion; -2 to $-3G_z$, severe facial congestion, throbbing headache, progressive blurring, graying, or occasionally reddening of vision after 5 s; $-5G_z$, limit of tolerance is 5 s, rarely reached by most subjects.

The effects of linear acceleration in the forward-backward directions $+G_x$ and $-G_x$ and lateral directions $+G_y$ and $-G_y$ are not as accentuated as in the vertical direction. The relatively higher G load that people can tolerate, especially in the forward direction $+G_x$, has been the basis for the use of this posture in space vehicles. Although symptoms of discomfort and pressure and breathing difficulty occur, G loads of 12 to $15G$ can be tolerated by some subjects, in some instances over 100 s [Chambers and Hitchcock, 9].

Voluntary Tolerance to Acceleration

The physiological symptoms brought about by acceleration of course have a pretty direct relationship to the levels that people are willing to tolerate voluntarily. The physiological effects of differential levels of linear acceleration in various directions mentioned above are paralleled by somewhat corresponding differences in average voluntary tolerance levels, such as shown in Figure 15-7.

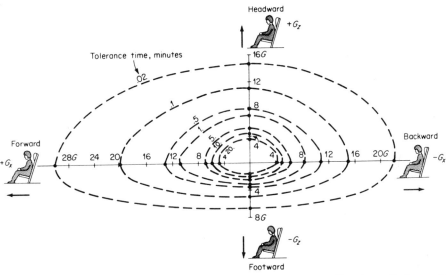

Figure 15-7 Average levels of linear acceleration, in different directions, that can be tolerated on a voluntary basis for specified periods of tome. Each curve shows the average G load that can be tolerated for the time indicated. The data points obtained were actually those on the axes; the lines as such are extrapolated from the data points to form the concentric figures. [Adapted from Chambers, 7, fig. 6.]

This figure shows the average levels for the various directions that can be tolerated for specified times. It can be seen that tolerance to footward acceleration $-G_z$ is least, followed by headward acceleration $+G_z$, with forward acceleration $+G_x$ being the most tolerable. Individual differences are of considerable magnitude; trained and highly motivated personnel frequently can endure substantially higher levels than the average.

Further indications of the differential effects on tolerance of acceleration forces operating in different directions are shown in Figure 15-8. Bypassing the

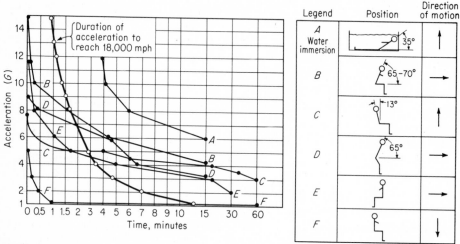

Figure 15-8 Comparison of acceleration tolerance of immersed and nonimmersed subjects in different postures. The angles indicated are not "back angles" as such, but rather are angles of the trunk relative to the direction of the motion that gives rise to the acceleration. [From Bondurant et al., 4, as adapted by Stapp, 35.]

water-immersion condition A for the moment, that figure shows the tolerance time for various G loads when people are oriented differently with respect to the direction of the acceleration forces (along with a line showing the duration required to reach the escape velocity of 18,000 mph in the case of space ships). Here, again, we see evidence of the limited tolerance to footward acceleration $-G_z$ as reflected by condition F. By comparison, we can see that water immersion (with welcomed breathing arrangements, of course) increases one's tolerance, particularly in a supine position with about a 35° angle of the trunk to the line of force. Under immersion conditions, the same acceleration forces affect the water as well as the person immersed in it. This has the effect of equalizing the forces over the parts of the body. Individuals so immersed can even move their body members under acceleration and can tolerate more acceleration than under any other known condition.

Performance Effects of Acceleration

Within the limits of physiological safety, acceleration can cause degradation in performance of certain kinds of human activities that might be required under acceleration conditions, such as in aircraft or space ships.[1]

[1]For reviews of the effects of acceleration on human performance, see Fraser [17].

Acceleration and Gross Body Movement As implied above, acceleration can make some simple physical movements difficult or impossible (and this effect, of course, carries over to the performance of certain physical tasks). The effects of vehicular (forward) acceleration $+G_x$ on gross body movements are shown in Figure 15-9. This figure shows the typical levels of G that are near the

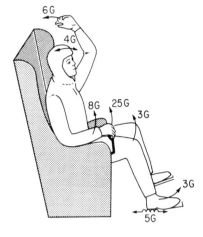

Figure 15-9 The G forces that are near the threshold of various body movements. For any given motion indicated, the movement is just possible at the G forces indicated; greater G forces usually would make it impossible to perform the act. [From Chambers and Brown, 8.]

threshold of ability to perform the acts indicated. Some of the implications for design decisions are fairly obvious; it would seem unwise, for example, to place the ejection-control lever of an airplane over the pilot's head.

Acceleration and Specific Activities Acceleration in its several variations (intensity, direction, etc.) has been found to be significantly related to performance on a varied spectrum of human activities and responses, such as: reading tasks; reaction time; some tracking and control tasks; and certain higher mental functions [Fraser, 17]. Although the effects on various types of performance depend somewhat on the direction and amount of acceleration, there is still a somewhat general pattern of performance degradation, as shown in Figure 15-10. This shows the average estimated degradation over several aspects of performance.

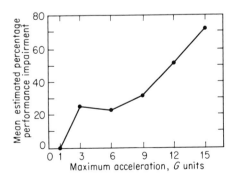

Figure 15-10 Average estimated performance impairment (across several aspects of performance) associated with acceleration forces. [From Chambers and Brown, 8.]

But while some types of human performance deteriorate under acceleration conditions, this is not necessarily true of all human performance under conditions of moderate acceleration. (Under near-tolerance or beyond-tolerance levels, all performance would deteriorate, of course.) For example, Brown [6] has demonstrated on a centrifuge that under certain conditions the accelerations which result from pilot control manipulations may be associated with improved performance. Whether such "improvement" effects would occur in actual flight, however, is probably unknown.

Protection from Acceleration Effects

Although minor doses of acceleration (as in most land vehicles and most commercial airplanes) pose no serious problem either in safety and welfare or in performance, the effects of higher levels (especially with long exposures as in space travel) usually require some protective measures. As indicated above, and as illustrated in Figure 15-8, certain postures make it possible to tolerate higher G loads than other postures. Immersion, although providing considerable protection, is generally not considered feasible in space flight because of the problem of lugging around a private swimming pool. Aside from the use of optimum postures, the presently available protection schemes include restraining devices, anti-G suits, and special body supports such as contour and net couches.

For high, sustained G loads in a forward direction $+G_x$, the use of contour couches (sometimes molded for the individual) are reasonably satisfactory. Such couches, however, do not provide adequate support for the rearward direction $-G_x$. For such purposes additional contraptions are required to provide adequate restraint, such as a restraint helmet and supporting face and chin pieces, and frontal supports (such as nylon netting) for the chest, torso, and body members; to further hem in the individual, an anti-G suit may be in order. Anti-G suits, however, generally have their greatest utility in connection with headward $+G_z$ acceleration, with some designs being more effective than others [Nicholson and Franks, 28]. There is no particularly effective scheme for providing protection against footward acceleration $-G_z$, so if you insist on being shot out of a cannon, have them shoot you out headward.

Deceleration and Impact

The normal deceleration of a vehicle imposes forces upon people that are essentially the same as those of actual acceleration, but in reverse. But acceleration usually is a gradual affair (except in unusual instances, such as with astronauts during blast-off), whereas deceleration can be extremely abrupt, especially in the case of vehicle accidents. When a vehicle hits a solid object or another vehicle head on, an unrestrained occupant will continue forward at his initial velocity until the body strikes some part of the interior of the vehicle, or is thrown clear from the vehicle. This is sometimes called the *second collision*. The deceleration of the occupant is a function of the deformation of the part of the interior which he hits, if any. To be protected from this "second" collision, the occupant must be "tied" to the vehicle (as with restraining devices) and decelerate with the vehicle (which may be through a distance of a couple of feet or more), or be provided with some energy-absorbing object which results in more gradual deceleration by increasing the travel distance of the occupant during deceleration (as by the use of air bags, collapsible steering wheels, etc.).

Restraining Device

An occupant can be "tied" to a vehicle by quite a few types of restraining devices, in automobiles and commercial aircraft these nearly always being seat (lap) belts. In some automobiles shoulder harnesses are used by themselves or in combination with seat belts, and in military aircraft more complex restraints are used, individually or in combination. A major deficiency of automobile seat belts is, or course, the possibility of head damage caused by the torso and head being propelled forward with the body hinged at the hips. Considerably greater protection is provided by lap-shoulder belts, which aid the occupant in "riding down" the vehicle as it impacts. A general comparison of some of these effects is shown in Figure 15-11. (However, there is some suspicion that, under impact

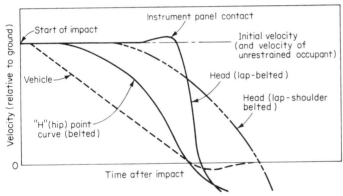

Figure 15-11 Velocity of vehicle occupants following impact with lap belt and shoulder-lap belt. The shoulder-lap restraint causes the upper torso and head to follow more closely the vehicle velocity change than does the seat belt alone. On impact, an unrestrained occupant would be propelled forward at the impact velocity until he hit some part of the interior. [From Cichowski and Silver, 11, fig. 1.]

conditions, shoulder belts might apply excessive force on the spine.) Some indication of the protection provided by seat belts comes from a survey in California [Tourin and Garrett, 36]. The use of seat belts reduced the incidence of fatalities by 35 percent, largely because they prevented the ejection of the individual. Although the seat-belt wearers were injured with the same frequency as nonwearers, the degree of injury was lower for those who used seat belts.

Another method of "tying" passengers to a vehicle is by the use of rear-facing seats. It has been estimated that properly designed aft-facing seats could protect occupants against as much as 40-G impacts with little or no injury [Eiband, 16]. Such seats would be particularly feasible in aircraft, trains, etc., but are not completely unrealistic for use in automobiles (if we could adapt ourselves to them). In the case of vehicles with forward-facing seats involved in rear-end crashes, the use of a full back support with headrest minimizes the possibility of whiplash.

Energy-absorbing Objects

The air bag seems destined to replace seat belts as the primary device for protection from vehicular head-on accidents. The air bag system is so designed that when the vehicle hits the back of another vehicle, the air bag is immediately inflated in front of the occupant and is then deflated. The effect is that of absorbing the energy of the occupant as he is catapulted forward upon the impact. In other words, the deceleration takes place over a longer travel distance than if the occupant were propelled forward at the vehicle velocity at the time of impact until he would strike a rigid object such as the windshield or the steering wheel. Figure 15-12 illustrates how the air bag functions upon impact.

In a serious crash the air bags automatically inflates in a split second...   protects...   and then deflates.

Figure 15-12 Illustration of the manner by which an air bag protects a passenger in a vehicle upon impact. The time between inflation and deflation permits a more gradual deceleration of the passenger. [Illustration through the courtesy of Allstate Insurance Company.]

WEIGHTLESSNESS

Since man has evolved in an earthbound environment the force of gravity has been his constant companion. Its constant presence has influenced our physiological makeup and is basic to all of our activities. It is only in space flight and in certain aspects of aircraft flight that the "natural" phenomenon of weightlessness or reduced gravity is experienced. For those few who do venture into outer space, there are two aspects of the weightless or reduced-weight state that are particularly important [Berry, 3]. The first of these is the absence of weight itself; the removal from the normal gravitational environment could be expected to have an impact on the human organism, such as on its physiological functioning and on perceptual-motor performance and/or sensory performance. In the second place the weightless condition is accompanied by a tractionless condition when moving and working.

Since the weightless condition does not exist in our earthbound environment, various schemes have been developed for creating such a condition, or simulating it, in order to study its effects. True weightlessness can be achieved within the earth's gravitational field by the use of aircraft in a parabolic flight maneuver. During the maneuver (up and down) a weightless condition can be achieved for upwards of a minute [Moran, 26].

Various forms of partial simulation of the weightless state include cable suspension rigs (which reduce the friction between the man and the surface on which he is operating), gimbals, air bearings, and water immersion (to achieve neutral bouyancy) [Deutsch, 15]. Although some such methods have provided some inklings about the effects of weightlessness, the most reliable basis for estimating the long-term effects is undoubtedly actual experience in orbital space flight.

Physiological Effects of Weightlessness

In summarizing the space flight experience of both the United States and the Soviet Union, Berry [3] indicates that although some physiological changes have been consistently noted, none of them have been permanently debilitating. Some of the temporary effects that have been observed include: aberrations in cardiac electrical activity; changes in the number of red and white blood cells; muscle tone; and loss of weight.

It should be noted that, to date, man's excursions into space have been for periods up to 24 days (the Soviet Salyut Soyuz 11 flight), so it is difficult to predict whether longer durations would, or would not, bring about more permanent physiological consequences.

Performance Effects of Weightlessness

Locomotion within a spacecraft adds a third dimension to our usual two-dimensional movements. Apparently no serious problems have occurred in normal locomotion within spacecraft [Berry, 3]. Although there has been serious consideration of generating artificial gravity in spacecraft (by rotating the entire space vehicle or station, or by having an on-board centrifuge), the experience of United States astronauts has so far suggested that this would not be necessary.

Performance outside a spacecraft in extravehicular activities (EVA) is a bit of a different story, since some astronauts have experienced considerable exhaustion in such activities. As Berry points out, this probably argues for careful consideration of work load in planning other EVA missions.

ILLUSIONS DURING MOTION

When man is in motion, he receives cues regarding his whereabouts and motion from sense organs, especially the semicircular canals, the vestibular sacs (the utricle and saccule), the eyes, the kinesthetic receptors, and the cutaneous senses. The interactions among these senses are quite intricate, and under some circumstances the sensory cues received are misleading and result in errors in perception and illusions, especially when certain cues are at odds with each other. Such conditions are perhaps most common (and most critical) in aircraft and space flight because of the unique factors and special circumstances that occur under such conditions [Clark and Graybiel, 12].

Disorientation (vertigo)

A rather common perceptual phenomenon in flight is associated with the sensation of orientation in space (such as sensing which direction is up). There are several manifestations of disorientation (vertigo), these generally resulting from stimulation of the vestibular and kinesthetic sense organs from acceleration and angular changes in direction. For example, especially during changes in direction and in velocity the sensation of the vertical is felt more in relationship to the aircraft than to the earth because of the manner in which the acceleration forces act on the vestibular organs. This effect is sometimes so strong that the sensation of tilt is grossly underestimated. This tendency to underestimate the amount of bank misleads the pilot into "perceiving" the plane as being oriented more vertically than it really is in such situations. If in such a case there is no external visual frame of reference (because of fog, darkness, clouds, etc.), the pilot also tends to perceive the ceiling of the plane as being vertical. Thus, both gravitational cues and visual cues can operate to "confirm" each other—and both can be wrong!

When facing forward in an aircraft, there is sometimes a sensation of backward tilt (or going into a climb) under conditions of forward acceleration, and of forward tilt (going into a dive) under conditions of forward deceleration. One can readily see how easy it would be under conditions of poor visibility for a pilot decelerating when landing to get the impression of going into a dive, and "correct" for it by overshooting the landing field. This emphasizes the importance, in building landing strips, of providing reliable visual cues (such as lights) to help pilots orient themselves visually when visibility is poor.

Other variations of the vertigo theme include sensations of climbing while turning, sensations of diving when recovering from a turn or following pull-out from a dive, the coriolis phenomenon (a loss of equilibrium that occurs when the pilot is rotating with the aircraft and moves his head out of the plane of rotation), and a sensation of reversed rotation that sometimes occurs after spinning [Roth, 31, chap. 7].

Visual Illusions

Certain illusions that sometimes occur during flight are essentially visual. One of these is autokinesis, which is manifested by a confusion with lights at night in which a fixed light may appear to move against its dark background. It has been reported that pilots have attempted to "join up" in formations with stars, buoys, and street lights which appear to be moving [Clark and Graybiel, 12]. Some pilots have experienced illusory horizons. This can be brought about, especially under conditions of poor visibility (such as over water, snow, and desert), when there is no clearly defined landmark. Such illusions can bring about errors in the estimation of altitude, distances, or directions. The oculogyral illusion is characterized by apparent movements of objects that actually are stationary. They occur following angular acceleration which stimulates the semicircular canals; the illusion is something of an aftereffect of their stimulation. Another type is the oculogravic illusion, which is induced by conflicting cues from the eyes and the vestibular sacs. It is characterized by the apparent displacement of objects in space as well as body displacement, the direction of the apparent displacement depending on the posture of the individual relative to the line of the force to which he is subject.

Discussion

The reduction of disorientation and illusions generally depends more on procedural practices and training than upon the engineering design of aircraft. Among such practices are the following [Clark and Graybiel, 12]: understanding the nature of various illusions and the circumstances under which they tend to occur, maintaining either instrument *or* contact flight, avoiding night aerobatics, shifting attention to different features of the environment, learning to depend upon the correct cue to orientation (such as using visual cues when possible and otherwise instruments), avoiding sudden accelerations and decelerations at night, and avoiding prolonged constant speed turns at night.

VEHICLE DRIVING

Very few among us will ever have the opportunity to experience the sensations of acceleration and of weightlessness such as those of the astronauts in spacecraft. And on the other hand, there are few among us for whom mobility in the family automobile is not part of our daily lives. The "system" of which a vehicle driver is a part consists of himself, his vehicle, the road, the signs, the surrounding traffic, and the physical environment, each of which abounds with human factors problems. Although we cannot cover the many human factors aspects associated with vehicle driving, we will touch on certain ones as illustrations.

Sensory and Perceptual Input in Driving

Various senses serve as input channels in driving, including vision, audition, the proprioceptors, the semicircular canals, and the vestibular sacs. Of these, vision is undoubtedly the most important.

Visual Search and Scan Patterns   The visual search and scan patterns of drivers have various implications in terms of road markings, placement of signs, and driver training. Such patterns typically have to be studied under somewhat controlled conditions. In one such study Mourant and Rockwell [27] used an eye camera to record the eye movements of drivers who were driving on an expressway at 50 mi/h. One of the implications from the study was that route familiarity plays a role in such patterns. Over unfamiliar routes the drivers typically "sampled" a wide area in front of them, but with increasing familiarity their eye movements tended to be confined to a smaller area. The investigators suggest that, since the unfamiliar driver has to rely on road signs for route guidance information, further investigations should deal with the relationship between density and sequencing of road signs with visual scanning patterns and the required control actions.

The results of this study also lend support to the hypothesis that peripheral vision is used primarily to monitor the road-edge markers. This points out that it is important to have good-quality and easily detected edge lines because of the poor visual acuity of the peripheral portions of the retina of the eye.

Response to Road Signs To serve their intended purposes road signs should be so designed and so located as to catch the driver's eye. In this regard some rather discouraging results of a survey are reported by Johansson and Rumar [22], who stopped and interviewed 1000 drivers to ascertain how many recalled having seen five specified types of road signs in a previous stretch of road. On the average, the signs were "recalled" by the motorists only 47 percent of the time; however, there were differences in the percents for the individual signs, as follows: speed limit (78 percent), "police control" sign (63 percent), road condition sign (55 percent), general warning (18 percent), and pedestrian crossing (17 percent).

Even when road signs are "seen" by drivers, they do not always affect the driver's driving behavior. This failure to be guided by road signs is illustrated by the results of an experiment dealing with driver behavior on curves [Ritchie, 30]. (Although an experimenter was in the car with the driver to instruct the driver on his route and to operate the mechanical recording equipment, the driver did not know the specific objective of the study.) One analysis that was made consisted of a comparison of the actual driving speed on curves which had "advisory" speeds with the advisory speeds given on the curve signs. By and large, the drivers tended to exceed the advisory speeds of 15 to 35 mi/h (24 to 56 km/h), but they did not exceed those of 45 and 50 mi/h (80 km/h).

Perceptual Style The visual input to drivers is of course influenced in part by the characteristics of the drivers as well as by the nature and location of objects within their environments. One of the personal characteristics that seems to be relevant is a perceptual style called *field dependence–field independence*. Field-independent people are better at distinguishing relevant from irrelevant cues in their environment than are those who are field-dependent. One indication of this perceptual style as related to driving behavior comes from a driving simulation study by Barrett and Thornton [2]. Figure 15-13 shows the relationship between perceptual style and deceleration rate of subjects when reacting to an emergency in the simulator (specifically, the figure of a child appearing on the "road" ahead).

There is some confirmation of the implications of this perceptual style as related to actual accidents from another study by Harano [20]. He found that drivers with accident records (three or more accidents in three years) tended to

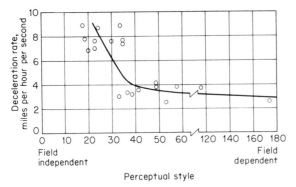

Figure 15-13 Relationship between perceptual style and deceleration rate of subjects reacting to an emergency situation in vehicle simulator. [From Barrett and Thornton, 2, fig. 3.]

be more field-dependent than did accident-free drivers. Presumably the field-dependent drivers have higher accident liability because they are more easily influenced by irrelevant cues when driving.

Information Processing

Given the information input to a vehicle driver (i.e., the various stimuli that impinge upon him, such as other vehicles, road signs, and the road), the driver "processes" the information somehow, leading up to his control responses.

Judgments of Vehicle Drivers   The inputs in part are used as the basis for judgments regarding the driving situation. For example, drivers certainly make some judgments about their own speed. Such judgments, however, are in part distorted by the phenomenon of adaptation. If a person has been exposed to one level of stimulus (in this instance, vehicular speed), his judgment of a different speed is influenced by the speed to which he is adapted. This influence is especially noticeable when changing from a high speed to a lower speed, the estimates of lower speeds usually being underestimates [Schmidt and Tiffin, 33]. Further, the longer the exposure to a high speed, the greater the underestimate of lower speeds. The effects of this would be for people who slow down (as on an exit ramp of a superhighway) to drive at a higher speed than they think they are driving.

Another type of judgment made by drivers is that of deciding whether or not to pass another vehicle. Such a judgment is made on the basis of the driver's perception of the speed of his car, that of the vehicle ahead, and that of an approaching vehicle, along with the relative distances between the vehicles. The inklings to date indicate that people tend to underestimate the time required for passing other cars. To add to the potential danger in passing, there is a further tendency on the part of some drivers to make less realistic judgments when passing vehicles traveling at high speeds than at low speeds [Gordon and Mast, 18]. One cannot provide protection from all of the harum-scarum drivers of the world, but a driver who is aware of the human tendency to underestimate the time required for passing can at least increase his own survival probabilities by *not* passing other vehicles if there is the slightest doubt about the time factor.

Vehicular Control

The actual control of vehicles is of course a function of the physical responses people make with the control mechanisms used (the steering wheel, brake, accelerator, etc.).

Reaction time One factor that differentiates between the quick and the dead in traffic is reaction time in emergency conditions. As discussed in Chapter 7, the reaction time of a person under "unexpected" conditions usually is longer than when a person is "expecting" a stimulus to which he is to respond. In one study Johansson and Rumar [40] measured brake reaction time of a sample of 321 drivers under an "anticipation" condition, and then adjusted the values (by a method we need not describe) to estimate the reaction time under a "surprise" condition. The median estimated brake reaction time was 0.9 s, with that for some subjects ranging up to 2.0 s. (As an aside, a 0.3-s increase in applying the brake when traveling at 62 mi/h (100 km/h) would increase the stopping distance by almost 32 ft (10 km)). The median of 0.9 s under the "surprise" condition can be compared to a median of 0.66 s under the "anticipation" condition.

A rather curious driver reaction pattern has been reported by Babarik [1] for a number of taxicab drivers who had had an abnormally high number of accidents in which they were struck from behind. In a laboratory test it was found that the pattern of some of these drivers was made up of a slow initiation time and a conpensatingly fast movement time, thus stopping their vehicles more abruptly than the drivers following them could duplicate.

Control Devices There are human factors implications with respect to all the control devices used in vehicles. One rather interesting (and promising) twist in vehicular control is represented by a combined brake-accelerator [Konz et al., 23]. The combination pedal has two fulcrums, such that when activated by the toe, the pedal serves as an accelerator, and when activated by the heel, it serves as a brake. Some rather persuasive evidence about this is shown in Figure 15-14. This shows the reaction times (i.e., "cycles") of 50 subjects when using a laboratory model of the dual-function pedal and 50 subjects when using the conventional systems with separate brake and accelerator. The figure shows the distribution of 500 reaction times (10 for each of the 50 subjects). The difference in mean values was 19 milliseconds (ms), this representing a potentially very significant difference in terms of stopping distance. Although the design of vehicular control devices is only one of several factors that influence driving behavior, the design of such devices in terms of human considerations (including any possible innovative features, such as the combined

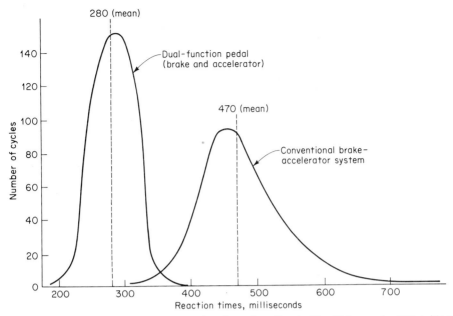

Figure 15-14 Distribution of 500 reaction times (i.e., "cycles")—10 for each of 50 subjects using a dual-function brake and accelerator pedal, and using a conventional brake-accelerator system. [Adapted from Konz et al., 23.]

brake-accelerator) can, to some degree, contribute to improved driving performance.

Prolonged Driving   The evidence about the effects of prolonged driving is rather skimpy, and what there is is somewhat conflicting. A study by Brown et al. [5], however, tends to confirm one's natural suspicion that prolonged driving is accompanied by some degradation in driving performance. In this study the subjects drove for 12 h (with breaks every 3 h), the first and last 3-h sessions being over a specified city route. A comparison of those two sessions indicated that the number of "risky" overtaking maneuvers increased 50 percent between these two session. Although this study was carried out in an "experimental" context, it is strongly suggestive of deteriorating performance over time in actual driving circumstances.

Road Characteristics

An example of still another relevant factor in vehicle driving is of course the nature of the road that is involved. The effects of road design on driving performance are illustrated, for example, by the reduced incidence of accidents on superhighways, which minimize the number of circumstances in which any given vehicle encounters other vehicles moving in other directions.

An example of a different kind of "effect" of road design characteristics is reported by Rutley and Mace [32], who rigged up electrodes to record the

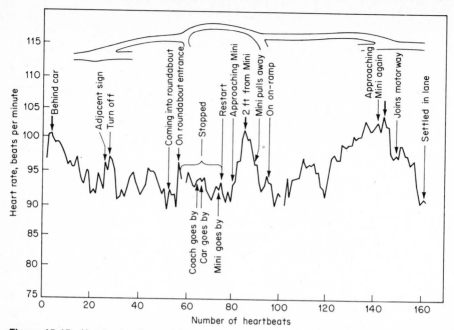

Figure 15-15 Heart rate of one driver on a section of a road near the London airport. The horizontal scale represents the time period during which 160 heart beats were recorded. The features of the road are sketched at the top, and the written entries indicate the "events" occurring during the time period. [From Rutley and Mace, 32, fig. 3.]

heart rates of a few drivers when driving on an approach to the London airport. An example of these recordings for one driver is shown in Figure 15-15. This figure shows the heart rate in relation to the road being traversed and the "events" that occurred when driving over the section of the road in question. One can consider such physiological responses on the part of the driver (in this instance, changes in heart rate) as indices of strain that reflect varying levels of stress that are generated by the driving situation.

DISCUSSION

The common denominator of the exotic experiences of astronauts in outer space, of the more mundane processes of moving around in automobiles or other vehicles, and of the operation of tractors and other related items of equipment is the fact that the human body is being moved somehow, e.g., accelerated, decelerated, transported, jostled, or shaken. Although some forms and degrees of such shuffling are not of any particular moment, other forms or degrees obviously have potentially adverse effects in terms of physiological, performance, or subjective criteria. The challenge to the human factors disciplines is that of so designing the physical systems or protective devices in question as to minimize such effects.

REFERENCES

1 Babarik, P.: Automobile accidents and driver reaction pattern, *Journal of Applied Psychology*, 1968, vol. 52, no. 1, pp. 49–54.

2 Barrett, G. V., and C. L. Thornton: Relationship between perceptual style and driver reaction to an emergency situation, *Journal of Applied Psychology*, 1968, vol. 52, no. 2, pp. 169–176.

3 Berry, C. A.: "Weightlessness," in *Bioastronautics data book* [38], chap. 8.

4 Bondurant, S., N. P. Clark, W. G. Blanchard, et al.: *Human tolerance to some of the accelerations anticipated in space flight*, WADC, TR 58–156, 1958.

5 Brown, I. D., A. H. Tickner, and D. C. B. Simmonds: Effect of prolonged driving on overtaking criteria, *Ergonomics*, 1970, vol. 13, no. 2, pp. 239–242.

6 Brown, J. L.: Acceleration and motor performance, *Human Factors*, November, 1960, vol. 2, no. 4, pp. 175–185.

7 Chambers, R. M.: "Operator performance in acceleration environments," in N. M. Burns, R. M. Chambers, and E. Hendler (eds.), *Unusual environments and human behavior*, The Free Press, New York, 1963, pp. 193–320.

8 Chambers, R. M., and J. L. Brown: *Acceleration*, paper presented at Symposium on Environmental Stress and Human Performance, American Psychological Association, September, 1959.

9 Chambers, R. M., and L. Hitchcock: "Effects of high *G* conditions on pilot performance," in *Proceedings of the National Meeting of Manned Space Flight*, April 30–May 2, 1962, Institute of the Aerospace Sciences, New York, 1962, pp. 204–227.

10 Chaney, R. E.: *Whole body vibration of standing subjects*, Boeing Company, Wichita, Kans., BOE-D3-6779, August, 1965.

11 Cichowski, W. G., and J. N. Silver: *Effective use of restraint systems in passenger cars*, paper presented at Automotive Engineering Congress, Detroit, Michigan, Jan. 8–12, 1968, SAE 680032.

12 Clark, B., and A. Graybiel: *Disorientation: a cause of pilot error*, USN School of Aviation Medicine, Research Project NM 001 110 100.39, Mar. 2, 1955.

13 Coermann, R. R.: Effect of vibration and noise on the human body, *Ringbuch der Luftfahrttechnic*, 1939, vol. 5, no. 1.

14 Cope, F. W.: Problems in human vibration engineering, *Ergonomics*, 1960, vol. 3, pp. 35–43.

15 Deutsch, S.: Preface to special issue on reduced-gravity simulation, *Human Factors*, 1969, vol. 5, no. 11, pp. 415–418.

16 Eiband, A. M.: *Human tolerance to rapidly applied acceleration*, NASA Memo 5-19-59E, 1959.

17 Fraser, T. M.: "Sustained linear acceleration," in *Bioastronautics data book* [38], chap. 4.

18 Gordon, D. A., and T. M. Mast: Drivers' judgments in overtaking and passing. *Human Factors*, 1970, vol. 12, no. 3, pp. 341–346.

19 Grether, W. F.: Vibration and human performance, *Human Factors*, 1971, vol. 13, no. 3, pp. 203–216.

20 Harano, R. M.: Relationship of field dependence and motor-vehicle accident involvement, *Perceptual and Motor Skills*, 1970, vol. 31, pp. 372–374.

21 Hornick. R. J.: "Vibration," in *Bioastronautics data book* [38], chap. 7.

22 Johansson, G., and Käre Rumar: Drivers and road signs: a preliminary investigation of the capacity of car drivers to get information from road signs, *Ergonomics*, 1966, vol. 9, no. 1, pp. 57–62.

23 Konz, S., N. Wadhera, S. Sathaye, and S. Chawla: Human factors considerations for a combined brake-accelerator pedal, *Ergonomics*, 1971, vol. 14, no. 2, pp. 279–292.

24 Lovesey, E. J.: The multi-axis vibration environment and man, *Ergonomics*, 1970, vol. 1, no. 5, pp. 258–261.

25 Magid, E. B., R. R. Coermann, and G. H. Ziegenruecker: Human tolerance to whole body sinusoidal vibration, *Aerospace Medicine*, November, 1960, vol. 31, pp. 915–924.

26 Moran, M. J.: Reduced-gravity human factors research with aircraft, *Human Factors*, 1969, vol. 11, no. 5, pp. 463–472.

27 Mourant, R. R., and T. H. Rockwell: Mapping eye-movement patterns to the visual scene in driving: An exploratory study, *Human Factors*, 1970, vol. 12, no. 1, pp. 81–87.

28 Nicholson, A. N., and W. R. Franks: "Devices for protection against positive (long axis) acceleration," in P. I. Altman and D. S. Dittmer (eds.), *Environmental biology*, AMRL, TR 66–194, November, 1966, pp. 259–260.

29 Radke, A. O.: *Vehicle vibration: man's new environment*, ASME, paper 57-A-54, Dec. 3, 1957.

30 Ritchie, M. L.: Choice of speed in driving through curves as a function of advisory speed and curve signs, *Human Factors*, 1972, vol. 14, no. 6, pp. 533–538.

31 Roth, E. M. (ed.): *Compendium of human responses to the aerospace environment*, NASA CR-1205, vol. 2 and 3, November, 1968.

32 Rutley, K. S., and D. G. W. Mace: Heart rate as a measure in road layout design, *Ergonomics*, 1972, vol. 15, no. 2, pp. 165–173.

33 Schmidt, F., and J. Tiffin: Distortion of drivers' estimates of automobile speed as a function of speed adaptation, *Journal of Applied Psychology*, 1969, vol. 53, no. 6, pp. 536–539.

34 Simons, A. K., A. O. Radke, and W. C. Oswald: *A study of truck ride characteristics in military vehicles*, Bostrom Research Laboratories, Milwaukee, Report 118, Mar. 15, 1956.

35 Stapp, J. P.: Acceleration: how great a problem? *Astronautics*, February, 1959, vol. 4, no. 2, pp. 38–39, 98–100.

36 Tourin, B., and J. W. Garrett: *Safety belt effectiveness in rural California automobile accidents*, Automotive Crash Injury Research of Cornell University, February, 1960.

37 Witkin, H. A.: The perception of the upright, *Scientific American*, February, 1959.

38 *Bioastronautics data book*, 2d ed., National Aeronautics and Space Administration, NASA SP-3006, U. S. Government Printing Office, Washington, 1973.

39 *Revised proposal to the secretariat: guide for the evaluation of human exposure to whole-body vibration*, ISO, Technical Committee 108 (mechanical vibration and shock), working group 7, ISO TC 108/WG 7, December, 1968.

40 Johansson, G., and K. Rumar: Drivers' brake reaction times, *Human Factors*, 1971, vol. 13. no. 1, pp. 23–27.

Chapter 16

The Living Environment:
The Physical Features

During its life span to date, the human factors discipline has been largely concerned with the design of equipment, facilities, and environments as related to work activities, as in the performance of military services, the operation of aircraft, certain manufacturing operations, the operation of computers, and space exploration. However, the basic approach of human factors is equally applicable to a wide spectrum of other areas—to the design of virtually all other man-made features or man-influenced features of our total life space—toward the possible improvement of the overall quality of human life. Despite the tremendous strides that have been made in technology, there are many aspects of human life that, on almost a worldwide basis, constrain the quality of life—aspects such as pollution of many varieties, health, crime, hunger, housing, congestion, lack of sanitation, lack of privacy, and unpleasant and unaesthetic surroundings [Ferguson, 11]. In fact, technology has contributed to the scope of some of these problems.

The quality of life is in large part a function of our involvement with a broad and ill-defined spectrum of various features or aspects of our total living environment. Such daily involvement includes our use of: buildings and their related facilities (houses, apartments, schools, offices, factories, stores, theaters, houseboats, etc.); the community proper, of whatever size and arrangement (including its aesthetic character, its facilities for entertainment, culture, and recreation); the transportation facilities; and the physical (ambient) environment (including the natural environment itself plus whatever we mortals do to it, such as polluting it and adding to its noise level). As Wells [29] points out, the environmental "unit" of an individual is never as simple as a single room or building, but rather is the composite of many interacting features of our total environment, such as depicted in Figure 16-1. All of these aspects of the living

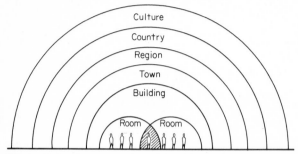

Figure 16-1   Representation of the different features of our total environment. The living environment of an individual is never as simple as a single room or building. (The shaded portion represents passage space between rooms.) [*From Wells, 29, fig. 1, p. 680.*]

environment have an impact, good or bad as the case may be, on human beings. One would hope that these features of our civilization would become the new frontier of human factors attention.

A wide-scale assault on these and other problems obviously would involve many disciplines, but there are, indeed, many facets of those that fall within the stock-in-trade expertise of people concerned with human factors. However, a couple of differences appear between the traditional areas of application of human factors and those related to the general living environment that probably would pose greater challenges to the human factors discipline. In the first place the "systems" of concern (e.g., cities and communities) are more amorphous and less well defined than pieces of hardware. In the second place, the relevant criteria for evaluation tend to be rather different. Instead of being primarily interested in system performance criteria (such as for a computer system) or physiological criteria (as in the case of astronauts), perhaps the dominant considerations in living environments tend to be more subjective, with a major focus on the fulfillment of human values and on the achievement of human satisfactions; such considerations, of course, have significant overtones in the social interactions among people.

DEPENDENT AND INDEPENDENT VARIABLES

It is not feasible within the remaining pages to delve into the living environment with its many human factors ramifications. But it may be useful to touch on some of the dimensions of the problem, especially certain of the dependent (i.e., criterion) variables and the independent (i.e., situational) variables. Beyond that, we shall bring in the results of a few relevant studies, but more for the purpose of illustration than for their substantive content as such.

Dependent Variables (Criteria)

If we were to ask a number of people to give their opinions about the standards (i.e., the criteria) by which they would evaluate the man-made features of their life space, we would of course receive a wide variety of responses, but such an assortment might fall into the following categories:

• Performance of activities (work in offices, factories, hospitals, etc.; preparing meals at home; working in home workshop; engaging in sports and games; etc.)
• Physical convenience (convenience of things that people use; proximity to places people want to go to)
• Convenient mobility (efficient mobility from place to place by public or private transportation, by foot, by bicycle, etc.)
• Physical and emotional health, and personal safety and security
• Physical comfort (temperature, seating comfort, avoidance of noise, etc.)
• Adequate physical space (adequate space relevant to the situation, such as at work, at home, or in travel; opportunity for privacy)
• Social interaction (opportunity for desired social contacts and interchange; individual and group interaction)
• Aesthetic values and preferences
• Fulfillment of personal values (opportunity for selection of activities and situations that fulfill one's own, individual values, such as recreation, entertainment, and culture)
• Financial considerations

Independent Variables

On the other side of the coin, one might likewise categorize the features of the total environment (i.e., the independent variables) that might impinge upon one or more of the above criteria (i.e., dependent variables). Such categories might include the following:

• Building design characteristics (structural characteristics such as size and arrangement of rooms, number and size of windows and doors, halls and passageways, and architectural style)
• Physical environment (nature and arrangement of furniture and other facilities, decor, etc.)
• Ambient environment (ambient outdoor environment, interior illumination, temperature control, noise control, etc.)
• Community (layout, arrangement, size, recreational and cultural facilities, shopping facilities, beauty and other aesthetic aspects, etc.)
• Services and related facilities (health services, transportation services and facilities, public services and related facilities, recreational facilities, etc.)

To put these two together, one could construct something of a matrix of the independent and dependent variables, such as in Table 16-1. In viewing this matrix it would be apparent that only certain of the cells would have meaningful implications, specifically those in which a particular characteristic of an in-

Table 16-1 Matrix of Possible Independent and Dependent Variables Relevant to Human Factors Aspects of Living Environment

| | Independent variables | | | | | | | | | | | | | | |
| Dependent variables (criteria) | Building characteristics | | | Physical environment characteristics | | | Ambient environment variables | | | Community characteristics | | | Services, related facilities | | |
	1	2	etc.	1	2	etc.	1	2	etc.	1	2	etc.	1	2	etc.
Performance of activities															
Physical convenience															
Convenient mobility															
Health and security															
Physical comfort															
Physical space															
Social interactions															
Aesthetic values															
Personal values															
Financial considerations															

dependent variable (such as, say, room arrangement) would have an *effect* on a specific criterion (such as convenient mobility).

Design Criteria

One could envision the possibility umpteen years from now when research might make it possible to set forth, for at least certain of the individual cells of this matrix, the specifications of that independent variable parameter that should be fulfilled in order to achieve a desired criterion value. These specifications would then become *design* criteria. Some such criteria, of course, are already available, as in the ambient environmental variables of illumination, noise, and temperature (discussed in earlier chapters). But, in the case of at least some of the cells, the current design criteria (if, in fact, any exist) are based on outmoded regulations (such as city building codes), on habit or rules of thumb that have somehow evolved, and on beliefs or hypotheses that may or may not have intrinsic validity.

One particularly complicating facet relating to possible design criteria is the fact of individual differences in value systems and preferences. People differ, for example, in their preferences for a degree of privacy, for group interaction at work or at home, for decor, for use of leisure time, and for ways of using their abode (house, room, penthouse, or igloo). Thus, there needs to be variability in living environments to provide for such individual differences.

In a very broad sense our living environment can be viewed as being comprised of the physical characteristics of our environment (i.e., houses, buildings, the physical layout of the town or city, etc.) and of the services (transportation services, health, recreation, and other services) that are available, although the distinction between these obviously is not very clear. In any event, in discussing the living environment, we will discuss these two aspects in separate chapters. The first of these will cover architectural and physical characteristics, and the second will cover various types of services.

ARCHITECTURAL DESIGN

An important ingredient in the living environments of people consists of the various buildings they use—homes, offices, factories, public buildings, schools, churches, etc. In this regard, there is an increasing awareness on the part of architects and others of the impingement of architectural design on the behavior and reactions and attitudes of people. For example, Wools and Canter [30] refer to the effect of the "meaning" of buildings on behavior. And these stirrings of interest have resulted in the coinage of the term *architectural psychology* and the development of research and academic programs in this area, such as at the University of Utah and the University of Strathclyde in Scotland.

People and Physical Space

The design and arrangement of buildings and related facilities and of furniture and other items within them define the physical space within which people live, and they can have a very distinct effect on people's behavior, comfort, emotions, and other subjective reactions. There are, however, different "contexts" within which people interact with respect to physical space.

Personal Space   One of these contexts is referred to as "personal space," this being the space immediately surrounding an individual, usually with invisible boundaries, within which "intruders" may not come [Sommer, 23, p. 26]. Such space is, in a sense, portable, in that the individual takes it with him wherever he goes. However, it has some of the flexibility of an accordion in that we will tolerate being closer to others in some circumstances (such as in a subway or at a football game) than in other circumstances (such as in a conversation at home or in an office). The amount of space also shrinks or expands depending on whether an invader is a close acquaintance or a stranger, and on the differential status of individuals (for example, a lowly subordinate of a high-ranking executive typically will remain farther away from the executive in the executive's offices than, let us say, an individual with more equal status). The amount of personal space is also a function of cultural background.

There are many manifestations of the existence of personal space, such as in seating oneself in a conversational group; in looking for a seat in a library, on an airplane, or in a theater (when it is usual for one to try to find a seat separated from others), or in "keeping one's distance" when walking along a sidewalk. People use various schemes (some rather ingenious) to define their space, such as glaring at a stranger who sits too close on a park bench (or even getting up and leaving, taking one's portable space with him), or by putting a coat or briefcase on a chair beside oneself in a public place.

Territoriality   Although personal space is "portable" (following a person like his shadow), territoriality tends to characterize a piece of "real estate" to which an individual makes some claim. (This is a very common phenomenon in the case of some animals. Intruders will be attacked by the resident.) Territoriality as related to human dwellings has been defined by Newman [20, p. 51] as "the capacity of the physical environment to create perceived zones of territorial influences." Newman points out that, by its very nature, a single-family house is its own statement of territorial claim in that it is on an integral piece of land that buffers the house from neighbors and from the public street. Shrubs, fences, walls, and gates sometimes serve as symbolic reinforcers of the house. The territoriality of semidetached or row houses is sometimes characterized by stoops, porches, or fenced-in front yards. In the case of apartments the territoriality depends very much upon the physical relationship of individual apartments to other areas. In many instances the door of the apartment separates the territory of the dwelling unit from the "public" area—the hallway.

Defensible Space   Closely related to the territoriality of the dwelling unit is the concept of defensible space, which is characterized by Newman [20, p. 3] as a model for residential environments which tends to inhibit crime by creating the physical expression of a social fabric that "defends" itself. It is, in essence, an area with features which form an environment in which latent territoriality and sense of community in the inhabitants can be translated into responsibility for ensuring a safe, productive, and well-maintained living space.

Discussion   The concepts of personal space, territoriality, and defensible space all represent reasonable human values that should be respected, and that therefore should be provided for in architectural design and in the arrangement of physical facilities and features within the living space of individuals.

Psychological Dimensions of Buildings

People tend to perceive virtually any stimulus object (such as other people, institutions, or news media) in terms of each of several "dimensions." In a similar fashion we tend to perceive buildings and environments in terms of certain dimensions, such as pleasantness or comfortableness.

In this regard Wools and Canter [30], following up some previous research by Canter [7], have identified certain of the dimensions that people generally reflect when appraising their physical environments. In their study they used a semantic differential scale [Osgood et al., 21] which consists of several or many pairs of opposing adjectives (such as *pleasant-unpleasant*) with a scale (usually a 7-point scale) between them that can be used to indicate a person's assessment of a stimulus object. In this study 49 pairs of adjectives were used, these being adjectives that could be considered as relevant in describing rooms in a building.

The subjects in this study (nonarchitectural students) were presented with 24 monochromatic line drawings such as the ones shown in Figure 16-2, and

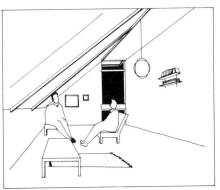
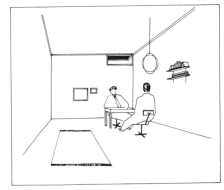

(*a*) Most friendly room   (*b*) Least friendly room

Figure 16-2   Examples of the drawings of rooms used in a study relating to the perceived dimensions of rooms. These examples are of rooms that were appraised to be most friendly and least friendly. [*Adapted from Wools and Canter, 30, figs. 4 and 5.*]

then "described" each room, using the 49 scales. The responses were then subjected to a statistical procedure (a principal-components analysis) for identifying the pairs of adjectives that tended to "go together" or to form groups. Eight such groupings were identified, these being referred to as *attitudinal* dimensions. Of these, three seemed to be of particular importance. These are listed below, along with an illustrative pair of adjectives:

Dimension	Illustrative adjectives
Activity	fast—slow
Harmony	clear—obscure
Friendliness	welcoming—unwelcoming

For any given room the responses of people to the several pairs of adjectives that form each dimension serve as the basis for deriving a score for the dimension as it applies to that room.

In a subsequent phase of the study, other subjects were asked to judge the drawings of eight rooms that had various combinations of windows, desk arrangements, and ceilings. Figure 16-2 shows the combinations that were judged to be most friendly and least friendly. The features which were associated with the most friendly room were: a sloping ceiling; a floor-to-ceiling window; and a grouping of chairs without an intervening table.

Although the above discussion deals with the subjective reactions of people to features of rooms as presented in the form of drawings, Wools and Canter [30] also report some "behavioral" differences on the part of persons being interviewed in two different rooms (an office storeroom and the office of a secretary, both of which had the same furniture). Thus, there is at least a hint that the perceptions of people of the features of rooms may have some behavioral counterparts.

What Kind of Office?

Arguments have been bandied about regarding the pros and cons of various types of offices—large and small, and landscaped and conventional. There have been a few studies dealing with this matter, although it must be said that the results of these have been somewhat ambiguous if not conflicting.

Office Size and Social Behavior   There are some indications that small-office environments are more conducive to the development of social affinity for others than large offices. This was reflected, for example, by the results of a sociometric study by Wells [28] in which office personnel in large and small offices were each asked to indicate their choices of individuals beside whom he/she would like to work. The data summarized below show the percents of the choices made by individuals (in open and small areas) of other persons who were in their own section or department:

	Percent of choices of members of own section	Percent of reciprocal choices within own section
Open areas	64	38
Small areas	81	66

Although there was greater internal cohesion among personnel working in the smaller areas than in the open areas, it should also be added that there were more isolates (individuals not chosen by anyone).

Office Size and Expressed Preference Although smaller offices tend to result in more work-group cohesion than open areas, there apparently is no widely pervasive preference for small offices. In one study Manning [15] asked a group of office workers in Great Britain to express their preferences for five different office arrangements varying in the degree of openness and found that the workers ranked the most closed arrangement first, with the most open last (and the others in between). On the other hand, Nemecek and Grandjean [19], on the basis of a survey of 519 workers in 15 large-space offices in Switzerland, found no such pattern of preferences. For example, one of the questions asked was: "Would you accept another job in a large-space office?" In reply to this 59 percent said "Yes" and 37 percent said "No," with 4 percent expressing no opinion. (It might be added that men judged large-space offices more favorably than women.)

As a further aspect of this study the workers were asked to indicate what they considered to be the advantages and disadvantages of large offices. The primary factors referred to were as follows (expressed as percents of all responses):

- *Advantages*: better communications (40 percent); personal contacts (28 percent); work flow, supervision, discipline (15 percent).
- *Disadvantages*: disturbances in concentration (69 percent); confidential conversations impossible (11 percent).

Landscaped Offices In recent years there has been a flurry of interest in the office landscape concept (Bürolandschaft), which originated in Germany [Brookes, 6]. Such an office consists of one large, open—but "landscaped"—area which is planned and designed about the organizational processes that are to take place within it. The people who work together are physically located together, the geometry of the layout reflecting the pattern of the work groups. The areas of the various work groups are separated by plants, low, movable screens, cabinets, shelves, etc., as shown in Figure 16-3.

One of the few field studies of the effects of landscaped offices was conducted by Brookes [6] with 120 employees of a major retail firm that had decided to build a new headquarters office. The study was actually a pilot study in which the existing space of three departments was converted into a landscaped office. In the original layout the departments were separated, with some of the managers and supervisors in private offices or in partitioned cubicles. The employees completed semantic differential questionnaires before the change and again nine months after the change, the comparisons of these being in terms of 13 dimensions (factors). The comparisons reflected general increases in judgments of aesthetic values but decreases in judgments of functional efficiency. Perceived increases in noise level, loss of privacy, and visual distractions were the chief causes of complaints. There were, however, some

Figure 16-3 Example of a landscaped office in which individual offices and work groups are separated by plants, screens, cabinets, and shelves. This office is in the Administration Services Building at Purdue University.

positive changes in expressed solidarity. Although the employees apparently sought a more sociable atmosphere, Brookes felt the landscaped design may have been too *avant garde* for them.

In summary, the investigator expressed it this way: "It looks better but it works worse," calling attention to the fact that the results of other such surveys have also failed to demonstrate significant benefits in efficiency from landscaped offices. Such results do not necessarily imply that landscaped offices should be avoided, but they do at least raise questions about the general desirability of such office arrangements.

Windows or No Windows?   Although windows are no longer necessary to provide light and ventilation, the question has arisen as to whether they have value such as in fulfilling what Manning [16] refers to as a "psychological need" for some "contact with the outside world" or for actual daylight. There are not many data bearing on this, but one study by Wells [27] sheds a bit of light on this topic. As one phase of his study he asked clerical workers to estimate, for the total illumination at their work stations, the percentage of that illumination that was from daylight (from the windows). He found that workers whose stations were quite removed from the windows markedly overestimated the amount of daylight at their stations. These results, along with the dominant opinion expressed that daylight is "better" for one's eyes than artificial light, add up to the impression that people generally seem to want "daylight" at their work (even though they seem to overestimate the amount they do have). At the same time, there are many people who work in windowless offices who presumably have not been driven up the walls because of that.

Discussion   Large or small offices? Landscaped offices? Windows or no windows? Although research data about these and other aspects of offices are still quite skimpy, the research that is available (such as that discussed above) gives an impression of ambiguity, inconsistency, and lack of support for at least certain expectations or hypotheses. In reflecting about this disturbing state of affairs one needs to keep in mind the fact that in part the measures of the effects of some design features consist of subjective reactions of people, such as preferences, attitudes, and aesthetic impressions. Overt manifestations of these in terms of behavioral criteria such as work performance are difficult to document, but at the same time the favorable disposition of workers to certain types of working situations indicates that they probably have some long-range hidden values. Further, it probably can be said that, although people might prefer a particular environmental situation, they have a fair quota of resiliency or adaptability that makes it possible for them to adjust to a variety of circumstances.

Private Dwelling Units

There are many aspects of dwelling units that have human factors implications. For our illustrative purposes let us consider room usage and certain features that are considered to influence the "difficulty" of various homemaking tasks.

Room Usage in Private Dwellings   In some housing studies, indexes are used that relate to the spatial adequacy of dwelling units. One of these is the number of *persons per room* (PPR), which is the simple ratio of the total number of occupants divided by the number of rooms. As pointed out by Black [5, p. 58], the upper limits of what are usually considered acceptable PPRs are somewhere around 1.00 or 1.20, with a national average of about 0.69. Another index that is sometimes used is the *square feet per person* (SFPP). As pointed out by Black, there are no United States norms, or standards, for the SFPP, although Chombart de Lauwe [8] has proposed the categories given below; the last column shows the percent in each category resulting from a survey of 121 houses in Salt Lake City [Black, 5, pp. 62–63].

SFPP	Category	Salt Lake City survey, percent
Less than 130	Poor housing	2
131–215	Adequate	21
Over 215	Very good	77

It might be added that, of the home owners surveyed, 7 out of 10 were satisfied with their present houses, and in the case of those who were not, house size had no apparent relation to their dissatisfaction.

Although Black's survey resulted in a correlation of $-.77$ between the PPR and SFPP values for the 121 houses,[1] he expresses the general opinion that the SFPP is the more discriminating of the two indexes (except in the case of overcrowded houses).

But, aside from deriving these gross indexes of spatial adequacy of the houses in the survey, Black was more concerned with usage of rooms, and in this connection used a questionnaire in which home owners reported the frequency of use of various rooms for such purposes as studying, TV viewing, family activities, and seeking privacy. Summaries of the responses to three such questions are given below, specifically questions regarding rooms in which reading usually was done.

		Type of reading	
Space	Books, percent	Magazines, percent	Newspapers, percent
Living room	50	62	50
Recreation room	24	25	21
Bedroom	24	12	2
Kitchen-dining area	5	7	32
Other (and "never")	6	3	2

Data on room usage could ultimately aid in designing dwellings which would be more useful to the occupant.

Housing Features Associated with "Difficulty" of Homemaking Tasks   As one phase of a broader study dealing with the "difficulty" of homemaking tasks by Steidl [25], 208 housewives were asked to rate the housing and equipment factors that they preceived as being associated with the difficulty of various tasks around the home, these being separated into the so-called "high-cognitive" tasks (those involving primarily mental activities) and "low-cognitive" tasks (those involving primarily physical activities). A number of these features were associated with the design features of the house and of the associated equipment, as shown in Table 16-2. Although some of these factors are associated with food preparation (such as the level of work surfaces), many of them deal with many other features of the house, a number of which are a function of the design characteristics.

Multifamily Housing

Although we cannot elaborate on all of the many human factors aspects of multifamily housing, an overview of the results of one survey dealing with such

[1]The negative correlation is brought about by the fact that PPR values *decrease* with house size, whereas the SFPP values *increase*.

Table 16-2 Housing and Equipment Factors Considered to Make Homemaking Tasks "Less Difficult" or "More Difficult"

Feature or factor	Satisfactory (That make work less difficult)		Unsatisfactory (That make work more difficult)	
	High-cognitive tasks	Low-cognitive tasks	High-cognitive tasks	Low-cognitive tasks
Work surface	28	17	80	21
Storage space	32	16	58	28
Quality of equipment, supplies	105	109	91	69
Availability of equipment, supplies	80	50	39	39
Location of equipment, task	24	23	22	28
Furniture, furnishings	15	35	22	32
Amount of space, no. of rooms	92	76	135	53
Arrangement of rooms	56	23	25	10
Temperature, light, ventilation, sound, safety	35	18	40	24
Quality of housing structure, age	15	24	20	21
Other	24	14	48	36
Total	506	405	580	361

Source: Steidl, table 3, p. 476.

housing may serve to illustrate the human implications of at least certain design features. The survey in question is one reported by Becker [2] that was sponsored by the Center for Urban Development Research (UDC) of Cornell University. Data were collected from public housing developments (three high-rise and four low-rise public housing developments) in urban and suburban areas throughout New York State. The data were obtained by interviews with 257 residents, questionnaire checklists from 591 residents, systematic observation (nearly 100 observation periods), and interviews with managers of the developments.

A few of the findings of the survey are given below (in very abbreviated and simplified fashion):

• Low-rise developments were preferred slightly to high-rise developments ("satisfaction" averaged 93 percent for the low-rise as compared with 87 percent for the high-rise).
• The exterior appearance of a development was "very important" to most residents (67 percent). Variation in the shape, pattern, and form of buildings which increased their "individuality" were much appreciated; straight, rectilinear, and symmetrical forms were strongly disliked.
• Residents liked lobbies that were visually pleasing (in the survey, the lobby of only one of the seven developments received a very high rating from the residents).
• Many variations of room size and arrangement were considered to be equally satisfactory by the residents, but (as would be expected) satisfaction tended to be greater with larger rooms.
• More residents preferred a separate dining area (39 percent) or separate dining area and large eat-in kitchen (28 percent) to combined living/dining area (19 percent) or living room with large eat-in kitchen (18 percent). A couple of these arrangements are illustrated in Figure 16-4 along with a couple of proposed designs that would permit several options for arrangement of eating and living space. Although most residents preferred a separate dining area, 83 percent of the residents said they would not be willing to give up any or all of their living room space for a separate dining area or an all-purpose room. In other words, residents presumably found living space so essential (whatever the size) that they could not conceive of reducing it.

Aside from presenting the results of the opinion survey, Becker [2] also depicts illustrative contrasting designs of certain specific features of multifamily housing developments, in each instance illustrating characteristics that can be considered as positive or negative in terms of human factors considerations. One of these features is the "tot" lot (a playground for small children), as illustrated in Figure 16-5. The arrangement at the right has a clear superiority in terms of ease of access and of increased opportunity for surveillance from the adjacent buildings. The other feature is a barrier around a development, as illustrated in Figure 16-6. The wire fence at the left gives an institutional impression, yet probably would not serve as a completely impenetrable barrier for someone intent on breaching it. On the other hand, the hedge-like fence at the right creates a psychological boundary and a feeling of enclosure, and would

(a) One of least preferred arrangements (in convenience of carrying food, food odors in living area, limited hobby space)

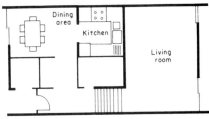

(b) More preferred arrangement (particularly because of separated eating and living areas)

(c) Proposed desirable arrangement

(d) More flexible proposed arrangement (with movable "wall" unit to permit flexibility)

Figure 16-4 Order of preference of residents of a couple of multifamily housing units regarding dining and eating arrangements, and a couple of proposed designs, one of which would permit several options for eating and arrangement of living space (this one having a movable wall unit which could be placed in various locations, such as at A or B). [*Adapted from Becker, 2, figs. 5a, 5b, 5c, and 5d.*]

tend to discourage strangers from entering while not communicating rejection to nonresidents.

Visitor Behavior in Museums

As another manifestation of the influence of the physical environment upon the behavior of people, Melton [17] reports the results of certain observational

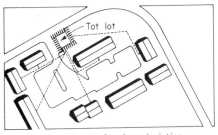

(a) Tot lot with negative characteristics

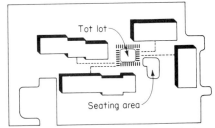

(b) Tot lot with more positive characteristics

Figure 16-5 Illustration of two "tot" lots (playgrounds for small children) in a multifamily housing development, one with negative and the other with positive characteristics. The one at the right provides more convenient access, better opportunity for surveillance from the adjacent buildings, and greater safety and avoids crossing the parking lot. [*Adapted from Becker, 2, figs. 9d and 9a.*]

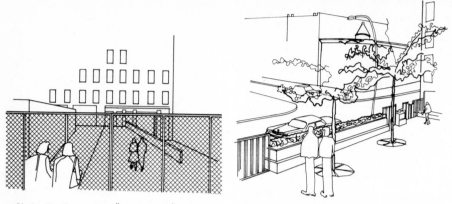

Chain-like fence gives "institutional" impression.   Hedge-like fence creates psychological boundary, with "soft" territorial definition.

Figure 16-6 Illustration of two types of barriers for a multifamily housing development. The one at the left gives an institutional feeling, whereas the one at the right has a more pleasing appearance yet still serves as a psychological boundary and gives a feeling of enclosure. [*Adapted from Becker, 2, figs. 7 and 8.*]

studies of the behavior of visitors in museums, in particular the patterns of movement and the time spent observing objects of art and other items, as related to room arrangement or mode of display. For example, more than 75 percent of the visitors in museums turn right on entering a gallery, this tendency probably accounting for the fact that the region directly to the right of the entrance of a room is the most effective display area. In turn, the region to the left of the entrance is the least effective display area. If the gallery has an exit directly across from the entrance, most of the visitors view only the objects along the side of the gallery toward which they first turn.

Another factor that influences viewing behavior is the degree of "crowding" of paintings on the wall. It appears, from Melton's surveys, that if the paintings are at least "moderately" crowded, the attention given to individual paintings is not further reduced if one crowds them closer together. However, below that "moderate" degree of crowding, further isolation of the paintings tends to encourage greater individual attention.

Still another environmental factor that presumably influences the behavior of museum visitors is the color of the room. In this regard, Srivastava and Peel [24] report a study carried out in a room of the museum of art at the University of Kansas, with Japanese paintings displayed on the walls. Paid (volunteer) subjects were informed that the experiment was intended to determine their reactions to the paintings (and this was done). But for the first half of the subjects (301) the room was painted light beige and had a corresponding rug, whereas for the second half the room was painted dark brown and had a matching rug. A special sensing device under the rug (*a hodometer*) made it possible to record each square foot of area over which the subjects walked, and a watch was used to measure the time spent in the room. Some of the comparisons of movements of the two groups are given:

Room	No. of steps	Area covered, ft^2	Time spent, s
Beige	42.7	9.0	38.1
Brown	46.2	17.9	26.4

It can be seen that the subjects in the brown room took more steps, covered more area (almost twice as much), and spent less time than those in the beige room. Although one might be hard pressed to explain why the differences in color affected the movements of people so markedly, one must conclude that the effect must be the indirect consequence of some subjective reaction to the environmental color.

Although one cannot as yet set forth a definitive set of guidelines for designing museums and for arranging objects of art within them, it is manifest that such factors do influence the behavior of museum visitors. As Melton points out [17], further studies are essential if existing museums are to make the best use of their galleries and if future museums are to have an architecture that is appropriate to the characteristics of the public it serves.

Color in Our Environment

There have been speculations about the emotional associations of color and about the general effects of color on human reactions. In this connection Birren [4] has made some generalizations regarding the effects of different hues, along with suggestions regarding the appropriate hues for particular types of environments. The empirical evidence about the effects of color in our environment, however, is very meager, and one needs to be very cautious in evaluating some of the generalized beliefs about the effects of color (even about those made below).

Thermal Effects One of the prevailing beliefs about color, for example, is that certain colors (especially red) are "warm" and that other colors (especially blue) are "cool." There has been no persuasive evidence that colors do influence the actual thermal sensations of people. In one rather well-controlled study, for example, Bennett and Rey [3] found no difference in the thermal sensations people expressed when in temperature-controlled rooms of various colors. They conclude that hue produces a "strictly intellectual effect, a belief that one is warmer or cooler but does not affect one's thermal comfort."

Subjective Reactions to Color Many years ago Eysenck [10] pulled together the results of many studies dealing with color preference based on over 21,000 judgments, and reported that the general order of preference was: (1) blue; (2) red; (3) green; (4) violet; (5) orange; and (6) yellow. Such an order, however, needs to be viewed with two qualifications. First, that there are distinct individual differences in the color preferences of different people (for example, one woman might like purple for a dress, while another woman would prefer blue).

In the second place, the "context," or the aspect of the environment in which the color is used, also can have an influence on the reactions of people to different colors. In this regard, for example, Acking and Küller [1] refer to some of their research in which five different perceptual dimensions (i.e., factors) had been identified on the basis of the combinations of words that people had used in describing various rooms. (The subjects had used a semantic differential scale consisting of many pairs of "opposite words" in describing different types of environments such as actual living rooms.) These perceptual dimensions, along with a single word that characterizes each dimension, are:

1   Personal evaluation —pleasant
2   Social evaluation   —expensive
3   Spatial enclosedness—open
4   Complexity   —motley
5   Unity   —of pure style

In a subsequent survey, 27 college students and 9 hospital patients and personnel used these words to describe three hospital rooms that were identical except for color, the colors being as follows:

Room	Hue	Lightness	Chromatic strength
A	green	low	medium
B	green	medium	high
C	white	high	low

The results indicated that the impressions of the three rooms differed on two of the five dimensions, as follows:

3   Spatial enclosedness (C being most "open," with B and A following in that order)
4   Complexity (with A and B being considered more "motley" than C)

Granting that these results are not profoundly earthshaking, they do reflect in quantitative terms the fact that people do have different perceptions of rooms of various colors, there being reasonable consistency across people in these perceptions.

Physiological Effects   On the basis of a review of various studies on the physiological effects of color, Acking and Küller [1] state that it seems to be beyond doubt that color has some direct physiological influence on people, such as that reflected by blood pressure, respiratory rate, and reaction time; however, the mechanisms that cause this are not known.

Behavior and Performance Effects of Color   There have been occasional cues indicating that, in certain specific circumstances, color does influence the nature of the behavior of people and even their level of performance. Mention was made above, for example, of the behavior of people in museum rooms of different colors. Another, rather dramatic, example was a report in *Time* magazine of a study by Ertel [9], director of Munich's Besellschaft für Rationelle Psychologie, dealing with the measurement of IQs (intelligence quotients) of 473 children. It was reported that when the children were tested in rooms they thought "beautiful" (light blue, yellow, yellow-green, or orange), their IQs were raised by 12 points, and when tested in "ugly" rooms (white, black, and brown), their IQs dropped 14 points. In a subsequent aspect of the study, two groups of children were studied over a period of months, an experimental group playing in "beautifully" colored rooms with "beautifully" colored building blocks, the other (a control) group playing in a conventional kindergarten. After 6 months the experimental group outstripped the control group by 15 IQ points and after 18 months by 25 points. In addition it was reported that the "positive social reactions" (i.e., friendly words, smiles) increased 53 percent with the experimental group and that negative reactions (irritating, hostile ones) declined 12 percent. One would want more supporting evidence about the possible effects of color on IQ scores before placing too much stock in such effects. If this turns out to be a valid phenomenon, it could of course be the result of the momentary effects on the attitude of children during the testing session of the color, or, conceivably, it could be that long-term exposure to "beautiful" colors might affect their basic attitudes or general mental development.

Discussion   In assessing the influence of color in our lives, one has the impression of being caught between Scylla (accepting the many common beliefs and pronouncements about the effects of color) and Charybdis (rejecting such beliefs and notions, and assuming that color is of only nominal consequence in human life—except possibly for the fate of paint stores). Such an in-between frame of reference actually seems to be warranted, but it must be acknowledged that hard data about the various aspects of color are still limited.

URBAN COMMUNITIES

The burgeoning problems of urban centers undoubtedly comprise one of the major challenges of current life. The many facets of these problems leave few inhabitants unscathed. A partial inventory would include problems associated with health, recreation, mobility, segregation, education, congestion, physical housing, crime, and loss of individuality. The current manifestations of these problems lend some validity to the forebodings of Ralph Waldo Emerson and Henry Thoreau, who viewed with deep misgivings the encroachment of civilization on human life, especially in the form of large population centers. The tremendous population growth, however, makes it inevitable that many people must live in close proximity to others (and thus requires the existence of urban centers); accepting this inevitability, however, it is proposed that one should operate on the hypothesis that, by proper design, urban centers can be created which might make it possible to achieve the fulfillment of a wide spectrum of reasonable human values—perhaps even those that Emerson and Thoreau, and maybe those that you and I, might esteem.

Following the discussion at the beginning of this chapter, we shall here try to view this problem in something of a human factors frame of reference, with the particular intent of trying to see how different community characteristics (i.e., independent variables) influence human behaviors and reactions (i.e., dependent variables, or criteria). For this purpose, the basic interrelationships of Table 16-1 might be relevant.

The manifestations of the "quality of life" (i.e., the criteria) can be expressed in various forms and at various levels of abstraction. For example, Ittelson et al. [13, p. 254] refer to the "pathology" of various living circumstances, this being a reflection of the incidence of such undesirable conditions as illness and crime. In turn, Lynch [in Holland, 12, pp. 120–171] proposes what he refers to as desirable formal goals or "performance characteristics" of cities, these being: (1) accessibility; (2) adequacy of facilities (houses, schools, recreation, etc.); (3) congruence (the fit of the system, the coordination of the parts in operation); (4) diversity; (5) adaptability; (6) legibility (a term something akin to the "image" of a city); (7) safety; (8) stress (achieving a balance in physiological and psychological stress, neither too much nor too little); and (9) efficiency. One can recognize most of these as variants of the criteria given in Table 16-1.

On the other hand, some of the relevant independent variables (i.e., the characteristics of living circumstances) that have been said to influence the quality of living include crowding, density, environmental overload, and urban "pattern." The possible effects of different variations or levels of these and other independent variables can of course be on the plus or minus side of the human-value ledger, but it must be stated that the nature of the effects of at least certain of such variables is not at all clear-cut.

Crowding and Density

Crowding relates to individual dwelling units, and usually is expressed as the ratio of the number of persons occupying a unit of living space (as a room). In turn, *density* refers to the number of people per unit of area, as per acre or hectare. In summarizing some of the research relating to the effects of crowding and density, Ittelson et al. [13, pp. 254–258] report varied and conflicting implications. Thus, although they point out that in certain surveys crowding has been found to be related to certain kinds of pathology and deviant behavior (as in city areas characterized by housing with very restricted living space per family), there have been certain surveys in which crowding was not related to such indices. And in certain instances in which such a relationship does exist, there is some question as to the extent to which the higher incidence of pathology in overcrowded dwelling units is a function of the overcrowding as such and the extent to which a function of the types of families which gravitate into crowded units. It has been suggested, for example, that socially disorganized families tend to live in crowded facilities, and that such families are more prone to a high incidence of pathological criteria. Further, there are indications of cultural differences in adjustability to crowding [Ittleson et al., 13, pp. 257, 258]. The evidence about the relationship between density and pathology is equally ambiguous.

Although crowding and density, individually or in combination, undoubtedly are contributing factors (i.e., causal factors) of various forms of pathological indicators or undesirable behaviors in some circumstances, this is by no means universally the case. And this leads us to the concept of overload.

The Overloaded Environment

In their discussion of cities, Ittelson et al. [13] make the point that cities are stimulating places in which to live, the stimulation arising from people, events, traffic, jobs, and perhaps the "pace" of life. When this stimulus information from the physical environment exceeds one's capacity to scan and process it, a possible consequence is cognitive overload. In the case of some persons this overload can have undesirable consequences, such as causing nervous breakdowns, causing one to "go off his rocker," causing other adjustment manifestations, or tilting the individual in the direction of undesirable behavior such as crime. The overload is of course the composite of all of the stimuli that impinge upon the person, and in this regard crowding and density can add their bit to the total, in some instances serving as the straw that breaks the camel's back. Thus, a tentative hypothesis relating to pathology in urban areas is one in which crowding and density serve as a couple of the several or many stimulus stressors that contribute to cognitive overload, which, in turn, serves to trigger any of several forms of pathology or other maladjustment.

Urban Patterns

Whether by intentional design or (more frequently) by fortuitous development, urban areas have some "pattern," this pattern being formed by the spatial and geographical arrangement and juxtaposition of the many elements that comprise the area, such as the physical structures, the facilities for the circulation of people (roads and transit systems), and the fixed facilities that draw upon or serve the population (stores, factories, hospitals, parks, theaters, etc.). With reference to these elements of spatial pattern, Lynch [in Rodwin, 22, chap. 6] points out that the most significant features of such patterns are the *grain* (the degree of "intimacy" with which houses, stores, and other facilities are intermixed), the *focal organization* (the interrelation of the "nodes" of concentration and interchange contrasted with the general background), and the *accessibility* (the general proximity, in terms of time, of all points in a region to a given kind of activity or facility).

Given these three parameters of the patterns of urban areas, any given area can be depicted graphically in such a manner as to reflect these features. And, in general terms, there are different types of patterns that can be identified (in the case of existing urban areas) or conceptualized for planning purposes.

The central point that we are leading up to is that differences in such urban patterns can have direct effects on the nature and quality of the living of people in urban areas of such patterns. The total spectrum of these effects probably is not now known. However, certain effects probably can be inferred from the representation of such structures, although we need to recognize such inferences as being substantially subjective. Lynch [in Rodwin, 22] describes five urban patterns and suggests certain of the possible effects of those patterns on the residents thereof. These patterns and their possible effects are summarized below.

- *Dispersed sheet.* Characteristics: low densities with substantial interstices of open land with dispersion of virtually all activities over a very broad area; no outstanding nodal points or terminals. Possible effects: convenient circulation of individual vehicles, dispersion of traffic loads, good accessibility, personal comfort, encouragement of local participation, negation of metropolitan character.

- *Galaxy.* Characteristics: numerous small community units, each with an internal peak of density, separated from the next by a zone of low or zero structural density. Possible effects: provides centers of activity, sharpened image of local community, accessibility to open country, concentration of city-wide activities, possibly monotonous similarity among communities but otherwise effects generally similar to those of dispersed sheet.

- *Core city.* Characteristics: highly compact community, virtually three-dimensional with many high-rise structures. Possible effects: dependence upon public transport, good accessibility, effects of high-density housing, spontaneous communication, high but possibly restricted privacy, possible discomfort from noise and poor climate, restriction of choice of habitat, strong sense of community identification, possible need for second (weekend) house in countryside.

- *Urban star.* Characteristics: dominant core with tongues of land radiating from the center, these having moderate density levels with secondary centers distributed along the main radials and transportation systems along the radials and between them. Possible effects: maintenance of metropolitan character with central-city facilities, opportunity for choice of habitat, varied access to services depending on individual location, possible transportation congestion along radials, and possible transportation problems in circumferential movements.

- *Ring.* Characteristics: like a doughnut, the "hole" having low density, with a surrounding high-density ring, with nodes of concentration for various community activities and functions; channels of flow systems consisting of a series of annular rings plus radial feeder lines. Possible effects: high accessibility to services and to open land, wide choice of habitat, variety and strong character of specialized centers, identity with individual centers, possible problem in controlling growth and maintaining open spaces.

We can see that certain criteria probably are intricately influenced (one way or the other) by the spatial pattern of the urban area, such as relative accessibility to various services, the type(s) of habitat available, availability of open spaces, identity with community, and metropolitan character. The patterns differ in the degree to which these and other criteria might be fulfilled. However, other kinds of criteria probably are unrelated to the urban pattern and depend more on other features of the community.

Urban Renewal and New Towns

Efforts to provide improved urban communities for people can consist of the renewal and rehabilitation of existing urban areas or the development of entirely new communities.

Urban Renewal   The characteristics of any existing urban areas frequently impose serious constraints on the nature and scope of any possible improvements. For example, the existing locations of employment opportunities (such as factories or offices) and of residential areas may preclude any major reduction in travel time to places of employment, or the spread-out (i.e., "dispersed-sheet") nature of an area may limit the feasibility of an efficient public transportation system, and the construction of a new highway or public transportation system may require the costly take-over of already built-up areas. At the same time, it is not particularly feasible to abandon existing cities (even though serious questions have been raised about the possible need of "saving"them). Therefore, there seems to be no reasonable alternative to the continuation of urban renewal efforts.

In this regard, however, there are major gaps in the possible guidelines for renewal which would result in the creation of what Mussen and Slyper [18] refer to as "the successful community," that is, the community which would provide increased opportunity for human satisfaction in the urban setting. True, one can take action to change certain specific community features with reasonable confidence that some particular aspects of living will be improved (such as in increased mobility or reduced crowding). The more challenging problem is that of providing for what Ittelson et al. [13, p. 382] refer to as the psychological implications inherent in changing people's city environment. Although these "psychological" implications undoubtedly arise in part from the physical characteristics of urban settings, they probably also arise in part from socially directed programs (such as in the case of certain Model Cities programs in the United States). However, the manner in which the physical characteristics of communities influence the "psychological" environment is not at all clear. In this regard Mussen and Slyper [18, p. 124] urge the crystallization of a "much more precise definition and measurement" of what constitutes "the successful community," and they cite the need for a "fuller understanding of desires for choice and variety by urban residents and the ingredients and mixes of ingredients that produce human satisfactions in the urban setting."

New Towns   In recent years a number of "new" towns have been created. These are communities that are planned from scratch, a major objective being that of creating communities that, if successful, would not be plagued by many of the problems of large cities. In a sense they represent an effort to "escape" those problems. Although the first "new" town was built in 1904 at Letchworth, England, it was not until after World War II that the idea took root. By 1968, 28 such communities had been created or were planned in England, with plans being made for others in Finland, Sweden, West Germany, Brazil, and India. In the United States, Reston (in Virginia) and Columbia (in Maryland) represent this development.

Question: How well have new towns fulfilled their expectations? In discussing the new towns, Ittelson et al. [13, pp. 384–387] express the opinion that such communities generally provide a cultural setting for promoting man's diverse occupational interests, artistic self-expression, and leisure-time hobbies, and they indicate that concerts, arts exhibits, and community activities typically flourish in most such communities.

On the other side of the coin, Ittelson et al. [13] cite a number of criticisms that have been leveled at the new towns. For example, although most such towns were intended to be classless, some towns have become havens for the middle and upper classes. In addition, it has been charged that the lack of diversity and architectural variety produces a rather bland, even an impoverished, environment lacking in novelty. For example, Izumi [14] reports that in connection with the development of Brasilia (the new capital of Brazil) a temporary town sprang up adjacent to Brasilia for workmen and their families. It was to this shantytown that some of the residents of Brasilia went for excitement at the end of the day. Izumi stated that Brasilia had "visual order" but lacked "the varieties of psychic rhythms that we need to experience in any environment." Thus, the new towns of the current world seem to comprise a mixed bag, with both positive and negative features. Although they do avoid some of the ailments of many metropolitan areas, they do not represent—in their present form—the utopia that mankind seems to seek in his living space.

Discussion

It is probable that utopia in one's living space is a will-o'-the wisp, an unattainable goal. In other words, it is doubtful if man's living space in the communities he builds can ever provide for the broad-scale fulfillment of the various criteria that are relevant to each of us—physical and mental health and welfare, aesthetic values, opportunity for social interchange or privacy, recreation, entertainment, culture, convenience, mobility, safety and security, psychological identity with the community, or whatever. Although perfection in such affairs is not a realistic possibility, we should never cease trying to achieve it in the matters of rehabilitating our existing urban areas or creating new ones. The problems that would be involved in such efforts (especially rehabilitation of certain existing metropolitan areas) could easily cause one to throw up his hands in utter despair, especially because of the economic and political aspects. In this connection it is the contention of many knowledgeable people that the magnitude of the problem would require a major national effort [Myrdal, Dyckman, and others, in Warner, 26].

But aside from the rehabilitation of parts of existing communities, every day additional elements of the fabric of communities are being constructed, such as new houses, schools, public and private buildings, streets, and highways. Probably a substantial portion of such construction perpetuates the errors of the past and will only amplify the total problem of developing communities that would be fairly optimum in the human values that community living should fulfill.

Although guidelines for achieving certain goals are relatively manifest, those for achieving other goals are still fragmentary or need to be developed. This is especially the case with some of the more subjective aspects, and it is these which must be considered as items of unfinished business for the behavioral scientists.

REFERENCES

1 Acking, C. A., and R. Küller: The perception of an interior as a function of its colour, *Ergonomics*, 1972, vol. 15, no. 6, pp. 645–654.

2 Becker, F. D.: *Design for living: The resident's view of multi-family living*, Center for Urban Development Research, Cornell University, Ithaca, N.Y., May, 1974.

3 Bennett, C. A. and P. Rey: What's so hot about red? *Human Factors*, 1972, vol. 14, no. 2, pp. 149–154.

4 Birren, F.: *Color psychology and color therapy*, University Books, Inc., New Hyde Park, N.Y., 1965.

5 Black, J. C.: *Uses made of spaces in owner-occupied houses*, Ph.D. thesis, University of Utah, Salt Lake City, April, 1968.

6 Brookes, M. J.: Office landscape: does it work? *Applied Ergonomics*, 1972, vol. 3, no. 4, pp. 224–236.

7 Canter, D.: *The study of meaning in architecture*, Building Performance Research Unit, University of Strathclyde, Glasgow, Scotland, GD/16/DC/A, Apr. 25, 1968.

8 Chombart de Lauwe, P.: The sociology of housing methods and prospects of research, *International Journal of Comparative Sociology*, March, 1961, vol. 2, no. 1, pp. 23–41.

9 Ertel, H.: *Blue is beautiful*, Article reported in *Time*, September, 1973.

10 Eysenck, H. J.: A critical and experimental study of color preferences, *American Journal of Psychology*, 1941, vol. 54, pp. 385–394.

11 Ferguson, D.: Ergonomics and the quality of living, *Applied Ergonomics*, 1972, vol. 3, no. 2, pp. 70–74.

12 Holland, L. B. (ed.): *Who designs America?* Anchor Books, Doubleday & Company, Inc., Garden City, N.Y., 1965.

13 Ittelson, W. H., H. M. Proshansky, L. G. Rivlin, and G. H. Winkel: *An introduction to environmental psychology*, Holt, Rinehart and Winston, Inc., New York, 1974.

14 Izumi, K.: Some psycho-social aspects of environmental design, mimeographed paper, 1969. (Discussed in Ittleson et al. [13].

15 Manning, P. (ed.): *Office design: a study of environment*, Pilkington Research Unit, Department of Building Science, University of Liverpool, Liverpool, England, SfB (92):UDC 725.23, 1965.

16 Manning, P.: Windows, environment and people, *Interbuild/Arena*, October, 1967.

17 Melton, A. W.: Vistor behavior in museums: Some early research in environmental design, *Human Factors*, 1972, vol. 14, no. 5, pp. 393–403.

18 Mussen, I., and J.L. Slyper: "Urban conversion in the context of social improvement and development strategies for the towns of Israel" in R. Alterman with A. Kirschenbaum (eds.), *Urban renewal planning in Israel*, Center for Urban and Regional Studies, Technical Institute for Research and Development, Haifa, Israel, 1972.

19 Nemecek, J., and E. Grandjean: Results of an ergonomic investigation of large-space offices, *Human Factors*, 1973, vol. 15, no. 2, pp. 111–124.

20 Newman, O.: *Defensible space*, The Macmillan Company, New York, 1972.

21 Osgood, C. E., G. J. Suci, and P. H. Tannenbaum: *Measurement of meaning*, The University of Illinois Press, Urbana, 1957.

22 Rodwin, L. (ed.): *The future metropolis*, George Braziller, Inc., New York, 1960.

23 Sommer, R.: *Personal space: the behavioral basis of design*, Prentice-Hall, Inc., Englewood Cliffs, N. J., 1969.

24 Srivastava, R. K., and T. S. Peel: *Human movement as a function of color stimulation*, Environmental Research Foundation, Topeka, Kans., April, 1968.

25 Steidl, R. E.: Difficulty factors in homemaking tasks: Implications for environmental design, *Human Factors*, 1972, vol. 14, no. 5, pp. 471–482.

26 Warner, S. B., Jr. (ed.): *Planning for a nation of cities*, The M.I.T. Press, Cambridge, Mass., 1966.

27 Wells, B. W. P.: Subjective responses to the lighting installation in a modern office building and their design implications, *Building Science*, 1965, vol. 1, SfB:Ab7:UDC 628.9777, pp. 57–68.

28 Wells, B. W. P.: The psycho-social influence of building environment: sociometric findings in large and small office spaces, *Building Science*, 1965, vol. 1, SfB (92): UDC 301.151, pp. 153–165.

29 Wells, B. W. P.: Towards a definition of environmental studies: a psychologist's contribution. *The Architect's Journal*, Sept. 22, 1965, SfB:Ab1:UDC61, pp. 677–683.

30 Wools, R., and D. Canter: The effect of the meaning of buildings on behavior, *Applied Ergonomics*, 1970, vol. 1, no. 3, pp. 144–150.

Chapter 17

The Living Environment:
Services and Related Facilities

Although the "services" of the living environment cannot be neatly separated from the physical features of the environment (such as the buildings and the urban layout), we will for practical purposes treat these in separate chapters, this chapter dealing specifically with transportation systems, personal health, safety, and security, and certain aspects of recreational and athletic services.

PUBLIC TRANSPORTATION SYSTEMS

A dominant aspect of present-day living for many people is the need for transportation, either by public transportation systems or by private vehicles. Although private automobiles presently serve as the primary mode of transportation for millions of people, pressures are building up for greater dependence on public transportation systems, these pressures including depletion of petroleum as a source of energy, increasing traffic volume, parking problems, air and noise pollution, and costs associated with some of these and other factors. However, these pressure will not be effective in wooing people from their own automobiles unless public transportation is perceived as adequately serving their transportation needs. In this regard Hoag and Adams [10] state that urban growth and the development of the megalopolis are bringing about a re-examination of urban transportation systems in terms of their ability to satisfy human needs and preferences, adding the note that this would require a total assessment of the physical and behavioral characteristics of the user population as well as their economic, social, and aesthetic preferences.

Human Factors Aspects of Public Transportation

The human factors overtones of public transportation systems are manifest. In this regard Voorhees and Keith [26] state two fundamental objectives for human factors engineers in transit, as follows:

1 Make more people want to use the service
2 Gain support at the community level through a new positive image

It should be added that the system should indeed possess the qualities and features which are attributed to it in the image that is promulgated.

In a general sense it would seem that the legitimate requirements of a public transportation system—from the point of view of the potential user—would include convenience (being able to go where one wants to go when one wants to go), speed, physical safety, and comfort, along with some sense of psychological acceptance. The fulfillment of such requirements, however, would be brought about by the appropriate design of many specific features of such systems, such as the list set forth by Millar [14] in Table 17-1. That list, however, consists of the human factors aspects of transportation systems from the point of view of the user. It should also be kept in mind that such systems also should be designed with the operating and maintenance personnel in mind.

In considering the human factors aspects of transportation systems from the point of view of the users (or potential users), there appear to be two broad areas that deal with the suitability or relevance of design features for people. One of these concerns the suitability in terms of the convenience and mobility characteristics to the users; and the other to the more aesthetic, psychological, and subjective values and preferences of the users. The implications of these cut across the many specific human factors considerations given in Table 17-1. It is of course not feasible here to deal comprehensively with the many human factors aspects of transportation systems. The following discussion, then, will simply allude to some such aspects, with examples or data to illustrate a few of them.

Physical and Mobility Considerations of Users

By and large, it must be acknowledged that the designers of many existing public transportation systems have given only casual, incidental attention to the anthropometric and other physical characteristics and the physical mobility of potential users of such systems. In a general sense the sins of omission fall into two classes, namely: (1) failure to provide adequately for certain functional requirements; and (2) failure to provide adequately for certain groups of users or potential users.

Table 17-1   Classification of Human Factors Considerations in Urban Public Transportation.

Major Areas of Consideration	Illustrative Problems within Area
Vehicle and station interior	Space and Seating Entrance and exit Package and baggage handling Speed, acceleration, and vibration Noise Illumination Air quality: temperature, humidity, movement, and odors
Convenience and mobility	Time: trip time, waiting, walking, transfer, and terminals Operations: availability and frequency of service, schedule reliability Information: intelligibility of signs and messages related to routing schedule, locations, and vehicle recognition Problems for handicapped individuals
Safety and security	Accident risk: real and perceived Emergency provisions: breakdown recoverability Maintenance Crime and vandalism Health and sanitation
Social factors	Desirable and undesirable groupings Freedom to choose travel companions Personal space, privacy, and crowding
Psychological state	Boredom Self-esteem Perceived safety and security Anxiety and uncertainty Sense of personal control of conditions Aesthetics Comfort and ride quality
Environment-system interaction	Congestion Pollution: air, acoustic, and visual Harmony with surroundings

Source: Millar [14] as adapted by Hoag and Adams [10, table 1].

Functional Requirements   The providing of transportation to the public involves an assortment of related functions and activities, such as: procurement of tickets or tokens and collection of tickets, tokens, or fares; loading and off-loading of passengers; handling and storage of packages and other items; use of restrooms; and the movement of passengers within terminals or stations. All such functions have some implications in terms of the anthropometric and physical characteristics of people.

The time for loading and off-loading of passengers, for example, influences the "dwell time" of a system, that is, the time a vehicle (such as a bus) would have to stop at stations or locations, thus of course affecting total trip times. The dwell time itself is in part influenced by the physical features of the doors, platform, and mode of collection of fares (if collected on the vehicle, such as a bus). Some data on such times are reported by Bauer [1], with examples of average loading times being given in Table 17-2.

Table 17-2 Illustrations of Loading Times of Passengers on Various Transportation Vehicles

Feature	Bus	Streetcar	Subway
Location	Detroit	New Orleans	Toronto
Platform	Street level	Street level	Vehicle-floor level
Door	30"	50"	44"
Steps	3 steps	3 steps	No steps
Fare collection	On board	On board	Prepaid
Passengers/second	0.4	0.7	1.2
Seconds/passenger	2.7	1.4	0.8

Source: Bauer [1, Table 3, p. 16].

It might also be added that the off-loading times for various types of vehicles as reported by Bauer [1] range from 0.8 passengers per second (on a curb-level platform bus in Toronto) to 1.6 passengers per second on Toronto subways, there being a difference of 2 to 1 in this ratio.

In connection with turnstiles used to collect tokens or fares, the Massachusetts Bay Transportation Authority arranged for the redesign of the Arlington station in Boston. The original and redesigned stations are shown in Figure 17-1. The new design emphasizes passenger orientation and harmony.

Groups of Users   Although most passengers of public transportation systems are hale and hearty adults, there are some who are not, including the very young, the elderly, and the handicapped. The handicapped have been categorized by Gutman [7] as follows:

1 Aged who need physical assistance
2 Those with visual, auditory, or vocal deficiencies
3 Those with neural or coordinative problems such as palsy
4 Semiambulatory persons requiring crutches, canes, or braces, or who wear artificial legs, and persons with spastic, cardiac, or arthritic problems which limit walking and climbing
5 Nonambulatory persons who use wheelchairs

Hoag and Adams [10] on the basis of government data estimate that there are about 44 million individuals in the United States whose mobility is severely limited. Although many such people do not use public transportation, or use it only to a limited extent, transportation systems should be so designed as to make the use thereof reasonably convenient and suitable for them, including the performance of such activities as: moving through stations or terminals, including maneuvering through crowds; going up or down stairs,

Figure 17-1 Original (top) and redesigned (bottom) turnstiles of the Arlington subway station of the Massachusetts Bay Authority, Boston. [Photographs by Harvey Hacker and David Hirsch for Cambridge Seven Associates, Inc. Reprinted from *Design in transit*, published by the Institute of Contemporary Art, Boston.]

inclines, escalators, and elevators; going through doorways, gateways, and turnstiles; using restroom facilities; perceiving relevant visual cues such as signs and directions, or auditory cues such as announcements of departure times; and handling packages or baggage.

In addition, of course, the vehicle itself should be so designed as to accommodate the handicapped. In this regard the United States Department of Transportation has developed a procedure for evaluating the suitability of vehicles for the handicapped. In particular this procedure provides for the calculation of the percentages of handicapped who would be able to use a particular transportation mode with the "removal" of various combinations of "barriers." In the case of buses and trolleys these barriers are: sudden movement of the vehicle; riding while standing; rapid self-locomotion; movement in crowds; wait standing; taking short steps; rising from seat; aisle width (restricted); and walking long distances. If all of these "barriers" were removed, 99 percent of the handicapped could use the transportation mode in question, whereas if only the first four were removed, only 35 percent could use it. Although the removal of all such barriers could pose a serious design problem, solutions or partial solutions for at least some of them have been proposed [*Travel barriers*, 29]:

- *Sudden movement*: provide special bus lanes to control traffic; pad hard surfaces to reduce accidental injuries; use vertical floor-to-ceiling stanchions.
- *Crowds:* limit bus seating; use smaller buses with more frequent service; redesign fare turnstiles to eliminate push-bar and widen channel; use pressure mats to open fare gates when coin is deposited; use automatic doors at exits; improve coin receiver to eliminate precision movements; modify buses to lower entrance; use mechanical steps, or add ramps or lift; provide raised platform at bus stops; major redesign of buses.
- *In-vehicle barriers:* pad hard interior surfaces; provide vertical stanchions for all seats; reserve open spaces for wheelchairs; widen aisles.

Since the incorporation of some of the above in system design would be entirely feasible and would not involve major costs, the primary limiting factor in their implementation seems to be lack of awareness on the part of some designers of the desirability of such features for certain groups of users.

Values and Preferences of Users

Aside from providing a transportation system that is adapted to the physical and mobility characteristics of users or potential users, such systems should also be reasonably compatible with the values and preferences of people. In this regard the Research Laboratories of the General Motors Corporation has sponsored a few surveys of user preferences. In one of these studies [Golob, Canty, Gustafson, and Vitt, 5] interviews were conducted with, and questionnaires obtained from, 1603 adults in 1260 households. Two types of questionnaires were used. One of these was based on the paired-comparison method, the questionnaire including pairs of items such as the following:

A   Making a trip without changing vehicles
B   Easier entry and exit from the vehicle

For this purpose 32 "system characteristics" were used as the items in the questionnaire, these being formed into nine blocks. These 32 items (i.e., characteristics) generally depict something about the service that might be provided by a transportation system, such as A and B above and as listed in Figure 17-2. (More will be said about that figure later.) The items within these blocks were

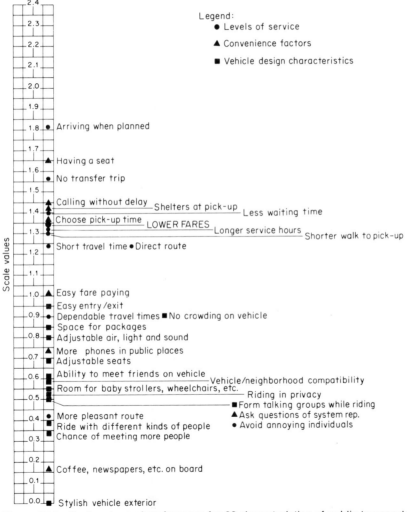

Figure 17-2   Scale values of preferences for 32 characteristics of public transportation systems based on the paired-comparison responses of 786 subjects. The different symbols represent three different groups of characteristics. [*Adapted from Golob et al., 5, figs. 6, 8a, 8b, and 8c.*]

formed into all possible pairs, there being 168 such pairs. A respondent indicated his preference for one of the items within each pair, and on the basis of the responses to all pairs within a block of pairs it is possible to derive a scale value for each item for that individual and the average of these scale values for the many respondents that can be considered as the scale value of the item for the sample of respondents.

Some of the results of the survey (based on responses from 786 individuals) are given in Figure 17-2, the figure showing the scale values of the 32 items. A word of caution is in order in evaluating the scale values, since the items were paired within each of nine blocks. In a strict sense, the scale values can be compared with each other only within each such block. Since one item ("lower fare") was common to all nine blocks, it did serve as a common link, but even so some caution in interpretation is in order. Recognizing this possible distortion of the scale values, the figure shows that "arriving when planned" comes out with the highest scale value, followed by "having a seat" and "no-transfer trip." These in turn are followed by a cluster of nine characteristics concerned mainly with the customers' time, fare, and shelters. Lower on the scale is a large cluster of 18 characteristics that are concerned primarily with interior design, aesthetic aspects of the trip, and passenger convenience. The items at the bottom of the scale are "coffee, newspapers, etc." on the vehicle, and "stylish vehicle exterior."

As an additional analysis, the characteristics were separated into three subgroupings: levels of service, convenience factors, and vehicle design characteristics. Figure 17-2 represents these groups by different symbols (a solid dot, a triangle, and a square).

Still another analysis consisted of the separation of the expressed preferences of each of several subgroups, such as the elderly, the handicapped, and those with low income. Without giving the details, it can be said that there were noticeable differences in the order of preference with respect to various characteristics between and among the various subgroups.

Although the results of this survey give an impression of the relative preferences of people for various characteristics (really the types of service) of transportation systems, such data do not reflect the reactions of people to specific types of systems or their relative preferences for alternative features of specific types of systems. Such preferences can be solicited in various ways, such as by the use of the paired-comparison method in which the various "items" characterize alternative features. An example of this approach is found in the results of a survey by Gustafson, Curd, and Golob [6], this example dealing with alternative methods of payment. Figure 17-3 shows the scale values (in this instance, it was found that the use of cash or change was the preferred method).

Variations in Urban Transportation Systems

Present urban transportation systems include buses, subways, trains, and rapid transportation systems, but various types of new systems are being proposed, a few examples of which are illustrated in Figure 17-4 [Canty and Sobey, 2].

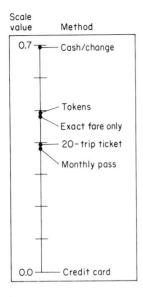

Figure 17-3 Scale values of preferences for various methods of fare collection, based on paired-comparison method. [*From Gustafson, Curd, and Golob, 6, fig. 5.*]

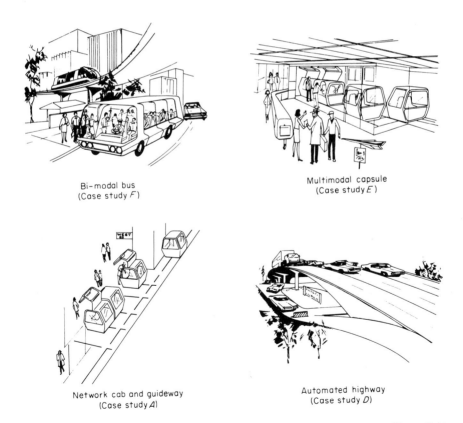

Bi-modal bus
(Case study *F*)

Multimodal capsule
(Case study *E*)

Network cab and guideway
(Case study *A*)

Automated highway
(Case study *D*)

Figure 17-4 Sketches of various possible urban transportation systems. [*From Canty and Sobey, 2, figs. 14, 11, 1, and 8.*]

The possible option of systems that might be considered for any given circumstance would of course be limited by technological and economic considerations. Within the range of such options, the choice should be based on the extent to which the various alternative systems fulfill the needs of the potential users.

Because of the variety of metropolitan areas, obviously no single transportation model would serve all purposes. At the same time, Canty and Sobey have at least characterized three basic classes of urban transportation services that would need to be provided for cities of various types. They are the following:

• *Central business district circulation (CBD)*. The primary purpose of this type of system would be to facilitate the movement of the people within a central business district (presumably of some size) once they have arrived in the general area via some other transportation mode. The primary need here would be for the transfer of people in small numbers to each of many specific areas within the central business district.

• *High-speed radial link systems*. Such systems would be intended to provide rapid service from residential areas to central business districts or other commercial and industrial areas within a community. Such systems could include bus systems, rail pallets (which combine the attributes of the automobile with the speed and capacity of rail systems), automated highways, and a bimodal bus (which could be operated electronically under fully automatic control, on a guideway, or under control on public streets).

• *Special-purpose links*. In certain instances special-purpose links are required, such as to and from airports, to provide direct service for passengers from and to a limited number of destination points.

It is not feasible to illustrate all of the variations of transportation systems that have been proposed, but Figure 17-4 does show four systems which reflect different variations of certain of those discussed by Canty and Sobey [2]. In particular, these include the following:

1 *Bimodal* (Case study F). Such a vehicle is an electrically propelled bus which could operate either as an automatically controlled, rail-like car on a private guideway, or as a bus operated in the usual fashion by drivers. It thus could serve for connecting central business districts and outlying residental and industrial areas.

2 *Multimodal capsule* (Case study E). This is one which would be optimum for communities to provide for transportation between the central business district in the city and the outlying residential and industrial areas, and between intercity homes and the increasing number of outlying job opportunities. However, it would probably be optimum in situations in which the population density is relatively low and in which the otherwise normal mode of travel would be the automobile. Operationally, it provides for electrically propelled buses which can operate as automatically controlled, raillike cars on a restricted right-of-way or by conventional bus-operational procedures on regular streets.

3 *Network cab* (Case study A). The network cab is intended primarily to facilitate the transfer of individuals within the central business district (CBD) in small vehicles accommodating perhaps two to four individuals.. Such vehicles would be supported on air cushions and would be propelled automatically, with the only passenger activity being that of indicating the particular stop at which the passenger would wish to get off.

4 *Automated Highway* (Case study D). The objective of an automated highway would be to move individual vehicles (especially automobiles) from outlying areas into more concentrated areas. The movement of the individual vehicles would be under the control of the system, although the operator would of course have the option of indicating his preference for entrance and exit to and from the system. In general, it would be expected that such a system would permit higher vehicle flow rates than those that can be maintained by individual control of vehicles, and a higher level of safety.

These illustrations, of course, reflect only a few of the many possible variations in modes of urban transportation that might be developed in the future to provide for varying requirements for transportation of people within urban areas.

PERSONAL HEALTH, SAFETY, AND SECURITY

The quality of life is in part a function of the risks to life and limb that are part and parcel of life. These include, for example, the risks to health from disease and related conditions, the risks of accidents, and the risks of criminal assault.

Health

Medical science and environmental conditions in at least many parts of the world have indeed reduced, and in some instances even eliminated, certain types of diseases. At the same time, as Ferguson points out [4], the human community still is subject to a state of "unhealth." Aside from many diseases and conditions that have always haunted man (including age), modern civilization is marked by increasing incidence of coronary heart conditions, respiratory diseases (such as emphysema), lung cancer and other forms of cancer, dental diseases, mental breakdowns, and other ailments which are partly of man's own doing. In part these are amplified by environmental conditions which we have brought about, such as pollution, slums, incorrect diet, use of drugs, and a frantic pace of life (which can contribute to mental breakdown and heart attacks). Some aspects of the environmental factors that contribute to this "unhealth" could possibly be reversed by some of the approaches of human factors in collaboration with other disciplines.

Accident Occurrence

The human factors discipline has for years been concerned with the problem of accidents, especially as related to the operation of vehicles and industrial equipment and to other occupationally related activities. However, as we all know (and can read in official reports), the incidence of accidents in other aspects of life is disturbingly high.

Accident Reduction Strategy What we call "accidents" involve the release of some form of energy. Although some "accidents" occur as the consequence of natural forces (such as hurricanes and lightning), others involve man-made items (such as automobiles or machines) or behaviors. In discussing such phenomena, Haddon [8] sets forth ten "countermeasure strategies" for reducing the human and economic losses that occur as the consequence of these phenomena. In their logical sequence, these are as follows:

1 Prevent the marshaling of the form of energy (as, preventing the generation of electrical energy).
2 Reduce the amount of energy marshaled (such as chemicals).
3 Prevent the release of the energy.
4 Modify the rate of "spatial distribution" of release of the energy (such as reducing the slope of ski trails for beginners).
5 Separate the energy in space or time from people or inanimate objects that could be damaged.
6 Separate the energy from susceptible personnel or objects by a barrier (such as shields or safety glasses).
7 Modify the contact surface (as having rounded corners or edges on objects).
8 Strengthen the susceptible structure (such as vaccines for people or the use of fire-resistant materials for physical structures).
9 Minimize the extent of loss or damage (as by immediate care for the injured, or the use of sprinkler systems in buildings).
10 Take rehabilitation measures following loss or damage.

Clearly, some of these would be impossible or impractical in all circumstances, but in any given situation there would be certain of these strategies that could be followed.

Accidents in the Residential Environment Referring specifically to the residential environment, Neutra and McFarland [15] call attention to the wide range of accidents that occur and point out that there are many ways to make our residential environments safer. Some solutions are fairly obvious and can be applied by residents themselves, whereas other solutions depend upon the design of physical facilities. A few examples are mentioned by Neutra and McFarland. For example, the injuries that occur as the consequence of people walking into glass doors have resulted in the development of new standards that specify wired glass or glass which is much stronger and which shatters into small, nonjagged pieces when broken. In a 164-unit high-rise apartment building constructed by the Toledo Housing Authority for the elderly and the handicapped, ramps and elevators were substituted for stairways, special guard rails were installed in bathrooms to prevent falls, and wall-to-wall carpeting was used to cushion unavoidable falls. And certain building codes now prevent the use of lead-based paints to reduce the possibility of lead poisoning.

Pedestrian Accidents   The use of overpasses and underpasses for pedestrians is clearly the most effective (albeit also the most costly) method of reducing pedestrian accidents. In fact, a survey by Great Britain's Road Research Laboratory [28] showed that these reduce the risk effectively to zero—*if* they are used. In this regard the study points out that it is imperative to plan them so they conform to the most convenient routes traveled. And oddly enough, it was found that underpasses will be used more than overpasses; overpasses must provide a faster route than surface travel if they are to attract a significant pedestrian load.

The effectiveness of town design in reducing accidents was reported by Lindstrom and Gunnerson [12] regarding two "new towns" in Sweden—Guldheden and Kortedala. In Guldheden the pedestrian and bicycle traffic was more separated by using underpasses for "secondary feeder streets" and had three-way junctions rather than four-way junctions where secondary streets met the main thoroughfares. Over a period of five years the rates of childhood pedestrian accidents were as follows:

Guldheden	0.49/1000 per year
Kortedala	1.44/1000 per year

The ratio of nearly one to three clearly reflects the difference in safety.

Health Services

As Rappaport [20] indicates, there is a profound need for the application of human factors technology to biomedical problems. Since there has been relatively little attention paid to the application of human factors technology to the broad spectrum of biomedical problems, a discussion of this must of necessity be more of the nature of mentioning problems than of presenting the results of completed research. A partial inventory of some of the specific areas of possible application would include the following, as suggested by Rappaport [20] and Ronco [22]: medical information systems; design of diagnostic and treatment instruments and devices; physical arrangement of hospitals and other health facilities in terms of efficiency of operations; the organization of services to patients; the psychological effects on the patients of confinement and lack of privacy; and the design and arrangement of patients' rooms, beds, and other facilities. A couple of examples of the research relating to health services will be summarized.

Operating Room Activities   One such example deals with an analysis of the activities of hospital personnel in the operating theater. In this study the investigators [Whitehead and Eldars, 27] found for example, that 38 percent of the working time was taken up by the movement of personnel into and out of various related areas. In the interest of developing an improved physical arrangement (to minimize movement time), an activity analysis was carried out to determine the frequency of movements between specific rooms and stations. Part of the results are shown in Figure 17-5. This figure represents what is real-

Total number of movements
(all areas)

488	1. Sterilizing room
677	2. Scrub-up room
1115	3. Antespace and nurses' station
711	4. Theatre No. 2
376	5. Anesthetic room No. 2
395	6. Emergency theatre
254	7. Workroom and clean supply
146	8. Sterile supply room

Movements between areas

16
13
182 123
111   4
111   13 16
51 52 13
85 113 13
9 31 49
56 15 10
1
13
3
26

Figure 17-5 Part of an association chart showing the number of movements between rooms and stations in and around an operating theater of a hospital. The cells in the intersections at the right show the number of movements between any pair of stations. (This example includes only a portion of the total chart.) [*Adapted from Whitehead and Eldars, 27.*]

ly a sequential link analysis. A reorganized arrangement based on this analysis was estimated to reduce movement time by one-fourth. That saving, reflected in reduced personnel or hours worked, would result in a salary saving of 8 percent.

Control of Surgically Induced Infections   One of the problems involved in surgery arises from airborne bacteria, which can cause infection in patients. An innovative approach to this problem was reported by Skidmore [24], the solution consisting of a complete system that comprised a transparent, collapsible, clean room, a laminar-flow air filter system, a full-bubble helmet with associated ventilation and communications, and special surgery gowns. The helmet is shown in Figure 17-6. This experimental system was put into actual use

Figure 17-6 Full-bubble helmet designed for use in an experimental surgical system for reducing wound contamination from airborne bacteria during surgery. [*From Skidmore, 24, fig. 2.*]

and an evaluation was made of the new system in comparison with the "regular" system in a total of 820 surgical cases. The use of the experimental system reduced wound contamination rates by 87 percent. A subsequent human factors analysis identified several additional areas of potential improvement.

Use of Mental Health Center Facilities   Another example deals with the behavior of patients in mental health centers as related to the "public" rooms and facilities. In this regard, Lipman [13] observed that in conventional large dayrooms the patients tended to occupy their "own" chairs in groupings that formed social cliques, with some accompanying hostility between the groups. In smaller rooms, however, the cliques could not be formed as readily since people were in closer proximity to each other and the groups could not separate. On the other hand, in the larger rooms the patients seemed to be more mentally alert and tended to participate more in social interaction, even though some such interaction tended to be more of an aggressive, hostile nature.

In connection with the accouterments of space (as contrasted with space itself), Taylor [25] relates some observations regarding the use of carpets on the floors of mental health centers. In one center, with carpeted floors, the patients tended to interpret their situation as one reflecting the feeling of "society" that they (the patients) were "worthwhile people." And in still another mental health facility (a psychiatric ward), Ittleson [11] was interested in observing the behavior of patients in certain areas. In a solarium, for instance, he found only "isolated" withdrawal types of behavior. The room had little furniture and no drapes, and was hot even in winter. The addition of drapes and furniture resulted in increased use of the room and, more importantly, use for more constructive activities. Because of the importance of social interchange in the recovery of some mental patients, the physical facilities should of course be such as to encourage such interchange.

Criminal Attack

The distressing rise in the incidence of various forms of crime is undoubtedly the consequence of a combination of many aspects of society, and the solution to the problem will require a many-faceted response, including certain aspects that have human factors implications.

Defensible Space   One such aspect relates to the concept of defensible space as discussed above. Newman [16] well documents the point that certain characteristics of building design and related physical layout (particularly in multifamily housing projects) combine to produce a "defensible space" environment, and that, in turn, other characteristics reinforce criminal behavior. One of the features that is associated with the incidence of crime is the number of families sharing a common hallway. The results of an analysis of crimes in a sample of public apartments in New York City, for example, revealed the following [Newman, 16, p. 69]:

	No. of apartments per hallway		
	2–5	6–8	9+
Crime rate/1000 population	4.5	8.7	8.3

In particular, "double-loaded" corridors (long central hallways with several apartments at both sides) seem to be conducive to high crime rates.

Newman [16, p. 50] sets forth four characteristics of defensible space, as follows:

1   The capacity of the physical environment to create perceived zones of territorial influence. Depending on the housing unit, territoriality can be enhanced by the use of fences, porches, gates, etc. (as discussed above), by having limited numbers of units opening into hallways (in the case of apartments), and by appropriate arrangement of outdoor play and sitting areas.

2   The capacity of the physical design to provide surveillance opportunities for residents and their agents. Surveillance of entrances and lobbies of apartment buildings can be facilitated, for example, by the geometry of the building, as illustrated in Figure 17-7. Since most crimes in apartment buildings

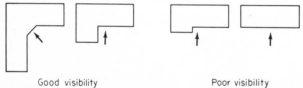

Good visibility   Poor visibility

Figure 17-7 Illustration of varying levels of visibility of entrance lobbies in apartment buildings. The designs at the left provide more opportunity for surveillance by residents than those at the right. [*Reprinted with permission of Macmillan Publishing Co. from Newman, 16, fig. 61, copyright 1972 by Oscar Newman.*]

occur in the visually deprived semipublic interiors of buildings (lobbies, halls, elevators, etc.), these areas should lend themselves to ready surveillance, as through the juxtaposition of windows with stairs and corridors and with the outside.

3   A geographical juxtaposition with "safe zones" of adjacent areas.

4   In the case of large-scale apartment developments, a perceptual image of uniqueness that minimizes any impression of isolation or of the "stigma" of some housing projects.

We cannot discuss here the many specific aspects of the design of residential environments that contribute to the creation of a sense of defensible space. However, Newman [16, pp. 39–49] makes a comparison of the experiences of two New York housing projects which illustrates the implications of design characteristics as related to crime rates. The two projects are equal in terms of population (6000 each) and in the social characteristics of the residents—in fact, they are side by side. But they differ markedly in their physical design. Brownsville consists of a spread-out assortment of primarily 6-story buildings, whereas Van Dyke Houses consists largely of 13- to 14-story high-rise buildings. Other features also are different with respect to a sense of teritoriality and better surveillance. For example, the Brownsville apartments have more entrances and only a few apartments clustering around each, whereas the Van Dyke Houses have the double-loaded corridors that are more characteristic of many high-rise buildings. The Brownsville apartments provide for a greater sense of territoriality and for better surveillance, thus being better designed in terms of defensible space considerations. A comparison of the number of crimes occurring in them indicated that the Van Dyke Houses had a crime rate that was about 60 percent higher than that of Brownsville.

Law Enforcement

One of the important ingredients in personal safety and security is of course the protection and assistance that are provided by law enforcement agencies. Here, again, there are many aspects of such programs that have human factors implications, such as: information systems (including computerized systems); protective devices; devices and procedures for restraining or immobilizing suspects without imposing injury on them; methods of identification of suspects; devices and procedures for control of groups of people; mechanical weapons and chemical agents suitable for varying levels of conditions; vehicles; and facilities for retention of suspects and/or those convicted of various crimes. As in many other areas of civil and community affairs, relatively little research has been carried out in the human factors aspects of law enforcement. We will here touch briefly on a couple such aspects, one dealing with police intelligence and the other with police vehicles.

The Use of Link Analysis in Police Intelligence   An important phase of police intelligence work deals with organized crime, which involves groups of individuals and/or organizations engaged in some form of illegal activity. Many police intelligence programs obviously incorporate provisions for collection and organization of information relevant to such individuals and/or organizations, such information consisting of a variety of types, such as investigation reports, arrest records, informant reports, surveillance reports, telephone call records, financial statements, newspaper articles, and public records of many kinds.

In an interesting systematic approach to the analysis of such police intelligence data, Harper and Harris [9] used a variation of the link analysis technique to "organize" such data. In this probing effort they used 78 law enforcement officers organized into 29 teams (20 three-person and 9 two-person teams) to develop link analyses between and among various individuals and/or organizations. The basic information (which had been collected from various sources and by various methods) consisted of 900 items of information on separate sheets of paper, with an accompanying cross index on cards.

The officers were asked to examine these materials and to develop link analyses that would reflect "strong" links (such as the relationship between a

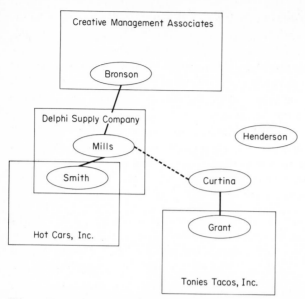

Figure 17-8 Example of link diagram of suspected criminal relationships between and among various individuals and/or organizations, as developed by teams of law enforcement officers from various sources of police intelligence data. Solid lines represent "strong" links and broken lines represent "weak" links. [*Adapted from Harper and Harris, 9, fig. 2.*]

father and son), "weak" links (such as a relationship that is suspected but unconfirmed or that requires some inferential judgment), or "no" link.

An example of the results of such a link analysis is given as Figure 17-8, this showing the evaluated relationships between and among certain individuals and/or organizations (with fictitious names). Although there were some differences in the link relationships developed by the various teams, the consistency among the various teams was generally quite substantial.

Harper and Harris [9] report that at least 20 local law enforcement agencies in California alone have reported successful use of link analysis of intelligence data, and report successful results of certain such applications, including the identification and subsequent destruction of a $2 million-per-year conspiracy engaged in the interlocking activities of prostitution, gambling, and robbery.

Police Patrol Vehicles As pointed out by Clark and Ludwig [3], most police patrol vehicles used at present are standard production models of commercial vehicles, with certain added items or specifications. For example, the "police package" specified for vehicles of the Detroit Police Department includes a special handling package (heavy-duty springs, shocks, and stabilizer bar), heavy-duty brakes, calibrated speedometer, large battery, heavy-duty drive train, and heavy-duty seat.

As Clark and Ludwig imply [3], it is probably not now feasible to specify what the optimum design characteristics of police patrol vehicles should be (this would depend upon extensive further investigation and research as related to actual usage). However, they do propose consideration of such features as: a front-seat-area "package" (including individually adjustable, scientifically designed seats and a center "console" to hold some of the patrolmen's equipment, such as flashlights and ticket books); a certain overhead console (controls for radio, siren, revolving light, etc.); a prisoner detention screen (which is used on some vehicles now); restraining devices for patrolmen other than seat belts (the air bag might be such a possibility); remote-control devices for control of traffic signals ahead; possible push bumpers; and a "roof package" (a possible substitute for the regular vehicle roof that would contain police lights, siren, radio, etc.). The problem for the human factors disciplines is that of determining through research the desirability of these and other features and the specific designs for them if they seem to be desirable.

RECREATION AND ATHLETICS

Various forms of recreation and athletics are becoming part and parcel of the lives of increasing numbers of people and undoubtedly will continue to attract new devotees, as either participants or observers. A partial listing would include football, baseball, basketball, hockey, tennis, handball, badminton, skiing, skating, camping, fishing, hiking, motorcycling, bicycling, snowmobiling, swimming, gymnastics, jogging, rifle practice, archery, gliding, sky diving, scuba diving, woodworking, boating, and sailing. Various human factors implications emerge from many of these activities.

The positive benefits from such activities include of course the enjoyment on the part of the participants and, in some cases, the observers, and the improvement in physical condition and mental well-being that frequently occurs after participation. On the other side of the coin, the unhappy consequences fall into two classes. In the first place, injuries and deaths are the unfortunate consequences of some such activities, and in the second place we are becoming aware of the ecological damage to our environment caused by certain such activities as snowmobiling, motorcycling across open terrain, camping, and other out-of-doors sports.

The human factors aspects enter the picture primarily in terms of the motivational factors that cause people to participate in various types of recreational sports and athletic activities, especially the risk-taking aspects; the decision processes involved in making choices of actions involved in such activities; the skill factors associated with performance; the design of equipment (particularly to minimize the possibility of personal injury or damage to the ecology); the selection and training of those who are to participate; the designation of, or restrictions relating to, the areas within which such activities should be carried out; and the establishment of practices and procedures which would be designed to minimize injuries or ecological damage.

To date there has been relatively little attention given to the human factors aspects of recreation and athletics, and we can here recap only a couple of the few investigations that have been carried out, to illustrate certain of the human factors implications of this area.

Ski Bindings

As one example let us consider various types of ski bindings as related to accident rates. Three types of ski bindings include: (1) a cable binding (which was developed many years ago); (2) a "two-release mode," a varient which retains the downward force at the heel but reduces or eliminates the amount of forward force on the toe; and (3) a "more than two-release mode" (> two-release mode), which allegedly takes into consideration the different nature of the forces encountered in normal skiing and in falls [Shealy et al., 23]. Without going into the specific features of these, it can be said that the incidence of accidents for these three types of bindings is different, and that there are accompanying differences in the nature of the injuries (upper extremities versus lower extremities) and in the skill of the skiers (beginners, intermediate, and advanced). These differences are shown in Figure 17-9, as based on data from

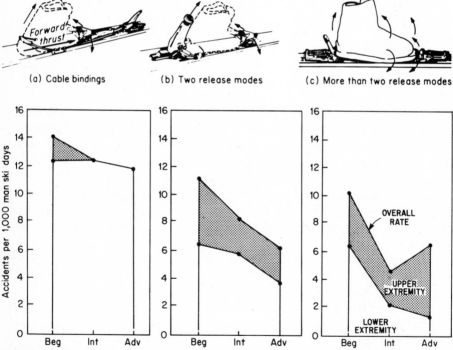

Figure 17-9 Comparison of incidence of injuries from skiing accidents with various types of ski bindings for beginners (Beg.), intermediates (Int.), and advanced (Adv.) skiers. [*Adapted from Shealy et al., 23, figs. 1 and 6.*]

Shealy et al. [23]. This shows differences in injury rates as related to experience. It is also apparent that the total number of injuries was influenced by the type of binding. Further, the type of injury was related to type of binding (the > two-release-mode bindings showing markedly fewer lower-extremity injuries than the other bindings, especially the cable bindings). (There are of course variations in many other types of sports equipment that also presumably are related to differential rates of injuries to participants.)

Snowmobiles

The increase in snowmobiling in recent years has taken on epidemic proportions in the northern states of the United States and in Canada and in certain other countries. The complaints about snowmobiles include the following, as reported by Rabideau [19]: injuries and fatalities (for example, 118 deaths in Canada in one year); some injuries that occur because of the G forces (up to 20 G) that result from "jumping over obstacles"; collision with other vehicles; noise (ranging up to 105 or 115 dB); damage to vegetation (especially destruction of small trees); damage to fences (some from intentional cutting); and damage to animal life.

In connection with injuries, for example, Rabideau reports that injuries to the spine are fairly common, these frequently occurring because of high-speed operation of snowmobiles over rough terrain. It was noted, for example [Roberts and Hubbard, 21], that vehicle landings after the negotiation of hard-packed snow ramps about 1 foot (30 cm) in height can result in the spine's receiving a short-period acceleration in excess of 20 G, a value which the U.S. Air Force considers a tolerance limit for acceleration on ejection from aircraft cockpits [Patrick and Mertz, 17]. Figure 17-10 shows a typical jumping maneuver that might result in such an injury.

Figure 17-10 Illustration of a snow-vehicle jumping maneuver common at speeds exceeding 20 mi/h on rough terrain. [*Courtesy of Kitchener-Waterloo Record, as presented by Rabideau, 19, fig. 4.*]

Figure 17-11   Posed view of how a snowmobile can adversely affect the ecology, by breaking down or otherwise damaging small trees. [*From Rabideau, 19, fig. 7.*]

As an illustration of the possible damage to the ecology, Figure 17-11 represents a posed view of a vehicle showing how a snowmobile body or track can break or otherwise damage small trees in wooded areas [Rabideau, 19, fig. 7].

Discussion

As indicated by Rabideau [18], the boom in recreation and sports has created millions of "novices," that is, individuals naïve with respect to the operating aspects of many kinds of equipment for camping, boating, motorized all-terrain travel, and a wide range of noncompetitive and competitive athletics. In many instances this has left the consumers or "users" of relevant equipment to their own devices—often trial and error—with respect to selection and purchase of equipment, acquisition of skill in its use, and even the development of procedures for operational utilization. Clearly, this whole area opens up a broad, new field for investigation and application by the human factors disciplines.

THE LIVING ENVIRONMENT: COMMENTS

The examples of research investigations relating to the living environment discussed in this and the previous chapter clearly represent only a modest portion of relevant human factors research relating to the living environment. But even so, this sample clearly represents a mixed bag. Human factors research in this area is admittedly limited, but that which has been carried out is very spotty, in that it deals with only limited aspects of the total possible range. Granting this, it has been the primary intent of these two chapters to focus attention on the wide range of human factors implications of the total living environment, on the premise that this area of research and application represents what it is hoped will be an area of increasing attention on the part of the human factors disciplines.

REFERENCES

1 Bauer, H. J.: *Public transportation and human factors engineering*, Research Laboratories, General Motors Corporation, Research Publication GMR-9982, Warren, Mich., Apr. 3, 1970.

2 Canty, E. T., and A. J. Sobey: *Case studies of seven new systems of urban transportation*, Research Laboratories, General Motors Corporation, Research Publication GMR-845, Warren, Mich., Jan. 13–17, 1969.

3 Clark, G. E., and H. G. Ludwig: Police patrol vehicles, *Human Factors*, 1970, vol. 12, no. 1, pp. 69–74.

4 Ferguson, D.: Ergonomics and the quality of living, *Applied Ergonomics*, vol. 3, no. 2, pp. 70–74.

5 Golob, T. F., E. T. Canty, R. L. Gustafson, and J. E. Vitt: Research Laboratories, General Motors Corporation, Research Publication GMR-1037, Warren, Mich., Oct. 26, 1970.

6 Gustafson, R. L., H. V. Curd, and T. F. Golob: *User preferences for a demand-responsive transportation system: A case study report*, General Motors Corporation, Research Laboratories, Research Publication GMR-1047, Warren, Mich., January, 1971.

7 Gutman, R.: *People and buildings*, The Macmillan Company, New York, 1972.

8 Haddon, W., Jr.: Energy damage and the ten countermeasure strategies, *Human Factors*, 1973, vol. 15, no. 4, pp. 335–366.

9 Harper, W. R., and D. H. Harris: The application of link analysis to police intelligence, *Human Factors*, 1975, vol. 17, no. 2, pp. 157–164.

10 Hoag, L. A., and S. K. Adams: Human factors in urban transportation systems, *Human Factors*, 1975, vol. 17, no. 2, pp. 119–131.

11 Ittleson, W. H.: "Environmental psychology of the psychiatric ward," in C. W. Taylor, R. Bailey, and C. H. W. Branch (eds.), *The second national conference on architectural psychology, May 26–28, 1966, Park City, Utah*, University of Utah, Salt Lake City, September, 1967.

12 Lindstrom, S., and S. Gunnarson: *Road safety for children in two residential areas*, Proceedings of the 1964 International Road Safety Congress, 1964.

13 Lipman, A.: Building design and social interaction, *The Architects' Journal*, Jan. 3, 1968, SfB (94):Aa3:UDC 725.56:301, pp. 23–30.

14 Millar, A. E. (ed.): *The human commotion, human factors in transportation* NASA-ASEE Report, Contract NGT 47-003-028, Langley Research Center and Old Dominion Research Foundation, 1972.

15 Neutra, R., and R. A. McFarland: Accident epidemiology and the design of the residential environment, *Human Factors*, 1972, vol. 14, no. 5, pp. 405–420.

16 Newman, O.: *Defensible space*, The Macmillan Company, New York, 1972.

17   Patrick, L. M., and R. J. Mertz: "Human tolerance to impact," in D. F. Heulke (ed.), *Human anatomy, impact injuries, and human tolerances*, No. 700195, Society of Automotive Engineers, Inc., 1970, p. 99.

18   Rabideau, G. F.: Overview (of special issue in athletics and recreational systems), *Human Factors*, 1974, vol. 16, no. 5, pp. 445–446.

19   Rabideau, G. F.: Human, machine, and environment aspects of snowmobile design and utilization, *Human Factors*, 1974, vol. 16, no. 5, pp. 481–494.

20   Rappaport, M.: Human factors applications in medicine, *Human Factors*, 1970, vol. 12, no. 1, pp. 25–35.

21   Roberts, V. L., and R. P. Hubbard: *Biomechanics and snowmobile injuries*, Highway Research Institute, University of Michigan, Ann Arbor, 1972.

22   Ronco, P. G.: Human factors applied to hospital patient care, *Human Factors*, 1972, vol. 14, no. 5, pp. 461–470.

23   Shealy, J. E., L. H. Geyer, and R. Hayden: Epidemiology of ski injuries: Effects of skill acquisition and release-binding accident rates, *Human Factors*, 1974, vol. 16, no. 5, pp. 459–473.

24   Skidmore, R. A.: Experimental system for control of surgically induced infections, *Human Factors*, vol. 17, no. 2, pp. 132–138.

25   Taylor, C. W.: *Architectural psychology: a pioneering program*, mimeographed paper, Department of Psychology, University of Utah, Salt Lake City, 1968.

26   Voorhees, A. M., and R. A. Keith: Urban transportation, *Consulting Engineer*, March, 1969, vol. 32, no. 3, pp. 130–136.

27   Whitehead, B., and M. Z. Eldars: An approach to the optimum layout of single-storey buildings, *The Architects' Journal*, June 17, 1964, SfB:Ba4:UDC 721.011, pp. 1373–1380.

28   *Research on road safety*, Road Research Laboratory, Her Majesty's Stationery Office, 1963.

29   *Travel barriers*, PB187-237, U.S. Department of Transportation, National Information Service, 1970.

Part
Six

Overview

Chapter 18

Application of Human Factors Data

Each of the previous chapters of this book has dealt with some particular aspect of human factors as related to the design of the things people use and of the environments in which people work and live. The intent of these discussions has been to demonstrate the importance of taking human factors into account in the design processes and to illustrate some of the research carried out in the various areas that could be applied in these processes. In this chapter we will take an overview of the application of human factors data to the design process.[1]

THE NATURE AND USE OF HUMAN FACTORS DATA

The human factors data that are available are based on varying combinations of research results, experience, and expert judgments. (Such data can be found in various types of sources, such as some of the references given in this book, especially those in Appendix C. Some illustrative data are given in this book.)

[1]For more thorough treatment the reader is referred to Meister [13] and Meister and Rabideau [14].

Types of Human Factors Data

The data relevant to various human factors design problems cover a wide span (with many gaps yet to be filled) and exist in many forms, such as the following:

- Common sense and experience (such as the designer has in his "storage," some of which may be valid, and some not)
- Comparative quantitative data (such as relative accuracy in reading two types of visual instruments)
- Sets of quantitative data (such as anthropometric measures of samples of people and error rates in performing various tasks)
- Principles (based on substantial experience and research, that provide guidelines for design, such as the principle of avoiding or minimizing glare when possible)
- Mathematical functions and equations (that describe certain basic relationships with human performance, such as in certain types of simulation models)
- Graphic representations (nomographs or other representations, such as tolerance to acceleration of various intensities and durations)
- Judgment of experts
- Design criteria (consisting of specifications for the many specific areas of application, such as displays, controls, work areas, escape hatches, and illumination)

Much of the research in human factors is directed toward the development of design criteria which can be used directly in the design of relevant items. Such design specifications sometimes are incorporated into the form of checklists; in the case of the military services, they are incorporated into military specifications called MIL-SPECS.

Evaluation of Human Factors Data

The application of relevant human factors data to some design problems is a fairly straightforward proposition and in some circumstances can even be done by computers. More generally, however, it is necessary to exercise judgment in this process—in particular, in evaluating the applicability to specific design problems of whatever data seem to have some potential relevance. In making such evaluations there are at least four considerations that should be taken into account.

Practical Significance   One of these deals with the practical significance of the application of some relevant human factors information. For example, although the time required to use control device A might be significantly less than for device B, the difference might be so slight, and of such nominal utility, that it would not be worthwhile to use device A, especially if other factors (such as cost) argued against it.

Extrapolation to Different Situations   Much of the available human factors data has been based on research findings or experience in certain specific settings. The second consideration in the evaluation of human factors data deals with possible extrapolation of such data to other settings. For example, could one assume that the reaction time of astronauts in a space capsule would be the same as in an earthbound environment? Referring specifically to laboratory studies, Chapanis [4] urges extreme caution in applying the results of such studies to real-world problems.

Perhaps three points might be made about these sobering reflections. In the first place, despite Chapanis's words of caution, there are some design problems for which available research data or experience are fully adequate to resolve the design problem. In the second place, in certain areas of investigation it may be possible to "simulate" the real world with sufficient fidelity to derive research findings that can be used with reasonable confidence. In the third place, there are undoubtedly circumstances in which one might extrapolate from data that are not uniquely appropriate to the design problem at hand, on the expectation that this would increase the odds of achieving a better design than if such data were not used. Clearly this strategy involves risks, and should be used only in the case of noncritical design problems.

Consideration of Risks   A third consideration in applying human factors data is associated with the seriousness of the risks in making bad guesses. Bad guesses in designing a manned capsule for shipping an astronaut to the moon obviously would be a matter of greater concern than those in designing a hat rack.

Consideration of Trade-off Function   Still a fourth consideration is trade-off values. Since it frequently is not possible to achieve an optimum in all possible criteria for a given design of a man-machine system, some give-and-take must be accepted. The possible payoff of one feature (as suggested by the results of research) may have to be sacrificed, at least in part, for some other more desirable payoff.

HUMAN FACTORS IN DESIGN PROCESSES

In the consideration of human factors in the design of equipment, facilities, and other physical items that people use, there are certain basic stages or processes that typically have to be carried out. If the item or system is very complex (such as a new aircraft or a petroleum refinery), these processes usually are highly organized and elaborate; whereas in the design of very simple items (such as a new type of egg beater or hedge trimmer), these processes may be very informal, and in some instances certain stages may be completely irrelevant.

A representation of certain of the human-factors-related functions in system design is given in Figure 18-1. This representation would apply in the case of a complex system, such as certain military systems. In the development of certain systems (especially in the military services) the human factors functions

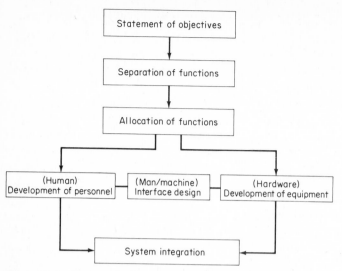

Figure 18-1 Representation of certain human-factors-related functions involved in systems design. [*From Applied ergonomics handbook, 15, fig. 15.5, p. 156.*]

extend into various personnel-related areas such as job design and task analysis, job specifications, selection criteria, training, and manpower planning.

We will touch on certain of the human-factors-related functions involved in design processes, recognizing that all of these would not apply on an across-the-board basis.

Specification of System Functions

Let us begin by assuming that the objectives and performance requirements of the system (or whatever is being designed) have been spelled out. In the case of at least certain systems (expecially complex ones), the fulfillment of the objectives will require that certain basic functions be performed. For example, a postal system requires mail collection, sorting by general area, transportation, sorting into local-area categories, and delivery. This function analysis should be concerned with *what* functions need to be performed to fulfill the objectives, and not with the *way* in which they are to be performed.

Allocation of Functions

Given functions that have to be performed, in some instances there may be an option as to whether any particular function should be allocated to a human being or to some physical (machine) component(s). In this process, the allocation of certain functions to human beings, and of others to physical components, is virtually predetermined by certain manifest considerations, such as obvious superiority of one over the other or economic considerations. Between these two extremes, however, may be a range of functions that are within the reasonable repertoire of both human beings and physical components. In this allocation process, however, we should keep in mind the qualms expressed by

Jones [9] that there are circumstances in which the notion of a one-to-one correspondence between functions and components may be severely strained. In such instances one might view the functions (on one side of the fence) and the components (on the other side) in more of a "collective" frame of reference than a one-to-one relationship.

Because of the importance of these decisions for some systems, one would hope that there would be available some guidelines to aid the system designer in allocating specific functions to human beings versus physical machine components. The most common types of guidelines that heretofore have been proposed consist of general statements about the kinds of things human beings can do better than machines and vice versa. As Chapanis [3] points out, however, such comparisons serve a useful function in only the most elementary kind of way. He points out that such generalizations are useful primarily in directing one's thinking toward man-machine problems and in reminding one of some of the general characteristics that men and machines have as system components. Aside from the potential practical utility of such comparisons for use in the allocation process, however, Jordan [10] and others raise the more basic question of the appropriate role of human beings in systems, especially as we continue the trend toward automation. Before discussing such views, however, let us present a set of such generalizations, first to demonstrate what such "lists" are like, and second to serve what Chapanis refers to as the "elementary" purposes for which they may have some utility.

Relative Capabilities of Human Beings and of Machines   The following generalizations about the relative capabilities of human beings and of machine components are drawn from various sources [Chapanis, 2, p. 543; Fitts, 6 and 7; Meister, 12; Meister and Rabideau, 15; and others], plus some additional items and variations on previously expressed themes.

Humans are generally *better* in their abilities to:

• Sense very low levels of certain kinds of stimuli: visual, auditory, tactual, olfactory, and taste.
• Detect stimuli against high-"noise"-level background, such as blips on cathode-ray-tube (CRT) displays with poor reception.
• Recognize patterns of complex stimuli which may vary from situation to situation, such as objects in aerial photographs and speech sounds.
• Sense unusual and unexpected events in the environment.
• Store (remember) large amounts of information over long periods of time (better for remembering principles and strategies than masses of detailed information).
• Retrieve pertinent information from storage (recall), frequently retrieving many related items of information; but reliability of recall is low.
• Draw upon varied experience in making decisions; adapt decisions to situational requirements; act in emergencies. (Does not require previous "programming" for all situations.)
• Select alternative modes of operation, if certain modes fail.
• Reason inductively, generalizing from observations.

- Apply principles to solutions of varied problems.
- Make subjective estimates and evaluations.
- Develop entirely new solutions.
- Concentrate on most important activities, when overload conditions require.
- Adapt physical response (within reason) to variations in operational requirements.

Machines generally are *better* in their abilities to:

- Sense stimuli that are outside man's normal range of sensitivity, such as x-rays, radar wavelengths, and ultrasonic vibrations.
- Apply deductive reasoning, such as recognizing stimuli as belonging to a general class (but the characteristics of the class need to be specified).
- Monitor for prespecified events, especially when infrequent (but machines cannot improvise in case of unanticipated types of events).
- Store coded information quickly and in substantial quantity (for example, large sets of numerical values can be stored very quickly).
- Retrieve coded information quickly and accurately when specifically requested (although specific instructions need to be provided on the type of information that is to be recalled).
- Process quantitative information following specified programs.
- Make rapid and consistent responses to input signals.
- Perform repetitive activities reliably.
- Exert considerable physical force in a highly controlled manner.
- Maintain performance over extended periods of time (machines typically do not "fatigue" as rapidly as humans).
- Count or measure physical quantities.
- Perform several programmed activities simultaneously.
- Maintain efficient operations under conditions of heavy load (men have relatively limited channel capacity).
- Maintain efficient operations under distractions.

In discussing such "lists" of relative advantages of men and machines, Jordan [10] boils the assortment down to a nub, as follows: "Men are flexible but cannot be depended upon to perform in a consistent manner, whereas machines can be depended upon to perform consistently but have no flexibility whatsoever."

Limitations of Man-Machine Comparisons   The previously implied limitations regarding the practical utility of general comparisons of human and machine capabilities stem from various factors. Some of these have been pointed up by Chapanis [3][2] and Corkindale [5] and are discussed in the following (with a fair portion of editorial license):

[2]Discussed by permission of the editor of *Occupational Psychology* (1965, vol. 39, no. 1, pp. 1–11), quarterly journal of the National Institute of Industrial Psychology, 14 Welbeck Street, London W1M 8DR.

1   General man-machine comparisons are not always applicable. Given some general superiority of men or machines, there are circumstances in which it would be inappropriate to apply the dictates of that generality. For example, the amazing computational abilities of computers do not imply that one should use a computer whenever computations are required.

2   Lack of adequate data on which to base function allocations.

3   Relative comparisons are subject to continual change. For example, machines are not very effective in pattern recognition, but this may change in the years to come.

4   It is not always important to provide for the "best" performance. For example, although human beings who serve as toll collectors on superhighways offer some advantage over mechanical collectors, the mechanized devices do the job well enough to be acceptable.

5   Function performance is not the only criterion. One also has to consider the trade-off values of other criteria such as availability, cost, weight, power, reliability, and cost of maintenance. For example, although remote-control devices for the family TV and garage are available, their cost probably has limited their widespread acceptance. At the present stage of affairs, there are very few systematic guidelines to follow in figuring out the relative trade-off values of various criteria.

6   Function allocation should take social and related values into account. The process of allocation of functions to men versus machine components directly predetermines the role of human beings in systems and thereby raises important questions of a social, cultural, economic, and even political nature. The basic roles of human beings in the production of the goods and services of the economy have a direct bearing upon such factors as job satisfaction, human motivation, and the value systems of individuals and of the culture. Since our culture places a premium on certain human values, the system should not require human work activities that are incompatible with such values. In this vein, Jordan [10] postulates the premise that men and machines should not be considered comparable, but rather should be considered complementary. Whether one would agree with his conclusion that the allocation concept becomes entirely meaningless, the fact does remain that decisions (if not allocations) need to be made concerning the relative roles of men and machines. In this context, although the objectives of most systems are not the entertainment of the operators (pinball machines and gambling devices excepted), it would seem, for example, that, within reason, the human work activities that are generated by a system preferably should provide the opportunity for reasonable intrinsic satisfaction to those who perform them.

A Strategy for Allocating Functions   The discussion above implies that there are no sets of clear-cut guidelines available for use in deciding what system functions should be performed by men and by machines. Rather, one needs to pursue a general strategy, bringing to bear at various phases the most adequate data that are available and exercising the best (well-informed) judgment possible. As proposed by Chapanis [3], this strategy should *not* be directed toward the allocation of functions as though each function were in vacuum-packed isolation from the others. Rather, the strategy should be directed toward making decisions about functions in such a manner as to enhance the operation of the system as a whole, and toward the creation of jobs that are interesting, motivating, and challenging to the human operator.

Design of Physical Components

Although the actual design of the physical components is dominantly an engineering chore, this phase represents a second stage at which human factors inputs usually are of considerable moment. The specific nature of the design decisions made during this phase can (if inappropriate in terms of human considerations) forever plague the user and cause decrement in system performance or, conversely (if appropriately designed), facilitate the user's use of the equipment and bring about better system performance. In a sense, the design of physical components embraces two phases, the development of a design *concept* (i.e., the basic conceptual formulation of the component) and the composition of design *detail* [Meister, 13, p. 273], the latter being an elaboration of the first.

The human factors design phase consists of tracking down sources of human factors data that are potentially relevant to the design problem and evaluating their possible applicability, taking into account the various considerations discussed earlier in this chapter. Such sources include various sets of design criteria, such as specifications.

Although the primary responsibility for this function lies with the designer (such as the engineer, the architect, and industrial designer), Meister [13, chapter VII] reports, on the basis of a few studies, that (unfortunately) design engineers typically do *not* consider human factors in their design procedures. He therefore urges that human factors specialists be brought into this process to give greater assurance that the design will be more compatible with human factors considerations.

Job Design

The design of equipment and other physical facilities which people use predetermines to some degree the nature of the jobs they perform.

Human Values in Job Design   As discussed above in the frame of reference of function allocation, there are certain philosophical considerations in the design of jobs. In this regard there has been considerable interest in the notion of job enlargement and job enrichment, this concern being based on the assumption that enlarged or enriched jobs generally bring about higher levels of job satisfaction. There are a number of variations on this theme, such as increasing the number of activities to be performed, giving the worker responsibility for inspection of the employee's own work, delegating responsibility for a complete unit (rather than for a specific part), providing opportunity to the worker to select the work methods to be used, job rotation, and placing greater responsibility on work groups for production processes. The primary objective of this approach focuses on the criterion of job satisfaction, with secondary emphasis on work efficiency or productivity.

On the other hand, the human factors approach has placed primary emphasis on efficiency or productivity, and it has then tended toward the creation of jobs that are more specialized and that require less skill than jobs that are enlarged. In discussing these somewhat disparate objectives, McCormick [11] emphasizes the fact of individual differences in value systems, indicating that some people do not like "enlarged" jobs.

It should be noted that the degree of possible conflict between these two approaches is not yet clear, and that there are as yet no clear-cut guidelines for designing jobs that would be compatible with the dual objectives of work efficiency and providing the opportunity for job satisfaction on the part of workers. Although we cannot here come forth with any pat solutions to this dilemma, it is timely to bring this issue to the attention of the human factors clan, with the hope that their efforts can be directed toward some merging of the objectives of the human factors and the job enlargement approaches.

Task Analysis   In the development of some systems, especially in the military services, some form of task analysis is carried out, leading toward two objectives. In the first place, task analysis as carried out during the design phase is intended to contribute to further improvement in the design. Such task analysis, being based on inferences from tentative designs about what the users of the equipment would have to do, can turn up design features that are incompatible with human characteristics of abilities. The second purpose of task analysis is to have a description of each job as it would in fact exist with the "final" design, such descriptions to be used for various purposes, such as the development of personnel specifications, manpower requirements, and training programs. Although there are variations in the methods and specific information provided by different task analysis approaches, a basic process is common to all, namely, that of inferring, from the design of the equipment, what human tasks will be required when the system is completed; this is a process of *predicting* the nature of future tasks. Some such inferences are quite evident; others are less so. It should also be added, however, that some tasks are not necessarily implied by the equipment characteristics, since some tasks come into being by reason of other considerations, such as decisions regarding safety practices and philosophies regarding maintenance procedures.

An example of a task analysis scheme is given below. This scheme provides for the following task-related information, using a form with these items specified in the headings [Gael and Stackfleth, 8]:

Task number
Task name
Indicator (usually some type of display)
Response (behavior)
Feedback
Incidence

Time
Frequency
Performance Criteria
Time
Accuracy
Location of task
Criticality
Newness

Personnel Subsystem

We can see that, for at least some systems, one has to start worrying about *personnel* affairs during most of the design and development process, such affairs including task and function analysis, job design, operating procedures, personnel selection, training, the development of training aids, and manpower planning. Undoubtedly the most comprehensive program of this type is that of the U.S. Air Force [16], which is referred to as the *personnel subsystem* (PSS). (In fact, the term is also being used somewhat outside the Air Force.) As developed by the Air Force, this program embodies a systematic set of procedures and guidelines to be followed during the development of at least major systems, for which trained personnel must be available at the time the final system is produced. Although many (perhaps most) systems do not require an elaborate personnel subsystem in this sense, there may be somewhat corresponding functions that require attention during system development for the benefit of the ultimate user of the system. Such functions might include actual training of people (such as mechanics for a new model of automobile), and more likely the preparation of instructional manuals for self-training of users (such as how the householder can sharpen the blade of his new lawnmower).

Human Factors Evaluation

Evaluation in the context of system development has been defined as the measurement of system-development products (hardware, procedures, and personnel) to verify that they will do what they are supposed to do [Meister and Rabideau, 14, p. 13], and, in turn, human factors evaluation is the examination of these products to ensure the adequacy of attributes that have implications for human performance. Actually, almost every decision during the design of a system includes some evaluation, such as deciding whether to use a visual signal or an auditory signal. Although many such evaluative decisions need to be made as part of the ongoing development cycle, for most systems certain systematic procedures should be followed to evaluate the system-development products (hardware, procedures, and personnel). Our interests are, of course, focused on the human factors evaluation. We shall mention at least a few such evaluation processes.[3]

---

[3]For a thorough treatment of human factors evaluation in system development, the reader is referred to Meister and Rabideau [14].

Experimental Procedures The testing of a system or component (either actual or simulated) is essentially an experiment and therefore requires the use of procedures in which appropriate experimental practices should be followed. While it is not feasible here to describe or discuss such practices, various texts deal with this topic.[4] While conventional experiments involve the manipulation of experimental conditions (and while some tests also involve such manipulation), in other test situations there is no variation in such conditions; in such cases the purpose may be to evaluate performance under a single specified condition, usually to compare with some predetermined performance requirement standard. It should be pointed out, however, that in any event some measure of performance (a criterion) is essential.

Test Conditions The conditions under which the evaluation test is carried out should usually simulate as closely as possible those in which the system is ultimately to be used. These include physical environment, procedures, numbers of personnel, and, where possible, even conditions of stress.

Subjects The subjects used in the tests should be the same types of individuals that are expected to use the system, taking into account considerations of aptitude and training.

Adequate Number of Repetitions As in any experiment, repeated observations, or trials, are required in order to give a reliable indication of performance. Where feasible, there should be enough replications to provide the basis for appropriate tests of statistical significance.

HUMAN FACTORS DESIGN CONSIDERATIONS

It is probably fairly obvious that the application of human factors data to design processes does not (at least yet) lend itself to the formulation of a completely routine, objective set of procedures and solutions. However, a systematic consideration of the human factors aspects of a system usually would at least focus attention on features that should be designed with human beings in mind. In this connection, therefore, it might be useful to characterize at least some of the considerations that are in order in approaching a design problem. These considerations will be presented in the form of a series of questions (with occasional supplementary comments). Three points should be made about the implications of these considerations. In the first place, some of these would not be pertinent in some systems; in turn, this is not intended as an all-inclusive assortment of considerations. In the second place, the design of any given item or facility should above all be focused on the ultimate users— whether this be the public or some segment of the population. The diverse segments of the population (in terms of age, physical condition, sex, etc.) can

---

[4]A particularly useful source is Chapanis [1]. There are also several experimental psychology tests available.

differ markedly on virtually every human attribute—sensory and perceptual skills, cognitive abilities, physical dimensions, psychomotor skills, communication skills, etc.—each of which has distinct human factors implications. Thus, the questions given below should be viewed as they would be relevant to the ultimate user population. In the third place (and as indicated frequently before), the fulfillment of one objective may of necessity be at the cost of another.

In a general sense these questions may serve as "reminders" of some of the human factors considerations that should be taken into account in the design process.

1 What are the functions that need to be carried out to fulfill the system objective?

2 If there are any reasonable options available, which of these should be performed by human beings?

3 For a given function, what information external to the individual is required? Of such information, what information can be adequately received directly from the environment, and what information should be presented through the use of displays?

4 For information to be presented by displays, what sensory modality should be used? Consideration should be given to the relative advantages and disadvantages of the various sensory modalities for receiving the type of information in question.

5 For any given type of information, what type of display should be used? The display generally should provide the information when and where it is needed. These considerations may take into account the general type of display, the stimulus dimension and codes to be used, and the specific features of the display. The display should provide for adequate sensory discrimination of the minimum differences that are required.

6 Are the various visual displays arranged for optimum use?

7 Are the information inputs collectively within the reasonable bounds of human information-receiving capacities?

8 Do the various information sources avoid excessive time-sharing?

9 Are the decision-making and adaptive abilities of human beings appropriately utilized?

10 Are the decisions to be made at any given time within the reasonable capability limits of human beings?

11 Granting that aspects of some systems will be automated, is the basic *control* of the system that of the individual?

12 When physical control is to be exercised by an individual, what type of control device should be used?

13 Is each control device easily identifiable?

14 Is the operation of each control device compatible with any corresponding display, and with common human response tendencies?

15 Are the operational requirements of any given control (as well as of the controls generally) within reasonable bounds? The requirements for force, speed, precision, etc., should be within limits of virtually all persons who are to use the system. The man-machine dynamics should so capitalize on human abilities that, in operation, the devices meet the specified system requirements.

16 Are the control devices arranged conveniently and for reasonably optimum use?

17 If there is a communication network, will the communication flow avoid overburdening the individuals involved?

18 Are the various tasks to be done grouped appropriately into *jobs*?

19 Do the tasks which require time-sharing avoid overburdening any individual or the system? Particular attention needs to be given to the possibility of overburdening in emergencies.

20 Is there provision for adequate redundancy in the system, especially of critical functions? Redundancy can be provided in the form of backup or parallel components (either men or machines).

21 Are the jobs of such a nature that the personnel to perform them can be trained to do them?

22 If so, is the training period expected to be within reasonable time limits?

23 Do the work aids and training complement each other?

24 If training simulators are used, do they achieve a reasonable balance between transfer of training and costs?

25 Is the work space suitable for use by the range of individuals who will use the facility?

26 Are the environmental conditions (temperature, illumination, noise, etc.) such that they permit satisfactory levels of human performance and provide for the physical well-being of individuals?

27 In any evaluation or test of the system (or components) does the system performance meet the desired performance requirements?

28 Does the system in its entirety provide reasonable opportunity for the individual(s) involved to experience some form and degree of self-fulfillment and to fulfill some of the human values that we should all like to have the opportunity to fulfill in our daily lives?

29 Does the system in its entirety contribute generally to the fulfillment of reasonable human values? In the case of systems with indentifiable outputs of goods and services, this consideration would apply to those goods and services. In the case of systems that relate to our life space and everyday living, this consideration would apply to the potential fulfillment of those human values that are within the reasonable bounds of our civilization.

In the resolution of these and other kinds of human factors considerations, one should draw upon whatever relevant information is available. This information can be of different types, including principles that have been developed through experience or research, sets of normative data (such as frequency distributions of, say, body size), sets of factual data of a probability nature (such as percentage of signals that are detected under specified conditions), mathematical formulas, tentative theories of behavior, hypotheses that have been suggested by research investigations, and even the general knowledge acquired through everyday experience.

With respect to information that would have to be generated through research (as opposed to experience), while there is very comprehensive information available in certain areas of knowledge, in others it is pretty skimpy, and there are some areas in which one draws virtually a complete blank. Where adequate information is not available, the opinion is expressed that considered judgments based on partial information will, in the long run, result in better design decisions than those that are pulled out of a hat. But let us reinforce the point that such judgments usually should be made by those whose professional training and experience put them in the class of *experts*, whether in the field of night vision, physical anthropometry, hearing disorders, perception, heat stress, acceleration, learning, decision making, or otherwise.

We should here, again, mention the almost inevitability of having to trade off certain advantages for others. The balancing out of advantages and disadvantages generally needs to take into account various types of considerations—engineering feasibility, human considerations, economic considerations, and others. Granting that there probably are few guidelines to follow, nonetheless the general objective of this horse trading is fairly clear. This basically goes back to the stated or implied system objectives and the accompanying performance requirements. In other words, any trade-offs should be made on the basis of the considerations of their relative effects in terms of system objectives.

THE CLOUDED CRYSTAL BALL

To date, application of human factors principles and data has been concentrated on the design of military equipment, aircraft, electronics equipment, communication equipment, and some items of transportation equipment, although there has been some attention to human factors in other industries. But what of the future?

In reflecting about this, a distinction should be made between the *potential* areas of application and the areas in which applications *actually* will take place. The potential appears to embrace a wide spectrum, including some areas in which some research and/or application has been carried out. In these, the common denominator is the fact that there are obvious human factors aspects associated with them, in terms of more effective use or in terms of some aspect of human welfare (physical welfare, satisfaction, social values, or what one might refer to as the "quality of life").

Let us touch on some such areas, realizing that this is not intended as a complete inventory. Certainly the whole spectrum of the living environment discussed in the last two chapters would be high on the list. In addition we could suggest some of the following: many types of consumer products, such as household appliances, toys, and recreational and sports equipment; vehicles of various kinds, including what possibly will be new forms of transportation systems; industrial processes of many kinds; undersea operations involving recovery of metals and other materials; the raising of seafood in the oceans; agricultural operations; mining; and energy production.

In effect, we would like to propose that virtually any kind of equipment, facility, gadget, "thing," process, service, or environment that is created or influenced by man—presently in existence or to be created—be viewed as a prime candidate for human factors attention. Such systematic concern should of course be directed toward the identification of any existing or potential human factors problems, with the objective of designing the product or conditions for more effective human use and the improvement of the quality of life.

A FINAL WORD

It has not been the presumptuous intent in this volume to compile and organize a tremendous amount of available pertinent information relating to human attributes and behavior. Rather, the intent has focused around certain more modest objectives. These include the presentation of at least some information about certain human characteristics (such as sensory and motor processes) that might contribute to greater understanding of human performance and behavior.

In addition, it has been the intent to illustrate, by example, some of the research that has been carried out that reflects the implications of design features for the effectiveness with which people use the man-made facilities of our civilization and for the quality of life. Although the examples discussed are intended to be only illustrative, in many instances they deal with relationships that may have fairly general utility. It has also been the intent to present discussions of at least some techniques and methods, such as those relating to the conduct of human factors studies, and to discuss some aspects of the application of human factors data and principles to practical problems.

But perhaps the more important underlying objective has been to develop an increased sensitivity to, or awareness of, the many human aspects of the systems and situations that abound in our current civilization. Such awareness is at least the first prerequisite to the subsequent processes of creating those systems and situations in which human talents can be most effectively utilized in the furtherance of human welfare.

REFERENCES

1 Chapanis, A.: *Research techniques in human engineering.* The Johns Hopkins Press, Baltimore, 1959.
2 Chapanis, A.: "Human engineering," in C. D. Flagle, W. H. Huggins, and R. H. Roy (eds.), *Operations research and systems engineering,* chap. 19, pp. 534–582, The John Hopkins Press, Baltimore, 1960.
3 Chapanis, A.: On the allocation of functions between men and machines, *Occupational Psychology,* 1965, vol. 39, pp. 1–11.

4 Chapanis, A.: The relevance of laboratory studies to practical situations, *Ergonomics*, 1967, vol. 10, no. 5, pp. 557–577.

5 Corkindale, K. G.: Man-machine allocation in military systems, *Ergonomics*, 1967, vol. 10, no. 2, pp. 161–166.

6 Fitts, P. M. (ed.): *Human engineering for an effective air-navigation and traffic-control system*, NRC, Washington, D. C., 1951.

7 Fitts, P. M.: Functions of men in complex systems, *Aerospace Engineering*, 1962, vol. 21, no. 1, pp. 34–39.

8 Gael, S., and E. D. Stackfleth: *A data reduction technique applied to the development of qualitative personnel requirements information (QPRI)—Keysort card system*, USAF, WADD, Technical Note 60–133, 1960.

9 Jones, J. C.: The designing of man-machine systems, *Ergonomics*, 1967, vol. 10, no. 2, pp. 101–111.

10 Jordan, N.: Allocation of functions between man and machines in automated systems, *Journal of Applied Psychology*, 1963, vol. 47, no. 3, pp. 161–165.

11 McCormick, E. J.: "A human factors dilemma: Specialized or enlarged jobs?" Invited address, Human Factors Society Annual Meeting, Huntsville, Ala., Oct. 16, 1974.

12 Meister, D.: Methods of predicting human reliability in man-machine systems, *Human Factors*, 1964, vol. 6, no. 6, pp. 621–646.

13 Meister, D.: Human factors: theory and practice, John Wiley & Sons, Inc., New York, 1971.

14 Meister, D., and G. F. Rabideau: *Human factors evaluation in system development*, John Wiley & Sons, Inc., New York, 1965.

15 *Applied Ergonomics Handbook*, Part 1, "A first introduction"; chap. 15, "Systems design," *Applied Ergonomics*, 1971, vol. 2, no. 3, pp. 150–158.

16 *Personnel subsystems*, USAF, AFSC design handbook, ser. 1–0, General, AFSC DH 1–3, 1st ed., Jan. 1, 1969, Headquarters, AFSC.

List of Abbreviations

AFB	Air Force Base
AFHRL	Air Force Human Resources Laboratory
AFSC	Air Force Systems Command
AMD	Aerospace Medical Division, Air Force Systems Command
AMRL (of USAF)	Aerospace Medical Research Laboratory, Aerospace Medical Division, Air Force Systems Command
ANSI	American National Standards Institute
APA	American Psychological Association
ASD	Aeronautical Systems Division, Air Force Systems Command
ASHA	American Speech and Hearing Association
ASHRAE	American Society of Heating, Refrigerating, and Air-Conditioning Engineers
ASHVE	American Society of Heating and Ventilating Engineers (now ASHRAE)
ASME	American Society of Mechanical Engineers
FAA	Federal Aviation Administration
HFRB	Human Factors Research Branch, Adjutant Generals Research and Development Command, U.S. Army
IEEE	Institute of Electrical and Electronics Engineers, Inc.
IES	Illuminating Engineering Society

IRE	Institute of Radio Engineers
ISO	International Organization for Standardization
JSAS	Journal Supplement Abstract Service (of APA)
MRL (of USAF)	Medical Research Laboratory (see AMRL)
NAS	National Academy of Sciences
NASA	National Aeronautics and Space Administration
NAVTRADEVCEN	United States Naval Training Device Center (formerly Special Devices Center)
NEL	Navy Electronics Laboratory (USN)
NRC	National Research Council
NRL	Naval Research Laboratory (USN)
ONR	Office of Naval Research (USN)
OSHA	Occupational Safety and Health Administration
SAE	Society of Automotive Engineers
SDC	Special Devices Center (USN) (See NAVTRADEVCEN)
TDR	Technical Documentary Report (AMRL)
TR	Technical Report (term used by various organizations)
USA	United States Army
USAF	United States Air Force
USASI	United States of America Standards Institute (now ANSI)
USN	United States Navy
USPHS	United States Public Health Service
WADC	Wright Air Development Center, USAF (see AMRL and AFHRL)
WADD	Wright Air Development Division, USAF (see AMRL and AFHRL)

Appendix B

Control Devices

Table B-1 presents a brief evaluation of the operational characteristics of certain types of control devices.[1] Table B-2 presents a summary of recommendations regarding certain features of these types of control devices.[2] In the use of these and other recommendations, it should be kept in mind that the unique situation in which a control device is to be used and the purposes for which it is to be used can affect materially the appropriateness of a given type of control and can justify (or virtually require) variations from a set of general recommendations or from general practice based on research or experience. For further information regarding these, refer to the original sources given in the reports from which these are drawn.

COMMENTS REGARDING CONTROLS[3]

- *Hand push button:* Surface concave, or provide friction. Preferably audible click when activated. Elastic resistance plus slight sliding friction, starting low, building up rapidly, sudden drop. Minimize viscous damping and inertial resistance.
- *Foot push button:* Use elastic resistance, aided by static friction, to support

---

[1]Adapted largely from A. Chapanis, "Design of Controls," chap. 8 in H. P. Van Cott and R. G. Kinkade, *Human engineering guide to equipment design*, rev. ed., U.S. Government Printing Office, Washington, D. C., 1972.

[2]Adapted from ibid.

[3]Adapted from ibid.

Table B-1 Comparison of the Characteristics of Common Controls

Characteristic	Hand push button	Foot push button	Toggle switch	Rotary switch	Knob	Crank	Lever	Hand-wheel	Pedal
Space required	Small	Large	Small	Medium	Small–medium	Medium–large	Medium–large	Large	Large
Effectiveness of coding	Fair–good	Poor	Fair	Good	Good	Fair	Good	Fair	Poor
Ease of visual identification of control position	Poor*	Poor	Fair–good	Fair–good	Fair–good†	Poor‡	Fair–good	Poor–fair	Poor
Ease of non-visual identification of control position	Fair	Poor	Good	Fair–good	Poor–good	Poor‡	Poor–fair	Poor–fair	Poor–fair
Ease of check reading in array of like controls	Poor*	Poor	Good	Good	Good†	Poor‡	Good	Poor	Poor
Ease of operation in array of like controls	Good	Poor	Good	Poor	Poor	Poor	Good	Poor	Poor
Effectiveness in combined control	Good	Poor	Good	Fair	Good§	Poor	Good	Good	Poor

*Except when control is backlighted and light comes on when control is activated.
†Applicable only when control makes less than one rotation and when round knobs have pointer attached.
‡Assumes control makes more than one rotation.
§Effective primarily when mounted concentrically on one axis with other knobs.

foot. Resistance to start low, build up rapidly, drop suddenly. Minimize viscous damping and inertial resistance.

- *Toggle switch:* Use elastic resistance which builds up and then decreases as position is approached. Minimize frictional and inertial resistance.
- *Rotary selector switch:* Provide detent for each control position (setting). Use elastic resistance which builds up, then decreases as detent is approached. Minimum friction and inertial resistance. Separation of detents should be at least $\frac{1}{4}$ in.
- *Knob:* Preferably code by shape if used without vision. Kind of desirable resistance depends on performance requirements. Inertial resistance of little practical consequence, but may counteract harmful effects of friction and vice versa.
- *Crank:* Use when task involves two rotations or more. Friction (2 to 5 lb) reduces effects of jolting but degrades constant-speed rotation at slow or moderate speeds. Inertial resistance aids performance for small cranks and low rates. Grip handle should rotate.
- *Lever:* Provide elbow support for large adjustment, forearm support for small hand movements, wrist support for finger movements. Limit movement to 90°.
- *Handwheel:* For small movements, minimize inertia. Indentations in grip rim to aid holding. Displacement usually should not exceed ±60° from normal. For displacements less than 120°, only two sections need be provided, each of which is at least 6 in long. Rim should have frictional resistance.
- *Pedal:* Pedal should return to null position when force is removed; hence, elastic resistance should be provided. Pedals operated by entire leg should have 2 to 4 in displacement, except for automobile-brake type, for which 2 to 3 in of travel may be added. Displacement of 3 to 4 in or more should have resistance of 10 lb or more. Pedals operated by ankle action should have maximum travel of $2\frac{1}{2}$ in.

Table B-2 Summary of Selected Data Regarding Design Recommendations for Control Devices

	Size, in		Displacement		Resistance	
	Minimum	Maximum	Minimum	Maximum	Minimum	Maximum
Hand push button						
Fingertip operation	$1/2$	None	$1/8$ in	15 in	10 oz	40 oz
Foot push button	$1/2$	None				
Normal operation			$1/2$ in			
Wearing boots			1 in			
Ankle flexion only				$2^1/_2$ in		
Leg movement				4 in		
Will *not* rest on control					4 lb	20 lb
May rest on control					10 lb	20 lb
Toggle switch			30°	120°	10 oz	40 oz
Control tip diameter	$1/8$	1				
Lever arm length	$1/2$	2				
Rotary selector switch					10 oz	40 oz
Length	1	3				
Width	$1/2$	1				
Depth	$1/2$	1				
Visual positioning			15°	40°*		
Nonvisual positioning			30°	40°*		
Knob, continuous adjustment†						
Finger-thumb						
Depth	$1/2$	1				
Diameter	$3/8$	4				
Hand/palm, diameter	$1^1/_2$	3				$4^1/_2$–6 in/oz
Crank†						
For light loads, radius	$1/2$	$4^1/_2$			2 lb	5 lb
For heavy loads, radius	$1/2$	20			5 lb	10 lb
Rapid, steady turning						
<3-5 in radius					2 lb	5 lb
5-8 in radius					5 lb	10 lb
>8-in radius					?	?
For precise settings					$2^1/_2$ lb	8 lb

	Minimum	Maximum	Displacement	Resistance, min	Resistance, max
Lever†§					
Fore-aft (one hand)			14 in	12 oz	32 oz
Lateral (one hand)			38 in	2 lb	20–100 lb
Finger grasp, diam.	1/2	3			
Hand grasp, diam.	1 1/2	3	90°–120°		
Handwheel‡					
Diameter	7	21		5 lb	30 lb‡
Rim thickness	3/4	2			
Pedal					
Length	3 1/2				
Width	1				
Normal use			1/2 in		
Heavy boots			1 in		
Ankle flexion			2 1/2 in		
Leg movement			7 in		
Will *not* rest on control				4 lb	10 lb
May rest on control				10 lb	180 lb

*When special requirements demand large separations, maximum should be 90°.
†Displacement of knobs, cranks, and handwheels should be determined by desired control-display ratio.
‡For two-handed operation, maximum resistance of handwheel can be up to 50 lb.
§Length depends on situation, including mechanical advantage required. For long movements, longer levers are desirable (so movement is more linear).

Selected References

This appendix includes a list of selected books and journals that deal with human factors engineering. With a few exceptions, the books listed are general references; for references dealing with topics that are specific to the topics of the various chapters, the reader is referred to the lists following the individual chapters.

BOOKS

1 Bennett, E., J. Degan, and J. Spiegel (eds.): *Human factors in technology*, McGraw-Hill Book Company, New York, 1963.
2 Burns, N., R. Chambers, and E. Hendler: *Unusual environments and human behavior*, The Free Press, New York, 1963.
3 Chapanis, A.: *Research techniques in human engineering*, The Johns Hopkins Press, Baltimore, 1959.
4 Chapanis, A.: *Man-machine engineering*, Wadsworth Publishing Company, Inc., Belmont, Calif., 1965.
5 Damon, A., H. W. Stoudt, and R. A. McFarland: *The human body in equipment design*, Harvard University Press, Cambridge, Mass., 1966.
6 DeGreene, K. B.: *Systems psychology*, McGraw-Hill Book Company, New York, 1972.
7 Edholm, O. G.: *The biology of work*, World University Library, McGraw-Hill Book Company, New York, 1967.
8 Forbes, T. W. (ed.): *Human factors in highway traffic safety research*, John Wiley & Sons, Inc., New York, 1972.

9   Gagné, R. M. (ed.): *Psychological principles in system development*, Holt, Rinehart and Winston, Inc., New York, 1962.

10   Hanrahan, J. S., and D. Bushnell: *Space biology*, Science Editions, John Wiley & Sons, Inc., New York, 1961.

11   Harris, D. H., and F. B. Chaney: *Human factors in quality assurance*, John Wiley & Sons, Inc., New York, 1969.

12   Howell, W. C., and I. L. Goldstein: *Engineering psychology: current perspectives in research*, Meredith Corporation, New York, 1971.

13   McFarland, R. A., et al.: *Human factors in the design of highway transport equipment*, Harvard School of Public Health, Boston, Mass., 1953.

14   McFarland, R. A., R. C. Moore, and A. B. Warren: *Human variables in motor vehicle accidents*, Harvard School of Public Health, Boston, Mass., 1955.

15   McFarland, R. A., and A. I. Moseley: *Human factors in highway transportation safety*, Harvard School of Public Health, Boston, Mass., 1954.

16   Meister, D.: *Human factors: Theory and practice*, John Wiley & Sons, Inc., New York, 1971.

17   Meister, D., and G. F. Rabideau: *Human factors evaluation in system development*, John Wiley & Sons, Inc., New York, 1965.

18   Murrell, K. F. H.: *Human performance in industry*, Reinhold Publishing Corporation, New York, 1965.

19   Nadler, G.: *Work design*, Richard D. Irwin, Inc., Homewood, Ill., 1963.

20   National Aeronautics and Space Administration: *Bioastronautics data book*, NASA SP-3006. J. F. Parker, Jr., and V. R. West (managing eds.). U.S. Government Printing Office, Washington, D.C., 1973.

21   Parsons, H. M.: *Man-machine system experiments*, The Johns Hopkins Press, Baltimore, 1972.

22   Poulton, E.: *Environment and human efficiency*, Charles C Thomas, Springfield, Ill., 1970.

23   Sells, S. B., and C. A. Berry (eds.): *Human factors in jet and space travel: a medical psychological analysis*, The Ronald Press Company, New York, 1961.

24   Sinaiko, H. W.: *Selected papers on human factors in the design and use of control systems*, Dover Publications, Inc., New York, 1961.

25   Spector, W. S.: *Handbook of biological data*, USAF, WADC, TR 56–273, October, 1956.

26   Van Cott, H. P., and R. G. Kinkade: *Human engineering guide to equipment design*, rev. ed., U.S. Government Printing Office, Washington, D.C., 1972.

27   Woodson, W. E., and D. W. Conover: *Human engineering guide for equipment designers*, 2d ed., University of California Press, Berkeley, 1964.

28   *Personnel subsystems*, USAF, AFSC design handbook, ser. 1–0, General, AFSC DH 1–3, Jan. 1, 1969, Headquarters, AFSC.

29   *Studies of human vigilance*, Human Factors Research, Inc., Goleta, Calif., January, 1968.

JOURNALS

Applied Ergonomics, IPC House, Guilford, Surrey, England.

Ergonomics, Taylor & Francis, Ltd., London.

Human Factors (Journal of the Human Factors Society, Santa Monica, Calif.), Johns Hopkins Press, Baltimore.

Name Index

Subject Index